CHEMISTRY: WITH SELECTED PRINCIPLES OF PHYSICS

CHEMISTRY:
WITH SELECTED PRINCIPLES OF PHYSICS

ROSEMARY KENNELLY, M.A.
Associate Professor and Science Coordinator
Graduate School of Nursing
New York Medical College
Flower and Fifth Avenue Hospital
and
Science Instructor
Catholic Medical Center
Diocese of Brooklyn and Queens
School of Nursing

RAYMOND E. NEAL, S.B.
Late Associate Professor of Chemistry
Simmons College

SECOND EDITION

McGRAW-HILL BOOK COMPANY
A Blakiston Publication

New York   St. Louis   San Francisco   Düsseldorf   Johannesburg
Kuala Lumpur   London   Mexico   Montreal   New Delhi   Panama
Rio de Janeiro   Singapore   Sydney   Toronto

CHEMISTRY: WITH SELECTED PRINCIPLES OF PHYSICS
Copyright © 1962, 1971 by McGraw-Hill, Inc. All rights reserved. Printed in the United States of America. No part of this publication may be reproduced, stored in a retrieval system, or transmitted, in any form or by any means, electronic, mechanical, photocopying, recording, or otherwise, without the prior written permission of the publisher.

Library of Congress Catalog Card Number 76-140957
07-034060-9

4567890VHVH798

This book was set in Helvetica by Black Dot, Inc., and printed on permanent paper and bound by Von Hoffmann Press, Inc. The designer was Richard Paul Kluga; the drawings were done by John Cordes, J. & R. Technical Services, Inc. The editors were Joseph J. Brehm and Barry Benjamin. Matt Martino supervised production.

CONTENTS

	Preface		vii
chapter 1	THE SHAPE OF THINGS		1
chapter 2	SOME FUNDAMENTAL CONCEPTS		7
chapter 3	ATOMIC STRUCTURE AND VALENCE		13
chapter 4	STATES OF MATTER		23
chapter 5	PHYSICAL AND CHEMICAL CHANGE		31
chapter 6	ENERGY		39
chapter 7	FLUIDS		49
chapter 8	IMPORTANT GASES		71
chapter 9	IMPORTANT LIQUIDS		83
chapter 10	LIQUID MIXTURES		91
chapter 11	ELECTROLYTES		103
chapter 12	ELECTROLYTES IN THE BODY		123
chapter 13	ORGANIC CHEMISTRY		137
chapter 14	IMPORTANT ORGANIC COMPOUNDS		163
chapter 15	CARBOHYDRATES		175
chapter 16	LIPIDS		185
chapter 17	PROTEINS		195
chapter 18	METABOLIC REGULATORS		207
chapter 19	BLOOD		235
chapter 20	METABOLIC WASTES		241
chapter 21	MECHANICS		251
chapter 22	WAVES		259
chapter 23	IONIZING RADIATIONS		269
	Suggested References		285
	Glossary		286
	Index		291

PREFACE

This second edition is offered to students in the health professions as a more complete and up-to-date presentation of important chemical and physical principles fundamental to an understanding of life processes. Biochemical reactions have been expanded to include oxidative processes involved in the source of energy for living. Since the role of enzymes in diagnosis and treatment is becoming more important day by day, these marvelous proteins are considered in some detail. A brief discussion of the nucleic acids has also been included. As a result of valuable suggestions from students, portions of the text have been rewritten in the interests of simplification and clarification.

As in the first edition, the emphasis is on application of basic science principles to situations arising in working with patients. A sound understanding of the chemistry of electrolytes is fundamental to an understanding of fluid and electrolyte balance and replacement fluid therapy. A knowledge of gas and water pressure is basic to any understanding of the mysteries of suctioning apparatus.

The first six chapters contain the background material necessary for an understanding of any of the subsequent chapters. By using this design, it is easy to correlate the subsequent chapters with other areas in the curriculum. Thus having completed the first six chapters, the instructor may choose the order of presentation of the rest of the material.

Equation writing has been deemphasized except in those cases where equations serve a major role in clarification as in oxidations, chemical digestion, and electrolytes and electrolyte balance. Technical and mathematical material has been kept to a minimum and, when included, has been translated into terms and explanations that can be readily understood. As far as possible, examples, applications, and illustrative materials have been selected from situations in the health fields.

Rosemary Kennelly

CHEMISTRY: WITH SELECTED PRINCIPLES OF PHYSICS

chapter one THE SHAPE OF THINGS

As scientific instruments, tools, and measuring devices become more refined and precise, scientists are able to probe deeper and deeper into the composition and structure of matter, coming ever closer to a scientific explanation of life and life processes. The elucidation of the structures of DNA (deoxyribonucleic acid) and RNA (ribonucleic acid), their role in heredity, the structure and methods of synthesis of enzymes, and the application of this knowledge to the study of health and disease have opened vast areas of research, promising ever greater comprehension of the relationships between chemical transformations in health and disease.

On a fundamental level, the old distinctions between chemistry, biology, and physics have broken down. While we may still say that chemistry is the study of matter and its transformations, biology is the study of life, and physics is the study dealing chiefly with the interactions of matter and energy and the transformation of energy, increasingly studies are being done in molecular biology, biophysics, biochemistry, and biophysical chemistry. These interdisciplinary approaches have greatly expanded our knowledge and have emphasized the fact that we cannot study only matter or energy, but must consider also the ways substances are organized or arranged on the atomic level.

MATTER

Matter is anything that has mass. We recognize matter by its properties. One property of matter is inertia, namely, the ability of material to remain at rest or in uniform motion in a straight line unless acted upon by some external force. Mass enables an object to have weight in a gravitational field and imparts to the object its inertia. Weight is a measure of the gravitational attraction between the earth and an object. The weight of an object depends upon the mass of the object and its distance from the earth. We usually think of weight in relation to the earth but the law of gravity applies to the attraction between any two objects.[1] As the distance between an object and the earth increases, the gravitational pull or weight decreases. Thus we have the problem of weightlessness faced by astronauts. The mass, however, remains unchanged.

Substances

Any sample of matter has in addition to inertia, other properties. These are classified

---

[1] The law of gravity states that *the gravitational force is equal to the product of the masses of the bodies divided by the square of the distance between them.*

$$F = \frac{M_1 M_2}{D^2}$$

as physical and chemical properties. Physical properties can be detected or measured without altering the composition of the sample. They include color, odor, state, solubility, boiling point, freezing point, etc. Chemical properties, on the other hand, cannot be observed without altering the composition of the material. Combustibility is a chemical property. If a sample of material does burn, you no longer have your original starting material, but the products of the combustion.

If a sample of material has definite, recognizable, unvarying properties, it is classified as a pure substance. Gold, mercury, table salt, and glucose are examples of pure substances. Two or more substances mixed together can give a variety of properties depending upon the amounts of each substance present. These combinations are, logically enough, called mixtures. Milk, mayonnaise, flour, tincture of iodine, and sea water are examples of mixtures.

Pure substances are subdivided into elementary substances and compound substances, or simply elements and compounds.

Elements

An element is a substance that cannot be decomposed into simpler substances or built up from them. This definition is satisfactory for the present. However, radioactive elements are continually decomposing, forming daughter elements, and many of these continue to decompose until a stable element is formed. New elements can also be made from other elements in the cyclotron, betatron, or as the result of atomic detonations. We can say that elements are simple or fundamental substances composed of only one kind of atoms.

It is hardly safe today to state the number of elements. Through the study of the structure of atoms, it is possible to say with certainty that there are at least 103 elements. At room temperature, 2 of these are liquids: mercury and bromine; 11 are gases: hydrogen, helium, nitrogen, oxygen, fluorine, neon, chlorine, argon, krypton, xenon, and radon. The others are solids. Of the elements, only about 20 are normally found in the human body.

The elements are divided into two classes, the metals and the nonmetals. The metals are the more common. We meet them in our daily experiences: the iron of the beds, the mercury of thermometers, and the silver of silverware. The metals are characterized by the property of luster. This, obviously, can be seen only while observing the clean surface of the metal. Many of the metals are so active that the exposed surface soon becomes covered with a layer of oxide, which hides its true appearance. The luster of metallic sodium, for instance, can be seen only on a freshly cut surface. In some instances the oxide is transparent—e.g., aluminum oxide—and while this reduces the luster somewhat, its presence protects the metal from further action. The metals exhibit high conductivity of heat. Silver is notable in this regard, as everyone knows who has used a solid silver spoon in a cup of hot tea. They also have high electrical conductivity.

The nonmetals are not so well known. It is true that carbon and sulfur are familiar, but silicon and selenium are merely words to most people. Iodine, on the other hand, is well known because of the general use of tincture of iodine, an alcoholic solution of the nonmetal and potassium iodide.

Atoms

According to the theory postulated by John Dalton, an English schoolmater, elements are made up of atoms. He defined the atom as the smallest subdivision of an element that can exist. This means that as we continue to reduce the size of the particles of an element, we finally arrive at a submicroscopic, indivisible particle—the atom—which retains all the properties of the element. Today we know that further subdivision is possible, but we obtain particles that no longer possess

The Shape of Things

Table 1-1 APPROXIMATE COMPOSITION OF THE HUMAN BODY*

Element	Symbol	Per Cent	Element	Symbol	Per Cent
Oxygen	O	65.0	Iron	Fe	
Carbon	C	18.0	Iodine	I	
Hydrogen	H	10.0	Fluorine	F	
Nitrogen	N	3.0	Silicon	Si	
Calcium	Ca	2.0	Cobalt	Co	
Phosphorus	P	1.0	Zinc	Zn	Traces
Potassium	K	0.4	Copper	Cu	
Sulfur	S	0.3	Manganese	Mn	
Sodium	Na	0.2	Molybdenum	Mo	
Chlorine	Cl	0.2	Chromium	Cr	
Magnesium	Mg	0.1	Selenium	Se	

*Nursing students should memorize the symbols for the elements in the body.

the properties of the element. These particles are called *subatomic particles,* the most familiar of which are protons, neutrons, and electrons.

Symbols

When Dalton postulated the atomic theory, he proposed a set of geometric figures to represent the atoms. Thus ○ represented oxygen; ☉, hydrogen; ●, carbon; ①, nitrogen; and ⊕, sulfur. These symbols were clumsy to use, and they easily gave way to the modern symbols of Berzelius, who used, as far as possible, the capitalized initial letter of the name of the element. Thus O became the symbol for oxygen; H, for hydrogen; and C, for carbon. In cases where two or more names had the same initial letter, two letters were used. In this way, Ca became the symbol for calcium; Cd, for cadmium; Co, for cobalt; and Cr, for chromium. The symbols, however, were to be used internationally, and the names of some elements begin with different letters in different languages; e.g., "iron" is *fer* in French and *eisen* in German. In such cases, the symbols were derived from their Latin names. The symbol for iron therefore became Fe from *ferrum,* that of silver became Ag from *argentum,* etc. Oddly enough, however, Na for sodium and K for potassium are still derived from the German names *natrium* and *kalium.* A symbol may represent the element, an atom of the element, or the atomic weight of the element. Atomic weights will be explained later. The symbol O can represent the element oxygen or 1 atom of oxygen or 16 grams of oxygen. A table showing the symbols of the elements will be found on the *inside front* cover.

Molecules

Two or more atoms can combine chemically to form a molecule. If the atoms are the same, we have a molecule of an elementary substance. This is the case with ordinary oxygen, which is composed of 2 atoms. If the combining atoms are from different elements, the resulting molecule represents a compound. Each molecule of carbon dioxide contains 1 atom of carbon and 2 atoms of oxygen. A molecule of glucose contains 6 carbon, 12 hydrogen, and 6 oxygen atoms.

Formulas

Just as the symbol represents 1 atom of an element, the formula signifies 1 molecule of an element or compound. The formula tells us what elements are contained in the molecule and how many atoms of each element are present. The formula for helium, He, indicates that there is 1 atom in the molecule. Helium along with the other inert gases, neon, argon, krypton, xenon, and radon, are monatomic because, being inert, their atoms will not react with those of other elements or with each other. The formulas for the elementary gases, oxygen, hydrogen, nitrogen, and chlorine are O_2, H_2, N_2, and Cl_2, indicating that there are 2 atoms per molecule. These gases are *diatomic*. The formula for carbon dioxide is CO_2 and for glucose, $C_6H_{12}O_6$.

Compounds

The second type of pure substance we call a compound. Compounds are composed of two or more different elements, chemically combined. When a chemical combination occurs, the starting materials of which the compound is made lose their original properties and the resulting compound can have radically different properties. Consider common table salt, sodium chloride (NaCl). Sodium is a soft, silvery metal and is extremely active. If a small piece of sodium is dropped on the surface of water in a dish or beaker, a violent reaction occurs. Hydrogen gas is given off and the piece of sodium is propelled about on the surface of the water. If it is held in one spot by some obstruction, it often bursts into flame and burns with a bright yellow color. This metal that acts so violent is part of table salt. The other part is the nonmetal chlorine, a greenish-yellow poison gas with a disagreeable odor. Yet not only does no violent reaction occur when sodium chloride is added to water, but sodium chloride is essential for life and a constituent of body fluids and tissues. Many compounds found in the body contain hundreds of atoms per molecule. These giant molecules or macromolecules include the proteins such as hemoglobin in red blood cells and the enzymes involved in all body functions. In these and other compounds, not only must the proper number and kinds of atoms be present in the molecule but these atoms must be arranged in a definite order and the molecule itself must have a definite shape. If there is a deviation in either of these conditions, the molecule cannot function normally and metabolic functions will be impaired. An example of this kind of impaired function is sickle cell anemia, a disease resulting from the abnormal arrangement of the atoms in the hemoglobin molecule.

We see that a knowledge of the arrangement of atoms in molecules is important and that different arrangements form compounds with different properties. Compounds formed from the same numbers and kinds of atoms but with different arrangements are called *isomers*. In the formation of compounds in living organisms, DNA and RNA control not only the composition of the new substances but also their shapes. We shall recognize the importance of the shape of things as we learn more about substances.

Mixtures

Unlike compounds, in which the constituent elements exist in definite proportion by weight, mixtures are highly variable. Mixtures may be composed of elements, as a mixture of iron filings and sulfur. Flour is essentially a mixture of compounds, namely, starch, protein, and salts. Lugol's solution, a medication that is taken internally, is a mixture of the element iodine, the compound potassium iodide, and the compound water. We may even have mixtures of mixtures, as we have in an eggnog.

Frequently mixtures may be quickly recognized because they are heterogeneous. Many mixtures, however, appear homogeneous to the naked eye. Almost anyone would suppose merely from looking at baking powder that it was homogeneous. However, if it is examined under the microscope at least three different substances will be evident.

Compounds have a fixed composition by weight. It is impossible to alter this composition and still have the same compound. On the other hand, it is impossible to have two mixtures just alike. It is true that it is possible to make two mixtures sufficiently alike to serve a practical purpose; e.g., a prescription filled by the pharmacist may be refilled and be practically the same as the original. Baking powder from one batch has the same practical effect as a sample from a previous mixing. Nevertheless, although they seem alike, they cannot be exactly the same. Further, the composition of a mixture, other than a solution, may vary over the widest range, and yet the substance is still a mixture.

Mixtures differ from compounds in that they may be separated into their components by physical means. If we examine a sample of coarse sand even with the naked eye, we find it made up of grains of different colors. If we use a reading glass, a pair of tweezers, sufficient time, and enough patience, we may separate this mixture into several piles of colored grains—surely a physical process.

We may make use of other physical processes. The components of a mixture of iron filings and sulfur have very different densities, and if the mixture is shaken with water in a test tube, the heavy iron settles to the bottom and the lighter sulfur rises to the surface. This same mixture might have been separated by making use of the magnetic property of iron. A horseshoe magnet held over the mixture would draw out the iron filings and leave the sulfur behind.

Often a mixture may be separated into its components by making use of differences in solubility. If a mixture of sugar and white sand is shaken with water, the soluble sugar dissolves and the insoluble sand settles to the bottom. Now, the solution may be separated from the sand by filtration or decantation, the water allowed to evaporate, and the crystals of sugar obtained.

Solutions present a problem to the beginner when he tries to decide whether they are mixtures or compounds. A solution appears homogeneous regardless of the magnification used in its examination. In this respect it resembles a compound. Quantitatively, however, although a given solution is uniform in composition down to the last drop, its composition may be varied by dilution while qualitatively the same homogeneous solution remains. In this it resembles a mixture. Finally, the solution may be separated into its components by a *physical process*, since usually the solvent may be evaporated, leaving the solute. This establishes the fact that a solution is a mixture but a homogeneous mixture.

The properties of the components of a mixture are still evident in the mixture. Since each component is still present as a separate substance, it continues to exhibit its individual characteristics even though it is in a mixture. The properties are additive, which means that the properties of the mixture are the sum of the properties of the components. This principle is well illustrated by a mixture of sugar and butter, the familiar hard sauce. Alone, the sugar is white, tastes sweet, and is in the form of hard crystals large enough to be felt in the mouth. The butter, on the other hand, is yellow and greasy and has a characteristic flavor. Let us mix the two and consider the properties of the mixture. The color is now a lighter yellow. The taste is sweet, but the flavor of the butter is also still apparent. The granular nature of the sugar is still felt, reduced somewhat by the greasiness of the butter. Thus it is seen that each component has contributed its own characteristics to the mixture.

Often a prescription calls for a mixture of drugs. Each one retains its identity and exercises its specific effect in the body, but together they produce the accumulated effect, the cure or relief intended.

Study Exercises

1. Which of the following are pure substances and which are mixtures? table salt, vinegar, baking soda, baking powder, air, milk, 14-carat gold, platinum, tincture of iodine, and boric acid.
2. What are the differences between a mixture and a compound?
3. Define the following: element, compound, isomer, symbol, and formula.
4. How could the following mixtures be separated into their components? (a) sugar and white sand, (b) iron filings and charcoal.
5. Compare the properties of the elements sodium and chlorine with the properties of the salt sodium chloride.
6. Mention three properties typical of metals.
7. Write the sumbols for sodium, sulfur, carbon, calcium, chlorine, iron, iodine, potassium, phosphorus, nitrogen, magnesium, manganese, molybdenum, fluorine, and zinc.

chapter two SOME FUNDAMENTAL CONCEPTS

Both chemistry and physics are exact sciences, in which measurement plays an important part. In medicine, too, measurements are important and tell much more than descriptive terms. If fluids are to be forced on a patient, which statement by the nurse is more useful? "The patient drank a good deal of water and took a little orange juice." "The patient drank 1,500 ml water and 500 ml orange juice." It is obviously more informative to report that a patient's temperature is 104.2°F than to say that he has a high fever.

THE METRIC SYSTEM

The metric system is used in all scientific work and, in all the civilized countries of the world except the United States and the British Commonwealth of Nations, for everyday measurements. In the United States, many of the newer drugs are being marketed with labels describing dosage in metric units. Perhaps the day will come soon when one will no longer need to know drachms, grains, fluid ounces, etc., because of the widespread use of the metric system, which has appropriate units to describe any quantity. The metric system is easier to understand and to use than other systems. To convert from one unit to another, we simply divide or multiply by 10 or 100 or 1,000.

The standard unit of length is the meter; of weight, the kilogram; and of time, the second. This system is known as the *mks system* (meter, kilogram, second system). A related system is called the *cgs system* (centimeter, gram, second system).

In measurements, one chooses a measuring unit appropriate to the size of the object or distance to be measured. In English units, to measure long distances, one would use the mile; for shorter distances, the yard; still shorter, the foot; and for still smaller sizes, the inch or fraction of an inch. Other units such as leagues, fathoms, spans, and hands are unfamiliar to most of us. The metric system has appropriate units for measuring all kinds of things, including very small objects such as viruses and molecules.

The meter, used to measure length or distance, is equal to 39.37 inches, slightly longer than a yard. To measure long distances, a unit 1,000 times larger than the meter is used. This is the kilometer, which is approximately 0.62 miles. Other units, the hectometer, 100 meters, and the decameter, 10 meters, are rarely used. Now we consider units smaller than the meter. The unit one-tenth the size of the meter is the decimeter. The prefix deci- means one-tenth. Hence, 10 decimeters equal 1 meter. The prefix centi- means one-hundredth, so 100 centimeters equal 1 meter, and the prefix milli- indicates

[1] Great Britain is in the process of converting to a basically metric system called the International System of Units or Système International, abbreviated S.I.

one-thousandth, so 1,000 millimeters equal 1 meter. The decimeter, which is rarely used in scientific measurements, is approximately equal to 3.94 inches, the centimeter is 0.39 inches, and the millimeter 0.04 inches. To measure the size of small things such as microbes, a unit 1,000 times smaller than a millimeter is used. This is the micron.[1] A millimicron is 1,000 times smaller than a micron. A unit used to measure the length of X rays and average-size molecules is one-tenth the size of a millimicron and is called the angstrom unit. We are now considering objects too small to be seen, even with the electron microscope. To summarize:

$$1,000 \text{ meters (m)} = 1 \text{ kilometer (km)}$$
$$10 \text{ decimeters (dm)} = 1 \text{ meter}$$
$$100 \text{ centimeters (cm)} = 1 \text{ meter}$$
$$1,000 \text{ millimeters (mm)} = 1 \text{ meter}$$
$$1,000,000 \text{ micra } (\mu) * = 1 \text{ meter}$$
or
$$1,000 \text{ micra} = 1 \text{ mm}$$
$$1,000,000 \text{ millimicra } (m\mu) = 1 \text{ mm}$$
$$10 \text{ angstrom units (Å)} = 1 \text{ m}\mu$$

With these relationships in mind, the systems for weights and volumes are simple. The unit of weight is the kilogram (kg), which consists of 1,000 grams (Gm) and is equal to 2.2 pounds.

$$10 \text{ decigrams (dg)} = 1 \text{ gram (Gm)}$$
$$100 \text{ centigrams (cg)} = 1 \text{ Gm}$$
$$1,000 \text{ milligrams (mg)} = 1 \text{ Gm}$$
$$1,000 \text{ micrograms } (\mu g) = 1 \text{ mg}$$

One gram is equal to 0.035 ounces.

In measuring volumes, only three units are commonly used: the kiloliter, the liter, and the milliliter. The liter equals 1.06 quarts.

$$1,000 \text{ liters (l)} = 1 \text{ kiloliter (kl)}$$
$$1,000 \text{ milliliters (ml)} = 1 \text{ liter}$$

Summarizing the units most commonly used in medicine:

[1] The micron is designated by the lower-case Greek letter mu which is written μ. Micra is often used as the plural of micron.

$$1 \text{ cm} = 0.01 \text{ meter}$$
$$1 \text{ mm} = 0.001 \text{ meter}$$
$$1 \ \mu = 0.001 \text{ mm}$$
$$1 \text{ m}\mu = 0.000,001 \text{ mm}$$
$$1 \text{ m}\mu = 0.001 \ \mu$$
$$1 \text{ Å} = 0.1 \text{ m}\mu$$
$$1 \text{ Å} = 0.000,000,1 \text{ mm } (1 \times 10^{-7})$$
$$1 \text{ Å} = 0.000,000,01 \text{ cm } (1 \times 10^{-8})$$

$$1,000 \text{ liters} = 1 \text{ kl}$$
$$1 \text{ ml} = 0.001 \text{ liter}$$
$$1,000 \text{ Gm} = 1 \text{ kg}$$
$$1 \text{ mg} = 0.001 \text{ Gm}$$

and also the meter, the liter, and the gram.

Area is measured in square units derived from linear units; e.g., square centimeters may be written sq cm or cm^2. The same procedure applies to cubic measurements; we have cubic centimeters, or cm^3, or the very common cc.

To recap:

$$100 \text{ cm} = 1 \text{ meter}$$
$$10 \text{ mm} = 1 \text{ cm}$$
$$1,000 \text{ mm} = 1 \text{ meter}$$
$$1,000 \text{ mg} = 1 \text{ Gm}$$
$$1,000 \text{ ml} = 1 \text{ liter}$$

In converting from one unit to another within a system, small units are changed into larger units by dividing by the number of small units in the larger unit. To change larger units into smaller units, multiply the larger unit by the number of smaller units that it contains. Since all the conversion factors are multiples of 10, this amounts to moving the decimal point to the right or to the left.

Problems

The barometric pressure is expressed as 760 mm of mercury. You wish to change this measurement to centimeters. Millimeter is the smaller unit. There are 10 mm in 1 cm. Dividing 760 mm by 10 gives us the answer, 76 cm. In a whole number there is no need for a decimal point, but one is understood to be at the right of the number, as 760.0. To divide

Some Fundamental Concepts

Table 2-1 METRIC UNITS

Weight	Length	Volume
Gram	Meter	Liter
1,000 Gm = 1 kg	100 cm = 1 meter	1,000 liters = 1 kl
1,000 mg = 1 Gm	1,000 mm = 1 meter	1,000 ml = 1 liter

*Chemists and physicists use the abbreviation g for gram, but in medicine, to avoid confusion between gram and grain (apothecaries' system), the abbreviation Gm is used.

by 10, one moves the decimal point one space to the left.

An individual is 169 cm tall. To change this measurement to meters, we realize that centimeter is the smaller unit and that there are 100 cm in 1 meter. Moving our imaginary decimal point 2 places to the left gives the answer, 1.69 meters.

A patient voids 1,250 ml of urine in a 24-hr period. Since there are 1,000 ml in 1 liter, move the imaginary decimal point three places to the left to find the volume in liters, 1.250 liters.

In changing from smaller units to larger units, divide. Move the decimal point the appropriate number of places to the left. Change 50 mg to Gm. Divide by 1,000. Move the decimal point three places to the left. Answer, .050 Gm. As a precaution and to alert the reader to the decimal, if a number starts with a decimal, a zero is placed before the decimal thus, 0.050.

If a medication order was written as .5 Gm and the decimal was blurred or indistinct, one might read this order as 5 Gm. This would be a gross error and the patient would receive 10 times the intended dose. By writing 0.5, the reader expects a decimal and looks for it. The possible error is avoided.

To change larger units to smaller ones, multiply by moving the decimal point (real or imagined) the appropriate number of places to the right.

A baby is 55 cm long. Change this measurement to millimeters. Imagine the decimal point as 55.0. There are 10 mm in 1 cm. Multiply by 10 by moving the decimal point one place to the right. Answer, 550 mm.

An object is 0.25 meter wide. To express as centimeters, multiply by the number of centimeters in a meter, 100. Moving the decimal point two places to the right gives us the answer, 25 cm.

Express 2.65 Gm as milligrams. There are 1,000 mg in 1 Gm. Multiply by 1,000. Move the decimal point three places to the right to get 2,650 mg.

In these problems, one can only use the same types of units. Grams cannot be changed to volume units without additional information.

To describe volume of solids, a unit called the cubic centimeter (cc) is used in place of the milliliter. This is an older measurement in science, which is gradually disappearing. One could go to a dairy and ask for a cubic foot of milk. If the clerk filled a container whose capacity was 1 cu ft, it would hold 1 cu ft milk. However, it is more acceptable to sell milk by the gallon than by the cubic foot.

Since the cubic centimeter and the milliliter are almost exactly the same size, these units are often used interchangeably.

When we start learning a foreign language, we think in our native tongue and translate; after a while we can think in the foreign language. When one first starts using the metric system, one will probably refer to English, or apothecaries', units. Here are some useful equivalents:

2.2 lb = 1 kg
1 lb = 454 Gm
1 in. = 2.5 cm
39.3 in. = 1 meter
1 cm = 0.39 in.

1 qt, or 32 fl oz ≅ 1 liter, or 1,000 ml[1]
1 pt, or 16 fl oz ≅ 500 ml
1 fl oz = 30 ml
1 oz ≅ 30 gm

Since dosages are sometimes prescribed in terms of body weight, particularly for infants, it may be necessary to change weights from English to metric units in order to compute the proper dosage. Since 2.2 lb = 1 kg, to change pounds to kilograms divide by 2.2 and to change kilograms to pounds, multiply by 2.2. For instance, an infant weighs 5½ lbs; 5.5 divided by 2.2 gives 2.5 kg.

ATOMIC WEIGHTS

When we realize that atoms are too small to be seen with the electron microscope, we also realize that an individual atom is too small to be weighed. It is however possible to determine relative weights. One could weigh x trillion atoms of hydrogen and x trillion atoms of helium and discover that helium is four times heavier than hydrogen. Relative weights of atoms are extremely useful and may be obtained with great accuracy. In expressing relative weights, it is necessary to set up a standard or basis of comparison. In the early days, hydrogen, the lightest element, was assigned a weight of 1 and was used as the standard or reference point from which the weights of the other elements were computed. In other words, if hydrogen has a weight of 1, then carbon, which is 12 times heavier than hydrogen, will have a weight of 12, and oxygen, which is 16 times heavier than hydrogen, will have a weight of 16, etc. Subsequently, oxygen 16 was used as the standard, but at present carbon, atomic weight 12, is the official standard against which the atomic weights are computed. These relative weights have no units; they are simply comparisons.

Having established relative weights, the next step was to translate these relative weights into actual weights. How much hydrogen will weigh one gram? The number of hydrogen atoms weighing one gram is 602,000,000,000,000,000,000,000 or 6.02×10^{23} atoms. We have now the gram-atomic weight of hydrogen, which is 1 gram. The gram-atomic weights of the other elements are the weights in grams of an equal number of the atoms of that element.

The symbol of an element can represent the name of the element, an atom of the element, or the atomic weight of the element. In the formula H_2O, H taken twice can mean 2 atoms of hydrogen or it can mean 2 Gm hydrogen; O can mean 1 atom of oxygen or 16 Gm oxygen.

MOLECULAR WEIGHTS

The molecular weight of a substance is the sum of the atomic weights of the atoms contained in the molecule. The molecular weight of carbon dioxide, CO_2, is 44, for the weight of 1 atom of carbon is 12, and the weight of 2 atoms of oxygen is 32 (2×16).

The molecular weight of water, H_2O, is 18. Hydrogen, weight 1 times 2, plus oxygen 16. Since the atomic weights are numbers that show the relative weights of the atoms referred to carbon 12, the molecular weight of a substance is a number showing the weight of the molecule compared to carbon 12.

The gram-molecular weight or the *mole* of a substance is the number of *grams* of a substance equal to the sum of its gram-atomic weights or to its molecular weight; the gram-molecular weight of CO_2 is 44 grams. The weight of one mole of water is 18 gm.

law of definite composition

Every time we analyze a given compound, regardless of its source, if it is a pure sample,

---

[1] The symbol ≅ means approximately equal.

we always find that it contains the same elements and in the same proportion by weight. That is, no matter how we obtain pure sodium chloride, table salt, whether it is formed from the direct union of sodium and chloride, mined from a salt mine, crystallized from brine, or obtained as a by-product in a chemical reaction, it will always contain sodium and chloride and the elements will be present in the ratio of 23 parts of sodium to 35.3 parts of chlorine. Furthermore, if sodium and chloride react directly, 23 Gm sodium will react exactly with 35.5 Gm chlorine to form 58.5 Gm sodium chloride. These observations are summarized in the law of definite composition which states that *in every pure sample of a compound, the elements are always present in a definite proportion by weight.* By consulting the table of atomic weights, one can see that these proportions are related to the atomic weights of the elements involved.

Study Exercises

1. Distinguish between mass and weight.
2. An infant was reported to be 38 cm long and having a weight of 3,000 Gm. Convert these figures into units of inches and pounds.
3. A student is 5 ft 7 in. tall and weighs 125 lb. Express these figures in metric units.
4. The average diameter of a human red blood cell is 7.5 μ, platelets are 2 to 3 μ in diameter, and the average diameter of pathogenic bacilli is 0.5 μ. Express these measurements as millimeters.
5. The "recommended daily dietary allowances" for the average female, age 18 to 22, for iodine, iron, and magnesium are, respectively, 100 μg, 18 mg, and 300 mg. Convert these figures to grams.
6. Define the following: atomic weight, molecular weight, gram-atomic weight, gram-molecular weight, and mole.
7. What is the law of definite composition?

chapter three ATOMIC STRUCTURE AND VALENCE

Work done with the cyclotron or "atom smasher" has indicated that the atom is composed of various smaller particles called *subatomic* or *elementary particles.* The proton, neutron, and electron had been known before this work was done, but at present we have, in addition, the positron, meson, neutrino, pion, and muon, each with its antiparticle, to name but a few. Since some of these particles have very transitory lives and since we can explain the reactions in which we are interested by the proton, neutron, and electron, we shall confine our description of atomic structure to these three.

SUBATOMIC PARTICLES

The *electron* consists of a unit charge of negative electricity and has a mass that is about 1,800 times lighter than an atom of hydrogen. The *proton* consists of a unit charge of positive electricity equal in amount to the charge on the electron. Since the hydrogen atom consists of 1 proton and 1 electron, it follows that the mass of the proton must be about equal to that of the hydrogen atom, or about 1,800 times that of the electron. The *neutron* is electrically neutral and has a mass practically equal to that of the proton.

An atom resembles a miniature solar system in that planetary electrons revolve around a heavy central core, the nucleus. The nucleus is composed of protons and neutrons and is therefore positively charged. The weight of the atom resides in the nucleus; the atomic weight, therefore, is equal to the sum of the masses of the protons and neutrons. The number of positive charges on the nucleus depends upon the number of protons and is equal to the atomic number. If the elements are arranged in the order of increasing number of protons in the nuclei, the ordinal number in the succession is the atomic number of the element. Planetary electrons, equal in number to the protons in the nucleus, revolve about the nucleus in orbits at various levels. The number of electrons in each level of the atom of a given element is dictated by the configuration of atoms of the inert gas that has an atomic number closest to that of the given element. Since the number of planetary electrons (negatively charged) is equal to the number of protons (positively charged) in the nucleus, the uncombined atom is electrically neutral. The chemical properties of the atom depend upon the planetary electrons, particularly those in the outermost level.

Figure 3–1 illustrates the structure of some representative atoms. Note that in each case, the number of *protons* is equal to the *atomic number,* and the *atomic weight* is equal to the *sum of the protons and neutrons*. In uncombined atoms, the number of planetary electrons is equal to the number of protons, which is, of course, the atomic

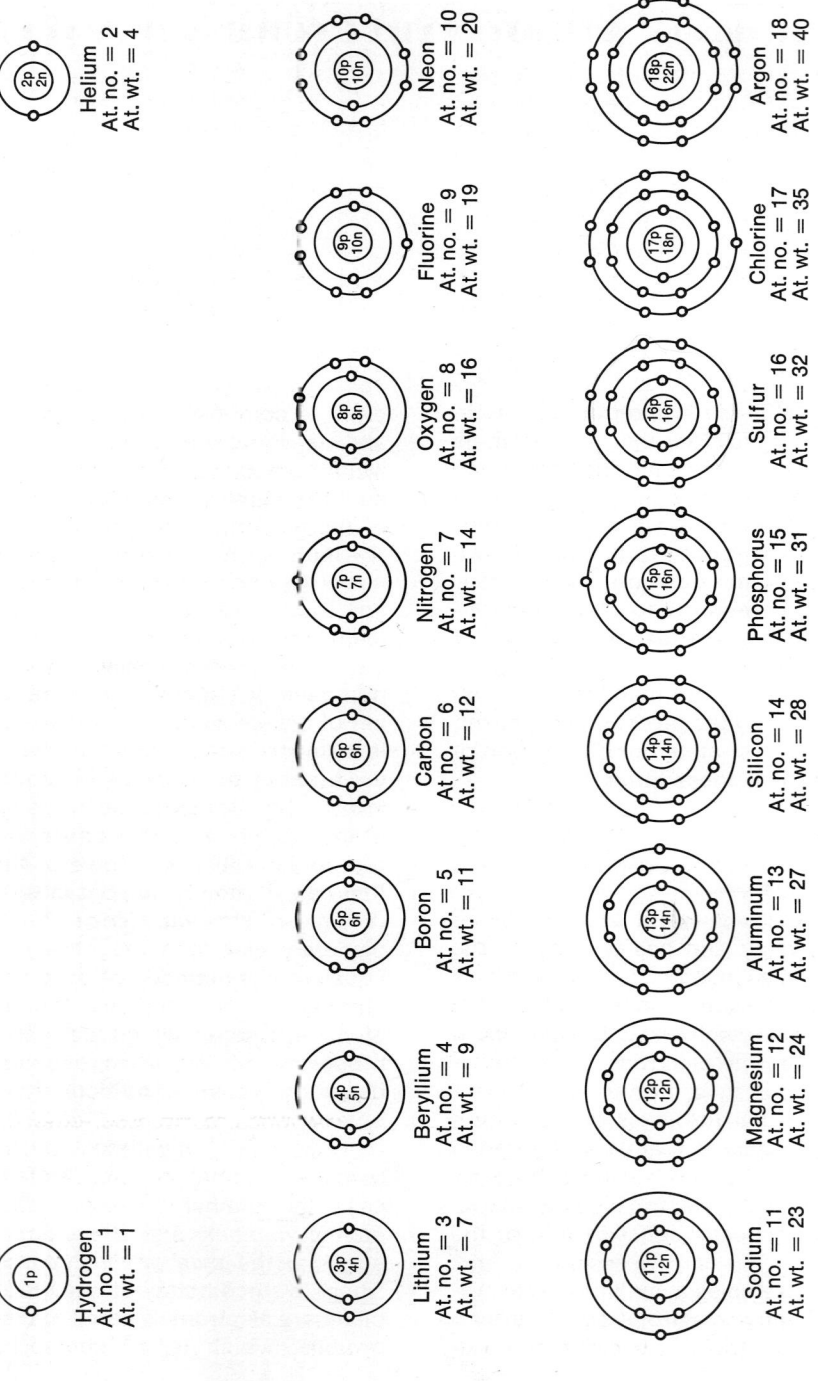

Fig. 3-1 The structure of some representative atoms.

Atomic Structure and Valence

number. In Fig. 3-1, the planetary electrons are represented schematically in orbital levels. However, each electron has its own path within its own level. Helium, the inert gas, atomic number 2, has 2 electrons in the first orbital level. Two electrons in the first orbital level is the stable configuration. All atoms following helium have 2 electrons in the first level. Electrons in excess of 2 are found in the second level until there are 8 electrons. Eight electrons in the second level is a stable configuration. The element with this arrangement is the inert gas neon. The third orbital level builds up until it also has 8 electrons, again forming an inert gas; in this case it is argon.

The orbital levels are designated by letters of the alphabet. The first level is called the K level with its maximum of 2 electrons; the second level, with its maximum of 8 electrons, is the L level, followed by M, N, etc. In chemical changes, electrons tend to become arranged like the outermost orbits of the inert gases.

ISOTOPES

A glance at the atomic weights of the elements will show that almost all of them have atomic weights that are not whole numbers. Why should this be if the atomic weight is the sum of all the protons and neutrons in the nucleus and the proton and neutron each has a weight of 1? Hydrogen has a weight of 1.008; chlorine has a weight of 35.45. The answer is that the sample contains atoms of the same element with at least two different weights.

Atoms that have the same atomic number belong to the same element even though the atomic weights may vary because of the different number of neutrons in the nucleus. An examination of Fig. 3-2 shows that the chlorine atoms have the same number of protons in the nuclei and therefore the same atomic number. This makes them isotopes. The number of neutrons, however, is different, giving

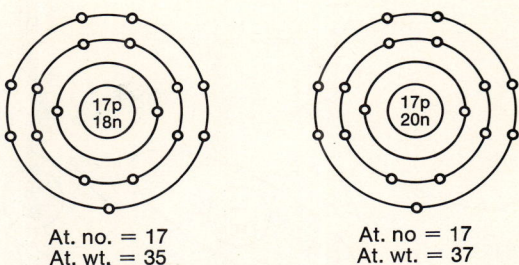

Fig. 3-2 Isotopes of chlorine.

the atoms masses of 35 and 37, respectively. Note, however, that the number of planetary electrons is 17 in each case, dictating identical chemical properties for the two isotopes.

The most common form of hydrogen has atomic weight 1 and atomic number 1, indicating 1 proton and 1 electron. So-called heavy hydrogen, deuterium, is found in heavy water and has an atomic number of 1 and atomic weight of 2.

Written symbolically, this would be $_1H^2$. The subscript number preceding the symbol is the atomic number, and the superscript to the right indicates the atomic weight. Atomic number 1 tells us that the atom contains 1 proton and 1 electron. The weight of 2 indicates 2 heavy particles in the nucleus, 1 of which we know is a proton. Therefore, the deuterium atom is composed of 1 proton and 1 neutron in the nucleus and 1 electron orbiting around it. Tritium, a hydrogen isotope formed in the atomic pile, is hydrogen with atomic number 1 and atomic weight 3. By the same reasoning, we conclude that there is 1 proton and 1 electron, which leaves 2 more heavy particles, 2 neutrons. (See Fig. 3-3.)

Isotopes, as we see, differ in weight. In addition, some are unstable and break up, emitting as they do radioactive particles or waves, which can be detected with appropriate instruments. For these two reasons isotopes are useful as tracers in reactions. Their paths can be followed and their final outcomes discovered. Let us emphasize the fact that isotopes have the same number and ar-

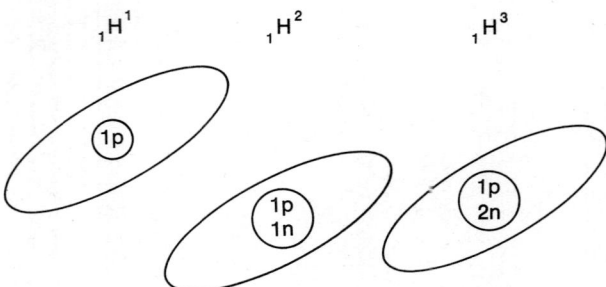

Fig. 3-3 Isotopes of hydrogen.

rangement of electrons and, therefore, their chemical reactions are, for all practical purposes, the same. One could feed to experimental animals a diet containing isotopes and then seek to answer such questions as: Does the bone formed during growth remain throughout adult life or is it constantly changing? Are deposits of fatty tissue, in an animal of constant weight, being used for metabolism? We have here a wonderful tool for the study of life processes.

VALENCE

Valence, very simply, is the ability of atoms to combine with other atoms to form molecules, or, expressed another way, the combining ability of an element. If an atom does not combine with other atoms, it has no valence. The inert gases—helium, neon, argon, krypton, xenon, and radon—are such. When atoms react, the planetary electrons are involved. The atom may lose one or more, gain one or more, or share one or more of its planetary electrons with other atoms. The <u>number of electrons involved is the valence humber.</u>

electrovalence

When a metal atom combines with a nonmetal, the metal atom loses electrons from its outermost level and thus acquires a positive charge equal to the number of electrons lost. The nonmetal gains the electrons, acquiring a corresponding negative charge. Stable electron arrangements are like those of the inert gases, i.e., 2 electrons in the first level and 8 in any other outermost level. Each atom attempts to attain the stable configuration of the nearest inert gas. Sodium has an atomic number of 11, and chlorine has an atomic number of 17. Sodium has 11 protons (+) and 11 electrons (−). These electrons are arranged with 2 electrons in the first level, 8 in the second level, and 1 in the outermost level.

$$\text{Na} \;(11p)\;\; 2)8)1)$$

The nearest inert gas is neon, atomic number 10, 10 protons, 10 electrons.

$$\text{Ne} \;(10p)\;\; 2)8)$$

To assume this configuration sodium could lose 1 electron from its outer level. The particle now has 11 positive charges and 10 electrons, giving it a total charge of +1. This charged particle is called *sodium ion*, and its valence is +1.

Chlorine, with atomic number 17, has 17 protons (+) and 17 electrons (−). The electron arrangement is 2 in the first level, 8 in the second, and 7 in the third. The nearest inert gas is argon, atomic number 18.

$$\text{Cl}\;(17p)\;\;2)8)7)\quad \text{A}\;(18p)\;\;2)8)8)$$

If chlorine picks up an electron to make

8 in its outer level, it then has 17 positive charges and 18 electrons or a total charge of −1. This charged particle is the chloride ion, valence −1.

By the same process, in the formation of magnesium sulfide, MgS, the magnesium atom may lose 2 of its electrons to become a magnesium ion, Mg^{++}, with a valence number of +2, and the sulfur atom could gain 2 electrons to become the sulfide ion, $S^=$, with a valence number of −2. In this connection it may be well to point out that when an atom with several electrons in its outer level loses 1 of these electrons, the particle becomes positively charged. Because of this residual positive charge, the next electron is lost with some difficulty. As the positive charge becomes greater, the electrons are lost with increasingly greater difficulty. This accounts for the paucity of ions with a high positive valence number. The same explanation in reverse accounts for the few ions with a high negative valence number.

Incidentally, any atom or group of atoms with a charge is called an ion. Those with positive charges are *cations*, and those with negative charges are *anions*.

When the metal sodium combines with the nonmetal chlorine, the electron lost by the sodium atom is gained by the chlorine atom, i.e., an electron is *transferred* from the metal atom to the nonmetal atom as shown:[1]

$$Na° + \cdot \ddot{C}\underset{\cdot\cdot}{l}\cdot \rightarrow Na^+ + {\,\cdot\,}\ddot{C}\underset{\cdot\cdot}{l}{:}^-$$

Since particles carrying opposite charges attract each other, these ions are held together by an electric force of attraction. This type of valence is called *ionic valence,* or *electrovalence*.

X-ray analysis of the crystals of sodium chloride shows the lattice of sodium and chloride ions as indicated in Fig. 3–4, wherein each sodium ion is surrounded by 6 chloride ions and each chloride ion by 6 sodium ions. It must follow, therefore, that there is no such thing as a molecule of sodium chloride. Instead, in a crystal of the salt, there are equal numbers of sodium and chloride ions. The formula NaCl is habitually used to represent the compound, but it should be clearly borne in mind that a molecule of an electrovalent compound does not exist. Such compounds are called *salts*. Sodium chloride has been used here as a representative salt, and the same ideas hold for potassium bromide, calcium chloride, or any other salt.

covalence

According to the theory advanced by G. N. Lewis, an atom may complete its quota of 8 electrons in its outer orbit by *sharing* electrons with another atom. In such a case 1 atom contributes 1 electron to the shared pair and the other atom does the same. If the atoms are the same, the shared pair of electrons is equidistant from the nuclei of both atoms; each atom may then be considered as gaining and losing half an elec-

Fig. 3–4 The crystal lattice of sodium chloride.

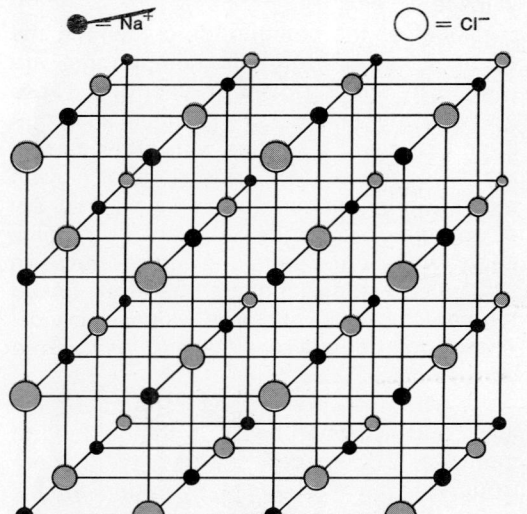

---

[1] In such electronic formulas, all but the valence electrons are omitted, and there is no actual difference in the electrons. They are depicted this way so that the student may better appreciate the source of each electron.

tron. The net result is that neither atom has acquired a charge. This is the sort of thing which occurs when 2 atoms of hydrogen form a molecule of hydrogen, H_2, or the diatomic chlorine molecule, Cl_2, is formed.

$$H:H \qquad :\!\ddot{C}l:\ddot{C}l:$$
Nonpolar

But if the atoms are not alike, the shared electron pair will be nearer one atom than the other, as in gaseous hydrogen chloride:

$$H:\ddot{C}l:$$
Polar

Here the chlorine atom, anxious for a complete outer orbit of 8 electrons, exerts a greater pull on the electron of the hydrogen atom than does the hydrogen nucleus. Therefore, it becomes more negative than the hydrogen. However, it is not a chloride ion, for the hydrogen atom has not relinquished its electron nor has the chlorine atom possessed the electron as would be the case in electrovalence. By the same token, the hydrogen atom has become more positive but has not formed an ion with a positive charge. Such molecules as gaseous hydrogen chloride are called *dipoles,* and the bond is called a *polar bond.* These are examples of covalence.

Most of the compounds containing the element carbon (organic compounds) are formed by sharing pairs of electrons. These are very stable bonds, and since a carbon atom can share electrons with other carbon atoms, complex chains and rings of carbon atoms can be built up, forming a great variety of organic compounds. Some carbon atoms share two or three pairs of electrons between 2 adjacent atoms and are therefore linked together by two or three valence bonds. These compounds are called *unsaturated compounds.*

Carbon has 4 electrons in its outermost level, and hydrogen has 1. In the compound methane, CH_4, carbon shares 1 of its electrons with each of the 4 hydrogen atoms, giving it effectively 8 electrons; each hydrogen atom, by sharing 1 electron with the carbon atoms, has effectively 2 electrons. Note that the numbers 2 and 8 are the arrangements in the inert gases helium and neon.

$$H\cdot \qquad \cdot\ddot{C}\cdot \qquad H:\overset{H}{\underset{H}{\ddot{C}}}:H$$
Hydrogen   **Carbon**   **Methane**
atom   **atom**

In the compound ethane, C_2H_6, 2 carbon atoms share a pair of electrons.

$$H:\overset{H}{\underset{H}{\ddot{C}}}:\overset{H}{\underset{H}{\ddot{C}}}:H$$

It can be seen that in covalent bonds a shared pair of electrons is a valence bond. Further, it can be observed that carbon has a valence number of 4 but that it is not + or −4, since electrons have been neither lost nor gained.

coordinate covalence

As a special case of covalence we may have an instance where the shared pair of electrons comes from 1 atom. This is called *coordinate covalence.* For example, in the molecule of ammonia in which nitrogen has valence 5

the atoms of hydrogen are linked to the nitrogen by covalent bonds, the shared pair of electrons being a joint proposition—but note the pair of unshared electrons on the nitrogen atom. Now, when ammonia reacts with hydrogen chloride, this unshared electron pair on the nitrogen atom has a greater attraction for the hydrogen nucleus than does its own electron, which is being shared with the chlorine. Therefore, giving up its own electron, the hydrogen nucleus (H^+) migrates to the nitrogen atom, forming a coordinate covalent bond at the pair of unshared electrons. This gives rise to the ammonium ion, NH_4^+, with 1 positive charge donated by the proton. Naturally, the chlorine atom, now in complete possession of the electron that originally belonged to the hydrogen atom, becomes a chloride ion, Cl^-.

$$H : \overset{H}{\underset{..}{N}} : H + H \cdot \overset{..}{\underset{..}{Cl}} \rightarrow \left[H : \overset{H}{\underset{H}{N}} \cdot H \right]^+ + : \overset{..}{\underset{..}{Cl}} :^-$$

Ammonium chloride, therefore, is a salt.

Dual Valence

Some of the metals form two kinds of ions with different valence numbers. Iron, for instance, exists in two valences, +2 and +3, giving rise to two sets of compounds. No confusion is encountered in the formulas because the valence, of course, is indicated in the formula, $\overset{+2}{Fe}Cl_2$ and $\overset{+3}{Fe}Cl_3$. In naming the compounds, the suffixes -ous and -ic are used to distinguish between the two series, -ous indicating the lower and -ic the higher valence. In the interest of euphony we do not say "ironous" and "ironic" but rather "ferrous" and "ferric." $FeCl_2$, then, is ferrous chloride and $FeCl_3$ is ferric chloride. However, when the common name is euphonious, we use it; e.g., we say "mercurous" and "mercuric." Most of the nonmetals form compounds in which the valence number varies, but when combined with a metal or hydrogen, the valence numbers are fixed.

Groups

As we proceed, we will become more and more conscious of the existence of certain groups of atoms that are very often used together, acting as a single unit. For instance, we notice HNO_3, $AgNO_3$, $Cu(NO_3)_2$, and $Al(NO_3)_3$, each of which contains the group (NO_3^-). Since the NO_3^- group acts as a unit, it is treated the same as the chlorine atom in HCl and therefore must have the valence number of -1. Similarly, the $SO_4^=$ group in H_2SO_4 would be like the oxygen in H_2O and have the valence number of -2. These groups of atoms are ions. Of course it is possible to determine the valence number of the group by obtaining the algebraic sum of the valence numbers of the atoms contained in the group, but this presupposes that the valence number of each of the atoms in the group is known. For instance, in the $SO_4^=$ group the 4 oxygen atoms each contribute 2 negative valences for a total of 8, and the sulfur atom with a valence number of $+6$ provides 6 positive valences. The algebraic sum of -8 and $+6$ is -2, which is the valence number of the group.

Valence of the Middle Element

In compounds containing three different kinds of atoms (ternary compounds), we often want to determine the valence number of the middle element in the formula. It should be obvious that if we know the valence number of any two of the elements, we may calculate that of the third. From the formula of sodium sulfate, glauber's salt, Na_2SO_4, we may calculate the valence of the sulfur if we know the valence number of sodium to be $+1$ and that of oxygen to be -2. First the product of the subscript 4 and the valence number of oxygen, -2, is -8. The product of the subscript 2 and the valence of sodium, $+1$, is $+2$. The difference between these products is 6, which is the valence number of sulfur in this compound.

In potassium permanganate, $KMnO_4$, what is the valence number of manganese? Potassium is $+1$, oxygen is -2, and there are 4 oxygen atoms, making -8. The difference is the valence of manganese, $+7$.

HYDROGEN BONDS

In addition to the bonds holding atoms together, there are attractive forces between molecules. In *polar compounds,* although the molecule itself is electrically neutral, there is an uneven distribution of positive and negative charges within the molecule itself. This means that one end of the molecule is more negative than the other, and the other end is therefore more positive. When such molecules come into close proximity to each other, they tend to orient themselves in space so that the positive end of one molecule is closer to the negative end of the other. Water is such a molecule. The electrons, although shared by the hydrogens and oxygen, are more attracted to the oxygen with its 8 protons than they are to the hydrogen. The oxygen part is the more negative part of the molecule; the hydrogen part, the more positive. Water molecules align themselves so that the oxygen part of one molecule is attracted to a hydrogen part of another molecule. This attractive force is called the *hydrogen bond.* It exists not only between hydrogen and oxygen but also between hydrogen and fluorine and hydrogen and nitrogen. Although these bonds are weak in comparison with valence bonds, they are very important in maintaining the shape in space of such substances as proteins and the nucleic acids, DNA and RNA.

Summary of Valence

a. If valence is the combining power of an element, an uncombined element is exhibiting no combining ability and therefore its valence is zero.
b. The valence of the metals and hydrogen is positive.
c. The nonmetals have a negative valence with metals and hydrogen, but a positive valence when combined with oxygen.
d. The common groups have a negative valence, except the ammonium group (NH_4^+), which is positive. Table 3-1 gives the valence numbers of the common elements and groups.

FORMULAS

A knowledge of valence numbers helps in the writing of formulas; conversely, a knowledge of formulas helps in determining valence numbers. It is to be hoped that the symbols and valences of the common elements will be committed to memory as a help in understanding formulas and equations.

Formulas are of two general types: molecular and graphic.

Molecular or Empirical. This type of formula tells us the elements of which a compound is composed and the number of atoms of each element present.

Water has the formula H_2O. It is composed of the elements hydrogen and oxygen and has 2 atoms of hydrogen and 1 atom of oxygen per molecule.

Hydrogen peroxide has the formula H_2O_2. We see that it also is composed of the elements hydrogen and oxygen, but in this case we have 2 atoms of hydrogen and 2 atoms of oxygen per molecule. As one would naturally expect, water and hydrogen peroxide are two entirely different substances with totally different properties.

Graphic or Structural. In addition to the information given by the molecular formulas, graphic formulas attempt to show, by diagraming the valence bonds, how the atoms are arranged in the molecule.

Let us consider H_2O and H_2O_2. We know that hydrogen has a valence of +1 and oxygen a valence of −2. Representing each valence bond by a dash, we have

Water, H_2O:
O—H
|
H

Hydrogen peroxide, H_2O_2:
O—O
| |
H H

Observe that each oxygen atom is attached by 2 bonds and each hydrogen atom by 1.

Writing the formulas for methane and

Atomic Structure and Valence

Table 3-1 VALENCE NUMBERS OF THE COMMON ELEMENTS AND GROUPS

Element or Group	Valence	Element or Group	Valence
Hydrogen (H)	+1	Chlorine (Cl)	−1
Potassium (K)	+1	Bromine (Br)	−1
Sodium (Na)	+1	Iodine (I)	−1
Silver (Ag)	+1	Oxygen (O)	−2
Calcium (Ca)	+2	Sulfur (S)	−2, +4, +6
Barium (Ba)	+2	Phosphorus (P)	−3, +3, +5
Magnesium (Mg)	+2	Nitrogen (N)	−3, +1 to +5
Zinc (Zn)	+2	Carbon (C)	4
Lead (Pb)	+2	Hydroxide (OH)	−1
Cuprous (Cu)	+1	Bicarbonate (HCO_3)	−1
Cupric (Cu)	+2	Nitrite (NO_2)	−1
Mercurous (Hg)	+1	Nitrate (NO_3)	−1
Mercuric (Hg)	+2	Chlorate (ClO_3)	−1
Ferrous (Fe)	+2	Sulfite (SO_3)	−2
Ferric (Fe)	+3	Sulfate (SO_4)	−2
Stannous (Sn)	+2	Carbonate (CO_3)	−2
Stannic (Sn)	+4	Phosphite (PO_3)	−3
Aluminum (Al)	+3	Phosphate (PO_4)	−3
Ammonium (NH_4)	+1		

ethane, we have a valence bond for each pair of electrons.

$$\underset{\text{Methane}}{\text{H}-\underset{\underset{\text{H}}{|}}{\overset{\overset{\text{H}}{|}}{\text{C}}}-\text{H}} \qquad \underset{\text{Ethane}}{\text{H}-\underset{\underset{\text{H}}{|}}{\overset{\overset{\text{H}}{|}}{\text{C}}}-\underset{\underset{\text{H}}{|}}{\overset{\overset{\text{H}}{|}}{\text{C}}}-\text{H}}$$

The formula C_2H_4 indicates the sharing of more than 1 valence bond between atoms.

$$\text{H}-\overset{\overset{\text{H}}{|}}{\text{C}}=\overset{\overset{\text{H}}{|}}{\text{C}}-\text{H}$$

Each carbon atom has a valence of 4 and each hydrogen atom of 1. This is the unsaturated compound ethylene, used as a general anesthetic.

In the compounds sodium carbonate, Na_2CO_3, washing soda, and sodium bicarbonate, $NaHCO_3$, baking soda, we would have the valence bonds distributed as designated by the structural formulas

$$\underset{\text{Sodium carbonate}}{\text{Na}-\text{O}-\underset{\underset{\text{Na}}{|}}{\overset{\overset{}{\|}}{\text{C}}}=\text{O}} \qquad \underset{\text{Sodium bicarbonate}}{\text{Na}-\text{O}-\underset{\underset{\text{H}}{|}}{\overset{\overset{}{\|}}{\text{C}}}=\text{O}}$$

Loss of Na^+ from sodium bicarbonate would leave bicarbonate ion $(HCO_3)^-$

$$-\text{O}-\underset{\underset{\text{H}}{|}}{\overset{\overset{\|}{\text{O}}}{\text{C}}}=\text{O}$$

valence −1.

Additional loss of H^+ would form carbonate ion $(CO_3)^=$

$$-\text{O}-\underset{\underset{}{|}}{\overset{\overset{\|}{\text{O}}}{\text{C}}}=\text{O}$$

valence −2.

Study Exercises

1. Define atom, electron, proton, atomic weight, atomic number, planetary electrons, isotope, neutron, valence, electrovalence, ion, cation, anion, covalence, valence number, molecular formula, graphic formula.
2. Indicate the atomic structure and the potential valence of each of the following elements as shown in the example:

Element	At. no.	At. wt.	Structure	Valence
Sodium	11	23	(12n, 11p) 2)8)1)	+1
Nitrogen	7	14		
Sulfur	16	32		
Calcium	20	40		
Oxygen	8	16		
Chlorine	17	35		

3. Which of the following are isotopes, as indicated by the composition given?

At. wt.	Protons	Neutrons	Isotopes
35	17	18	
38	18	20	
36	17	19	
37	17	20	
9	4	5	
10	5	5	
11	5	6	
12	6	6	

4. Draw the graphic formula for each of the following compounds: methane, CH_4; hydrogen chloride, HCl; carbonic acid, H_2CO_3; ethane, C_2H_6; nitric acid, HNO_3.
5. An advertisement states that a cooking oil is unsaturated. What does this mean?
6. Indicate the valence number and appropriate sign of the elements in the following binary compounds: HCl, H_2O, AgCl, NH_3, CuCl, H_2S, HBr, FeO, $RaCl_2$, $BaBr_2$.
7. What is the valence number of the middle element in each of the following formulas? $K_2Cr_2O_7$, $BaCrO_4$, KIO_3, HClO, $NaMnO_4$, $Na_2S_2O_3$, H_3BO_3, $KClO_4$.
8. Write the correct formula for each of the following compounds: mercuric chloride, sodium sulfide, calcium hydroxide, ammonium carbonate, and cupric sulfate.
9. Give the names of the compounds, whose formulas are: FeO, Fe_2O_3, MgO, $Mg(OH)_2$, $NaHCO_3$, K_2SO_4, $CaCl_2$, Na_3PO_4, Al_2O_3, NH_4Cl, CuO, Cu_2O, $MgSO_4$, $NaNO_2$.
10. The following compounds are used in medicine. Write their chemical formulas. Aluminum hydroxide, aluminum phosphate, barium sulfate, barium sulfide, barium sulfite, calcium phosphate, potassium permanganate, radium sulfate, sodium bicarbonate, sodium nitrite, sodium phosphate, sodium iodide, sodium bromide, sodium sulfate.

chapter four STATES OF MATTER

Matter can exist as solid, liquid, or gas. Most substances can be changed from one state to another by the addition or removal of heat or by changes in pressure.

Solids have a definite shape and a definite volume.

Ice exists in the form of ice crystals, which have definite shapes. If we melt 1 Gm ice, we will obtain 1 Gm water. The crystalline shape is, of course, lost; the water flows and assumes the shape of the container in which it is held. The volume of water is definite—slightly less than the volume of ice because water expands on freezing. Liquids have definite volume but no shape of their own.

Heating the liquid water can change the material to water vapor, a gas. How much water vapor have we? The amount of matter remains the same, 1 Gm, but the gas will be distributed throughout the whole container. If the container has a volume of 5 ml, the gas will occupy 5 ml. If the container has a volume of 10 ml, the gas will spread out to occupy the 10 ml of space. Gases have no definite shape or volume but expand to fill the container.

Both gases and liquids can flow and are therefore called *fluids*. Some substances that appear to be solids are actually fluids. Glass, for example, can flow very, very slowly and is properly described as a *supercooled liquid*.

THE KINETIC MOLECULAR THEORY

The *kinetic molecular theory* was proposed in an attempt to explain the observed behavior of gases. It can be modified to account for the behavior of liquids and solids as well. According to this theory, the following propositions are true:

a. Matter is made of molecules.
b. These molecules are discrete, separate particles. As uncounted grains of sand make a beach, so like numbers of molecules make a bit of matter.
c. Molecules possess energy. They are in constant motion, constantly vibrating.
d. The speed with which the molecules move is related to temperature. If the temperature is increased, the speed of the molecules increases. If the temperature is reduced, the molecules slow down.
e. Forces of attraction exist among molecules. These attractive forces exert greater influences on slowly moving molecules than on rapidly moving ones.[1]

---

[1] This principle may require a little thought. Consider an analogy: if a young man and a young lady from the same neighborhood walk to work in the same direction each day and if they both allow ample time for the journey, there may be opportunity for meeting and even conversation; perhaps eventually they would walk arm in arm. The attractive forces have a chance to work. If, however, they are both habitually late and run the distance, they may pass each other with a very brief greeting and be gone down the road.

f. Molecules, in their motion, collide with each other and with the walls of the container and bounce away with no loss of energy or speed. This accounts for the ability of gas molecules to diffuse in all directions.
g. Pressure is due to bombardment of molecules.

CHANGE OF STATE

solid to liquid (melting)

In the solid state, molecules vibrate around fixed positions and are relatively close together. Heating these molecules gives them greater energy and greater speed so that they overcome the attractive forces holding them in the crystalline position. Therefore, they can move about more freely. To break these bonds requires a final spurt of energy, which is called the *heat of fusion*. The heat of fusion of a substance is the number of calories of heat absorbed by 1 Gm of that substance in changing from the solid to the liquid state without changing its temperature.

It requires 79.7 cal heat to change 1 Gm ice at 0°C to 1 Gm water at 0°C. Roughly, the heat of fusion of water is 80 cal per Gm. If we had 10 Gm ice at 0°C and supplied 800 cal, the ice would change to 10 Gm water and the temperature would still be 0°C because only enough heat has been supplied to melt the ice. In an ice and water mixture, the addition of heat first causes all the ice to melt before the temperature can rise above 0°C. We see, then, that as ice melts, heat is absorbed.

When ice packs or ice bags are used to reduce inflammation, allay infection, lower temperature, etc., their effectiveness depends not so much on the application of a cold object as on the absorption of heat from body tissues. In deciding when to refill an ice bag, the nurse remembers that as long as any ice remains the temperature of the bag is 0°C, but once all the ice has melted the temperature of the water will rise. The melting point is the temperature at which a solid changes to a liquid (at 760 mm pressure).

liquid to solid (freezing or crystallization)

If we consider a liquid changing to a solid, we have the reverse of the above situation. The temperature at which a liquid changes to solid is the freezing point. The freezing point of water is 0°C (at 760 mm pressure). The melting point and freezing point are the same temperature. When a liquid is cooled, the speed of the molecules decreases; when the freezing point is reached, 79.7 cal heat is liberated for each gram of water that changes to ice. The heat absorbed in melting the solid is given off when the liquid freezes. Since the heat is temporarily hidden in the liquid, it is sometimes referred to as *latent heat*.

liquid to gas (evaporation)

To change a liquid to a gas, the forces of attraction between the molecules must be overcome. To do this, heat is required. This is called *heat of vaporization*, which is defined as the number of calories of heat absorbed by 1 Gm of a substance in changing from the liquid to the vapor state without any change in its temperature.

Water has an abnormally high heat of vaporization (540 cal per Gm). There are many applications of this high value for water. Much of the heat of the body is dissipated through the process of evaporation. This is demonstrated by the additional cooling effect of a sponge bath even though the bath water may be tepid. When giving a patient a bed bath, the nurse bathes a small area of the skin at a time to prevent too rapid a loss of body heat by evaporation. With debilitated patients, the nurse is particularly careful to prevent drafts while bathing the patient because so much heat may be lost

States of Matter

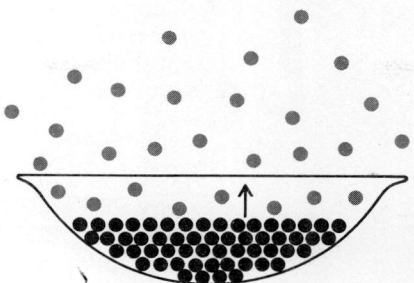

Fig. 4-1 Evaporation.

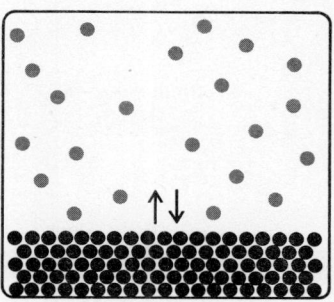

Fig. 4-2 Evaporation versus condensation.

from the body by the increased rate of evaporation in a moving stream of air that the patient may feel chilled. In the tropics water is stored in canvas bags or in porous earthenware jugs in the shade. The evaporation of the water that oozes through the container keeps the contents cool. Further, a wet sheet hung in a shady, breezy doorway cools the air as it passes by.

We may carry this idea of the cooling effect of evaporation still further. The more rapidly a liquid evaporates, the greater is the amount of heat required in the process; i.e., the greater will be the cooling effect. An alcohol bath is more efficient in reducing the temperature of a fevered patient than bathing with water. Ether spilled on the hands is noticeably cooler than alcohol. Ethyl chloride evaporates so rapidly, absorbing so much heat from the tissues, that it is used as a local anesthetic. Ammonia is normally a gas, and in the liquid state it would evaporate very rapidly; this is one of the reasons why liquid ammonia can be used as an artificial refrigerant.

gas to liquid (condensation)

Condensation is the reverse process of evaporation. The heat absorbed by a liquid changing to a vapor is released when the vapor changes to a liquid. Therefore, when 1 Gm water vapor changes to a liquid, it liberates 540 cal to the surroundings.

Boiling

evaporation versus condensation

Let us develop a picture of what is happening during evaporation. Consider an open dish containing a liquid as indicated in Fig. 4-1. Molecules of the liquid are moving about in random fashion, hitting each other and bouncing off, at an average rate that depends upon the temperature. Some of the molecules on the surface move out into the space above the water, but they do not possess enough energy to overcome the forces of molecular attraction, so they fall back into the liquid. Other molecules do succeed in getting beyond the field of molecular attraction, and they become free molecules in the atmosphere. Thus it is that water or any liquid evaporates spontaneously, a process familiar to everyone.

Now let us change the picture as shown in Fig. 4-2. Here we have a closed container instead of an open dish. This time the molecules in the gaseous state are not free as before. They are still moving as they were in the liquid state, but through greater distances. They have the same kinetic energy as those in the liquid state, for the temperature is the same. These molecules bombard the walls of the container and collide with each other. Some of them even go back into the liquid state (condense). The number of them returning to the liquid state increases

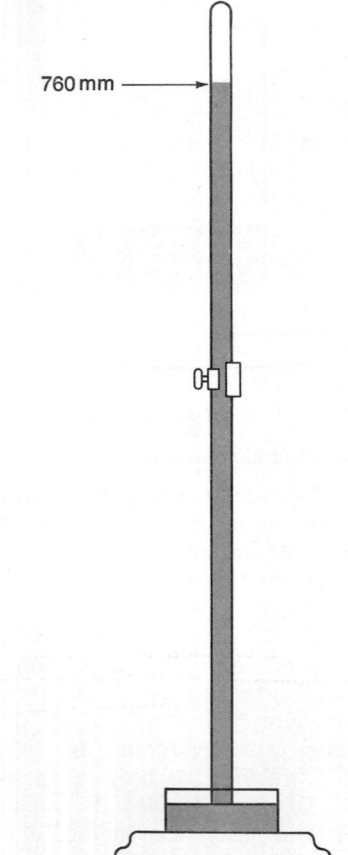

Fig. 4-3 A mercurial barometer.

as the concentration of molecules in the gaseous state increases, and eventually the rate of evaporation becomes equal to the rate of condensation at a constant temperature. We then say that we have reached a *dynamic equilibrium.* That is,

$$H_2O \text{ (liquid)} \underset{\text{condensation}}{\overset{\text{evaporation}}{\rightleftharpoons}} H_2O \text{ (vapor)}$$

As the temperature of a liquid is raised, the rate of evaporation per unit surface area increases; it is common knowledge that water evaporates faster in an open pan when the temperature is raised. In the closed container the same thing happens, but the rate of condensation also increases and again at a new temperature an equilibrium is established.

the barometer

A barometer is an instrument used for measuring the pressure of the atmosphere. One type of mercurial barometer consists of a glass tube, about a meter in length and closed at one end, filled completely with mercury and then inverted in a dish of mercury (Fig. 4-3). At sea level the mercury falls in the tube until its surface is about 76 cm above the surface of the mercury in the dish. The only thing above the mercury in the closed end of the tube is a very small amount of mercury vapor, which we will ignore. The pressure of the atmosphere on the surface of the mercury in the dish is supporting this column of mercury 76 cm high. That pressure is equal to the weight of the column on an area equal to the cross section of the tube. If the tube has a cross-sectional area of 1 sq cm and the column is 76 cm tall, the volume of the mercury is 76 cu cm. If the density of mercury is 13.6 Gm per cu cm, the column will weigh 1,033 Gm. The pressure on the surface of the mercury will be 1,033 Gm per sq cm.

This is the standard barometric pressure (76 cm, 760 mm, or 30 in. mercury). It was chosen as the standard because it is the average atmospheric pressure at sea level. The pressure of the atmosphere varies from day to day in one spot and, of course, will vary with the altitude. As we go higher in our atmosphere the concentration of the molecules becomes lower, and there are fewer bombardments on the surface of the mercury, which means lowered pressure.

what is boiling?

Boiling is the formation of bubbles of vapor *within* the liquid. These bubbles, being lighter than the liquid, naturally rise and burst on the surface. Boiling takes place when the

States of Matter

vapor pressure of the liquid becomes equal to the pressure of the surrounding atmosphere (Fig. 4–4). The temperature at which these two pressures become equal is known as the *boiling point* of the liquid at the specified pressure. In Fig. 4–5 a bubble of vapor is shown within the liquid. The pressure of the atmosphere is transmitted to this bubble through the liquid in all directions. In order for the bubble to exist, there must be pressure from within to offset the external pressure. This internal pressure is the vapor pressure of the liquid. So long as the bubble remains inflated, by definition the liquid will be boiling.

variation in the boiling point

Water boils at 100°C under standard conditions, for at that temperature its vapor pressure is 760 mm. As the barometric pressure changes, so will the boiling point. On top of a high mountain, water boils at a low temperature. At high altitudes, therefore, cooking by boiling is slow, if not impossible.

Water boils at an elevated temperature if the pressure of the gases above it is increased. In a pressure cooker or autoclave, steam pressure is produced within the container to create artificially an atmospheric pressure greater than normal. In this way

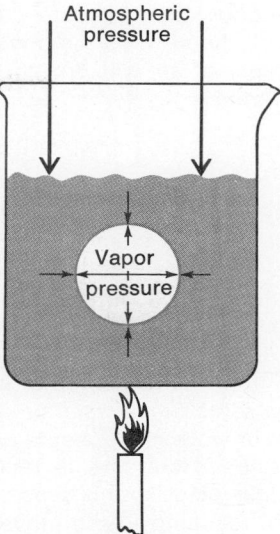

Fig. 4–5 Boiling.

temperatures as high as 120°C are easily attained. The increased temperature in the autoclave makes it a more effective sterilizer than an open steam sterilizer. Microorganisms are literally cooked and, therefore, killed more quickly at the higher temperature. The increased pressure does not kill the microorganisms. The increase in pressure causes an increase in the temperature of the steam; it is this higher temperatue that is effective.

Autoclaves are made with thick metal jackets to withstand the high pressure. They are equipped with safety valves to release the steam, should the pressure build up beyond the set point. If the safety valve becomes clogged and the steam pressure continues to build up, what might happen? Housewives who are careless in cleaning their pressure cookers or who overload them have discovered to their sorrow the answer to this question. Usually the safety valve or safety plug blows out. If this fails, the lid may blow off and the tremendous pressure exerted may distribute the food over the environment.

In packing the autoclave, objects should be placed in such a way that the steam can

Fig. 4–4 The vapor-pressure curve of water.

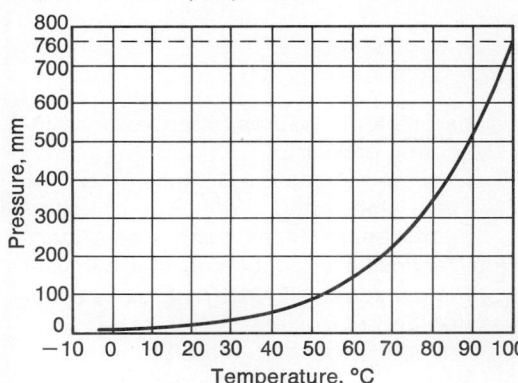

circulate around them. Before starting the autoclave, all the air must be driven out because pockets of air will prevent complete steam penetration.

putting change of state to work

Work in physics has a special meaning. Work involves motion or a change in velocity (acceleration). To cause motion or a change in velocity, energy must be transferred by the application of a force. *Force is defined as the cause of motion or acceleration of bodies.* It can be thought of as a push or a pull.

Work is force times distance. The greatest amount of work most people do involves climbing stairs. Here one is raising one's weight so many feet. In this sense, muscular contraction, resulting in the movement of a body part, is work, while sitting still and thinking is not. Taking notes is work: one *pushes* a pencil (force) across a page (distance).

As liquids change to gases, they expand, greatly increasing in volume. These expanding gases exert a force that can move a piston, a paddle, a wheel, etc. In the gasoline engine in the automobile, the gasoline air mixture changes, on ignition, to gases, primarily carbon dioxide, water vapor, and, unfortunately, some carbon monoxide. These gases push the pistons which are connected to the drive shaft. The drive shaft, a gear system, turns the wheels.

Whenever work is performed, energy is involved. In the gasoline engine, the source of this energy is the combustion of the gasoline. The ignition is so rapid, however, that there is incomplete combustion with the formation of carbon monoxide, a serious air pollutant. An indirect method of transforming the energy of fuels into work is to burn the fuel to produce steam which in turn can drive a turbine or engine which does the work. When automobiles were first developed, one type was driven by steam and was called the "Stanley Steamer." Although it may sound comical, it was a good system. If the public is truly interested in clean air, we may live to see steam-driven automobiles.

Work is done in the body, but not by change of state. The source of energy for work done by the body and in the body is the oxidation of food. Through this energy, liberated by chemical reactions, muscles contract and pull bones and tissues, food is moved along the digestive tract by contraction and relaxation of the walls of the gut, and contractions and relaxations of the heart muscles and arterial walls push or force blood through the body. In terms of physics, the body is working night and day.

In computing the amount of work done in metric units, force is expressed in dynes, distance in centimeters, and work in ergs. These are very small units. A more useful unit is the joule, which is 10,000,000 ergs.

Study Exercises

1. What are the characteristics of the solid, the liquid, and the gaseous states of matter?
2. List six principles of the kinetic molecular theory.
3. Define boiling, melting, freezing, evaporation, condensation, boiling point, pressure, vapor pressure, force, and work.
4. What is a barometer? Explain how the barometer measures variations in atmospheric pressure.
5. Denver, Colorado, is known as the "mile-high city." Why would instruments boiled in an open steam sterilizer need a longer boiling time in Denver than at sea level? Would the same situation prevail in Lima, Peru?
6. Water in a closed vessel cannot boil. Why?

States of Matter

7. Why are steam burns more severe than hot-water burns, both steam and water being at the same temperature?
8. Why is there a high degree of physical discomfort on warm, humid days?
9. Explain why an autoclave is a more efficient and effective sterilizer than an open steam sterilizer.
10. Foods cooked in pressure cookers require a shorter cooking time than foods cooked in open pots. Why is this true?

chapter five PHYSICAL AND CHEMICAL CHANGE

In a previous chapter it was stated that chemistry is the study of substances and the changes that they undergo. In order that we may recognize these changes when they occur, we must have some way of identifying substances. This is done by making use of their properties. To describe a substance completely two classes of properties, physical and chemical, must be used.

PHYSICAL PROPERTIES

Some of these properties, such as size and shape, are accidental and serve very little in the identification of the substance itself. More specific physical properties are *color, odor, taste, solubility,* and *density*. If the substance is a solid, its melting point is specific; if a liquid, its boiling point and freezing point are important; and if a gas, its critical temperature[1] is significant for that tells us how readily the gas may be liquefied. This in turn indicates which gases may not be used, for example, in artificial refrigeration.

Little trouble is encountered in the identification of the color of a substance. The solubility of a substance refers to its solubility in water unless some other solvent is specified. Its degree of solubility is expressed

[1] The critical temperature of a gas may be described as the temperature below which it must be cooled before it can be liquefied by increase of pressure.

by the terms "freely soluble," "slightly soluble," or "insoluble."

The other three of the five specific properties need some further consideration.

Human beings have a very delicate apparatus for detecting odors and for distinguishing between them. But we have an extremely limited vocabulary with which to describe what we smell. However, we are able to describe an odor in such a way that others understand even though we may be describing the feeling in the nose and throat: for example, we speak of a sharp, penetrating, or choking odor. Odors have been classified on the basis of the stimulation of only four kinds of nerves, each stimulated to relatively different degrees by any particular odor.

Taste is the property that is the most confusing to the uninformed. There are only four true tastes—sweetness, saltiness, sourness, and bitterness. Everything else that we enjoy along that line is odor. For instance, the taste of coffee is bitter; what is enjoyed about coffee is its odor. We should be very conscious of this from our national advertising today. Such expressions as "vacuum packed," "dated coffee," and "roaster to you" mean only one thing and that is the effort of the merchant to deliver to the customer a pound of coffee odor. To convince the skeptic of this explanation of taste, let him think back to his last bad head cold. Life was drab for he "could not taste a thing." Actually, one can taste as usual, but the nasal passages are clogged with mucus so that the odor of the food cannot penetrate to the nose. With the clearing of these passages, life takes on a rosy hue once more.

The taste buds are stimulated only by material in solution. Insoluble substances such as chalk and charcoal must therefore be tasteless. Strictly speaking, then, in predicting the taste of a substance we must think of what is present in solution.

Density is the relationship of the mass or weight of a substance or object to its volume. It is defined as *mass per unit volume* and is expressed by the equation.

$$\text{Density} = \frac{\text{mass or weight}}{\text{volume}} \quad \text{or} \quad D = \frac{m}{V}$$

If 10 ml of a substance weighs 20 Gm,

$$D = \frac{20}{10} = 2 \text{ Gm/ml}$$

Density is always expressed in terms of weight and volume. At 4°C water has a density of 1 Gm per ml. This means that 1 ml water weighs 1 Gm and 10 ml weighs 10 Gm, or 10 Gm would occupy a volume of 10 ml.

The density of mercury is 13.6 Gm per ml. Can 30 Gm mercury be contained in a vial whose volume is 10 ml?

$$D = \frac{m}{V}$$

$$13.6 = \frac{30 \text{ Gm}}{x \text{ ml}}$$

$$13.6x = 30$$

$$x = \frac{30}{13.6} = 2.2 \text{ ml}$$

Obviously yes, 30 Gm mercury would not even fill one-third of the vial's capacity.

Specific gravity is a figure resulting from the comparison of the density of a substance with the density of water or air. Liquids are compared to water, and gases are compared to air. The density of water is taken as 1.000 Gm per ml. Specific gravity readings have no units because they are a comparison or ratio. If we say that little Mary is 3 ft tall and her father is 6 ft tall, the ratio of these heights is 2, not 2 ft. Specific gravity is frequently used to describe the properties of body fluids. It is abbreviated sp. g. The specific gravity of urine, for example, usually falls within the range 1.015 to 1.030. A reading of 1.002 would indicate that the sample was almost like water, which would probably mean that the kidneys were not concentrating the urine. Such a condition occurs in diabetes insipidus.

CHEMICAL PROPERTIES

These properties cannot be determined by inspection. They are the characteristics of a substance that are brought out when it acts, i.e., enters into some chemical reaction. Obviously, no established set of chemical properties can be listed, for these properties depend upon a number of variables, such as whether the substance is an element or a compound or what its physical state is. A few chemical properties might be suggested. Is the substance combustible, or does it support combustion? Is it stable, or is it readily decomposed? Is it active or inactive? Does it react with water, and if so, what is formed? These questions may act as a guide in the future.

PHYSICAL CHANGE

It is convenient to classify all changes as either physical or chemical. In a physical change the accidental properties of the substance involved may be altered, or there may be a change in the physical state. However, there never is a change in the material composing the substance. For instance, the breaking of a pane of glass, by dropping it, certainly is a change but only in the size and shape of the pieces; the composition of the glass is the same. If we heat water and convert it into steam, there is a change in the physical state and the steam has different physical properties, but there has been no

Physical and Chemical Change

change in material for both steam and water have the same chemical composition. Further, if the steam is cooled to room temperature again, it condenses to liquid water. Physical changes have been called *temporary* changes because if we revert to the original conditions of temperature and pressure, the original properties are evident. All physical changes are not reversible, however, for by no stretch of the imagination may we expect the pieces of glass to revert to a pane when we pick them up.

CHEMICAL CHANGE

In a chemical change there is always a change of material; that is, the products of the change are always different from the reactants. There are many ways that we recognize a chemical change taking place; for example, a gas may be formed, heat may be evolved, or there may be an alteration in the color, odor, taste, or solubility. Other changes are possible, and it is not necessary that all of them occur in any one reaction. The result of heating cane sugar illustrates a chemical change very well. As it is heated in a test tube, the sugar melts and soon turns yellow, then brown. This is called caramelizing and is a common practice in cooking. As the heating continues, the color gets darker. Gases come off and may be ignited at the mouth of the tube. Finally only a black mass is left, which is one of the purest forms of carbon, sugar charcoal. The alterations that have taken place demonstrate that the change was chemical.

Types of Chemical Change

It is convenient to classify a chemical change under one of the following four types of change.

Combination or Synthesis. Two or more simple substances combine to form a more complex substance.

Hydrogen + oxygen → water
Carbon + oxygen → carbon dioxide
Carbon dioxide + water → carbonic acid

Decomposition or Analysis. A complex substance is broken down into simpler substances.

Water → hydrogen + oxygen
Carbonic acid → carbon dioxide + water

Replacement or Displacement. An element and a compound react so that the free element replaces an element in the compound.

Magnesium + hydrochloric acid →
 hydrogen + magnesium chloride
Potassium iodide + chlorine →
 iodine + potassium chloride

Double Replacement or Double Displacement. A reaction occurs between two compounds in which there is an exchange of the metals in the two compounds involved.

Sodium chloride + silver nitrate →
 silver chloride + sodium nitrate

Equations

Chemical changes may be expressed in the form of equations. An equation may be considered as an abbreviated statement of a chemical fact. Equations tell us not only the starting materials or reactants and the products but also the amounts of these substances involved in the change.

Before the discovery of atomic fission (as in the atomic bomb) and atomic fusion (as in the hydrogen bomb), the law of conservation of matter held without exception. This law states that we can neither make nor destroy matter. In both atomic fission and atomic fusion some matter is changed into tremendous amounts of energy, but for the usual chemical changes, the law of conservation of matter holds true. Equations describing chemical changes must balance; i.e., we must have the same number of each kind of atom at the end of the reaction as in the beginning.

Writing equations for the above reactions, we must know the formulas involved and we must be certain that the equations balance.

$$\text{Hydrogen} + \text{oxygen} \rightarrow \text{water}$$
$$H_2 + O_2 \rightarrow H_2O$$

The molecules of hydrogen and oxygen are diatomic, which is indicated by the subscript 2. If we start with 2 atoms of oxygen, we must account for 2 atoms of oxygen in our product. This would mean 2 molecules of water, $H_2O + H_2O$. We cannot write H_2O_2 because that represents a different compound, hydrogen peroxide. Our information so far would be written

$$H_2 + O_2 \rightarrow H_2O + H_2O$$

Instead of writing the formula for water twice, we can indicate 2 molecules of water by placing a 2 before the formula: $2H_2O$.

We now have 4 atoms of hydrogen in our product and must account for 4 atoms on the left-hand side of the equation. Since hydrogen is diatomic, this means 2 molecules of H_2.

or
$$H_2 + H_2 + O_2 \rightarrow H_2O + H_2O$$
$$2H_2 + O_2 \rightarrow 2H_2O$$

Checking through, we see that on the left-hand side of the equation we have 4 hydrogen and 2 oxygen atoms and on the right-hand side, 4 hydrogen and 2 oxygen atoms; the equation is balanced.

$$\text{Carbon} + \text{oxygen} \rightarrow \text{carbon dioxide}$$
$$C + O_2 \rightarrow CO_2$$

The equation is balanced as it stands.

$$\text{Carbon dioxide} + \text{water} \rightarrow \text{carbonic acid}$$
$$CO_2 + H_2O \rightarrow H_2CO_3$$

This one is also balanced.

$$\text{Water} \rightarrow \text{hydrogen} + \text{oxygen}$$
$$H_2O \rightarrow H_2 \uparrow + O_2 \uparrow$$

This is the reverse of the first equation.

If we have 2 atoms of oxygen in the product, we must have 2 molecules of water to supply them.

$$H_2O + H_2O \rightarrow H_2 + O_2$$

But now our 2 molecules of water give us 4 hydrogen atoms. Therefore,

$$H_2O + H_2O \rightarrow H_2 + H_2 + O_2$$
or
$$2H_2O \rightarrow 2H_2 \uparrow + O_2 \uparrow$$

The upward arrows indicate the formation of gases.

Carbonic acid breaks up to give us carbon dioxide and water.

$$H_2CO_3 \rightarrow CO_2 \uparrow + H_2O$$

Count up the atoms. This equation is balanced.

Magnesium reacts with hydrochloric acid to form hydrogen gas and the salt magnesium chloride

$$Mg + HCl \rightarrow H_2 + MgCl_2$$

To supply 2 hydrogen and 2 chlorine atoms in the product we will need 2 molecules of HCl.

$$Mg + HCl + HCl \rightarrow H_2 + MgCl_2$$
$$Mg + 2HCl \rightarrow H_2 + MgCl_2$$

$$\text{Potassium iodide} + \text{chlorine} \rightarrow \text{iodine}$$
$$KI + Cl_2 \rightarrow I_2$$
$$+ \text{potassium chloride}$$
$$KCl$$

To supply the 2 iodine atoms, we need 2 molecules of KI.

$$KI + KI + Cl_2 \rightarrow I_2 + KCl$$

Now we can team up the 2 potassium and 2 chlorine atoms into 2 molecules of potassium chloride

$$KI + KI + Cl_2 \rightarrow I_2 + KCl + KCl$$
or
$$2KI + Cl_2 \rightarrow I_2 + 2KCl$$

The eyes of newborn infants are rinsed with silver nitrate to destroy any gonococcus organisms that might be present. The eyes are then immediately flushed with a solution of salt water to stop any further action of the silver nitrate. In words, this reaction reads

Silver nitrate + **sodium chloride** →
  silver chloride + **sodium nitrate**

The equation says the same thing.

$$AgNO_3 + NaCl \rightarrow AgCl \downarrow + NaNO_3$$

The silver chloride formed is insoluble,

Physical and Chemical Change

Table 5-1 OXIDATION AND REDUCTION

Oxidation	Reduction
Loss of electrons	Gain of electrons
Increase in valence number	Decrease in valence number
Addition of oxygen	Loss of oxygen
Loss of hydrogen	Addition of hydrogen

as indicated in the equation by the downward arrow.

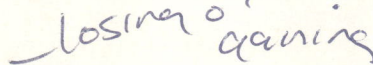
losing or gaining

Oxidation-Reduction or Redox

In the first three types of chemical changes mentioned above there is a shifting of electrons among atoms, resulting in a change of valence. Some atoms lose electrons; this is called *oxidation*. Some atoms gain electrons; this is called *reduction*. Both oxidation and reduction proceed simultaneously. Some atom loses, and some atom gains. Let us take the replacement reaction as an example.

$$Mg + 2HCl \rightarrow MgCl_2 + H_2 \uparrow$$

Free elements have not combined with other atoms, have neither lost nor gained electrons, and have zero valence. Magnesium has a valence of 0. Hydrogen in HCl has a valence of +1 and chlorine has a valence of −1. On the right-hand side of the equation magnesium has a valence of 2; chlorine, of −1; and hydrogen, now free, of 0. Since the valence of chlorine has not changed, it has simply served as a spectator and we can ignore it in this explanation.

$$Mg^0 \rightarrow Mg^{++}$$

Magnesium has atomic number 12: 12 positive charges (protons) and 12 negative charges (electrons). The loss of 2 electrons changes it from 0 valence to +2.

$$Mg^0 - 2e^- \rightarrow Mg^{++} \quad \text{oxidation}$$

Hydrogen has a valence here of +1.

Each hydrogen atom can pick up one electron from magnesium to form hydrogen with no charge.

$$H^+ + e^- \rightarrow H^0 \quad \text{reduction}$$

Putting these two reactions together,

$$Mg^0 - 2e^- \rightarrow Mg^{++}$$
$$2H^+ + 2e^- \rightarrow H_2^0$$

The number of electrons lost and gained is the same.

Oxidation and reduction can also be defined in other ways. As we see here, oxidation involves a mathematical increase in valence number and reduction, a mathematical decrease in valence number. Oxidation can also mean the addition of oxygen or the loss of hydrogen, and reduction can mean the loss of oxygen or the addition of hydrogen.

Substances that cause oxidation are called *oxidizing agents* or *oxidants*. Those that cause reduction are termed *reducing agents* or *reductants*. When an oxidation-reduction reaction occurs, one substance is oxidized while another is reduced. The substance that is oxidized is the reducing agent and the material that is reduced is the oxidizing agent. To put it another way, the reducing agent is oxidized and the oxidizing agent is reduced. In the reaction

$$2Ca + O_2 \rightarrow 2CaO$$

the oxidizing agent, oxygen, gains 2 electrons and is reduced, while calcium loses 2 electrons and is oxidized. Calcium is the reducing agent.

Many chemical reactions in metabolism are oxidation-reduction reactions. The oxidation of carbohydrates, fats, and amino acids in the tissues of the body are examples. Some commonly used antiseptics and disinfectants are effective because they cause oxidation-reduction reactions, which prevent the normal reactions of microorganisms, and the organisms die or their growth is greatly inhibited.

Potassium permanganate, $KMnO_4$, a deep-purple crystalline compound, liberates oxygen when in contact with organic matter. $KMnO_4$ is an oxidizing agent.

Hydrogen peroxide, H_2O_2, is used dissolved in water. In contact with tissues, bubbles of oxygen form, causing oxidation of the material present.

Some other oxidants used as antiseptics are sodium perborate, zinc peroxide, and sodium hypochlorite.

Agencies and Conditions Causing a Chemical Change

Heat is the most common form of energy bringing about a chemical change. Its action is illustrated by the heating of sugar, described above.

Electricity is another form of energy causing chemical changes. For instance, if a direct current is passed through water containing a small amount of sulfuric acid, hydrogen will appear at the negative electrode, the cathode, and oxygen at the positive electrode, the anode. Further, the volume of the hydrogen will be twice that of the oxygen, the same proportion that we find in water. This is an example of decomposition by electricity and is called *electrolysis.*

Light may cause a chemical change. A mixture of hydrogen and chlorine in equal volumes reacts so slowly in the dark that the action could be considered negligible. However, if the mixture is placed in sunlight or a beam of light is thrown on it, a violent explosion occurs. If the reaction is exothermic, i.e., one that evolves heat, the action of the light seems to be merely catalytic, but if the reaction is endothermic, i.e., absorbs heat, the light energy is stored up in the substances formed. Photosynthesis, the union of carbon dioxide and water in the presence of chlorophyll to form a carbohydrate, illustrates the storage of energy from the sun in these products. The most familiar action caused by light is the decomposition of silver bromide into metallic silver and bromine in photography.

Mechanical energy may bring about a chemical change. If we grind together crystals of mercuric chloride and potassium iodide, the red color of mercuric iodide, one of the products, soon develops. Incidentally, this experiment illustrates the fact that the speed of the reaction is affected by the size of the particles. If we mix solutions of these two salts, the red mercuric iodide is produced instantaneously. Here the substances are in their finest state of subdivision and the action is most rapid. In this connection also we might mention such mixtures as baking powder and effervescent medicinals. They remain unreactive until water is added. Steps must be taken, therefore, to keep them dry until the action is desired.

Catalysts are substances that affect the speed of a reaction but are not changed in the reaction. The catalyst does not cause a reaction but speeds it up or makes it possible for the reaction to take place at a lower temperature. Conversely, there are catalysts that slow up a reaction. These negative catalysts are called *inhibitors* and *preservatives.* Under normal conditions hydrogen peroxide, for example, will decompose to form water and oxygen at a rate that makes it useless long before it can be used up. However, if a small amount of acetanilid, a medicinal used as an antipyretic, is added, the hydrogen peroxide is much more stable. Again, a small amount of sodium benzoate is often added to sweet cider and to catchup as a preservative. Catalysts are specific; i.e., a catalyst that affects one reaction will not necessarily affect another. Finally, while it may be true in some cases that the catalyst takes part in the reaction and is regenerated at the end, the catalyst is not considered as one of the reacting substances, since it is not permanently changed. A catalyst is usually indicated in an equation for a chemical reaction by writing its name over the arrow.

Physical and Chemical Change

Liquid oil + hydrogen $\xrightarrow{\text{nickel}}$ solid fat

We shall continually meet catalysis throughout the course.

Metabolism

The chemical changes occuring in living things are called *metabolism*, or *metabolic changes*. Changes that result in the building up of substances needed by an organism—proteins, protoplasm, fats, catalysts, etc.—are called *anabolism*, or *anabolic changes*. Changes that result in breakdown or disintegration of tissue are called *catabolism*, or *catabolic changes*.

These changes can be classified according to the kinds of change we have discussed. Two particular classes of chemical change are so important in organisms that they have special names. One is *hydrolysis*. This is a chemical reaction with water; for example.

Table sugar (sucrose) + water → grape sugar (glucose) + fructose

This reaction usually forms two or more products.

The other important reaction involves building up larger molecules from smaller ones by removing molecules of water. This is a type of *dehydrolysis*.

x Amino acids → protein + x water

Study Exercises

1. What are the differences between physical and chemical change?
2. Distinguish between physical and chemical properties.
3. Define density, specific gravity, electrolysis, oxidation, reduction, and catalyst.
4. What are four agencies or sources of energy that may bring about a chemical change? Illustrate each by an example.
5. Which of the following bring about physical and which chemical changes? (*a*) sawing of wood; (*b*) rusting of iron; (*c*) burning of wood; (*d*) toasting of bread; (*e*) turning on an electric toaster; (*f*) making vinegar from cider; (*g*) souring of milk; (*h*) putting hydrogen peroxide on a cut; (*i*) writing with a "lead" pencil on paper; (*j*) making French dressing from olive oil and vinegar.
6. What are four types of chemical change?
7. Write a word equation to illustrate each type of chemical change.
8. Can a chemical change occur without an accompanying physical change? Give reasons for your answer.

chapter six ENERGY

With the possible exception of atomic energy (atomic fission and atomic fusion), all the energy on the earth comes from the sun. Energy is defined as *the power to do work*. Energy in operation is called *kinetic energy*, or energy of motion. Stored energy is potential or *latent energy* and can be used at a later date.

Green plants, in the presence of sunlight (energy), can manufacture, from carbon dioxide in the air and water in the soil, a variety of compounds. We think immediately of starches from potatoes and corn; sugars from grapes, beets, sugar cane, and fruits; oils from olives, soybeans, peanuts, and avocados. All these compounds contain stored energy from sunlight. When these substances are eaten by animals, they undergo chemical changes, forming carbon dioxide and water again and liberating energy. This energy in the body can be used to activate other chemical changes, to move muscles, and to supply heat to the body. Here it appears as kinetic energy. If the body does not need the energy at the time, it is not liberated but stored for future use as fat in adipose tissue. Fat represents potential energy.

KINDS OF ENERGY

Energy can exist in various forms: light, heat, sound, mechanical energy, electric energy, chemical energy, and magnetic energy. These forms of energy can be changed from one to another. It is interesting to trace such transformations. Suppose we consider the source of *heat* obtained from an electric heating pad. This heat comes from the resistance to the flow of *electricity* in the fine wires of the pad. The electricity came from a generator turned perhaps by a water wheel, which involves *mechanical energy*. The wheel was turned by water possessing *kinetic energy*, which came from the *potential energy* possessed by the water at a higher level. The water derived from the condensation of water vapor from the air in the form of rain; the water vapor was present in the air because of evaporation from lower bodies of water by the *heat* of the sun. Recent research has developed a practical method of changing heat directly to electricity without the intermediary steps of forming steam and turning a generator to make electricity. This discovery may, in a few years, change the present methods of generating electricity. In this chapter we discuss heat and electricity. Other forms of energy are considered in subsequent chapters.

einstein's equation

Einstein's equation revolutionized scientific thinking about the relationship of matter and energy. His equation states that *energy is equal to the mass (weight) times the speed of light squared.*

$$E = mc^2$$

The speed of light is 186,000 miles per sec. We can see that a very small amount of matter can be changed into a tremendous amount of energy. This reaction takes place at extremely high temperatures. It is thought to be the explanation of the energy of the sun. On earth, in atomic fission and atomic fusion reactions very small amounts of matter are converted into vast quantities of energy. Reactions occurring at lower temperatures do not cause a change of matter to energy and we say that, under ordinary conditions, *matter can neither be created nor destroyed* and *energy can neither be created nor destroyed*. These two statements are known as the *laws of conservation of matter and of energy.*

units of energy

Different units are used to describe the quantity of different kinds of energy; for example, electric energy is measured by watts and kilowatts, mechanical energy by horsepower, and heat by calories. Since energy can be changed from one form to another, it is often convenient to measure different kinds of energy in heat units.

Heat

Heat results from the random motion of molecules.

The *calorie* is a unit that measures the quantity or amount of heat; it is the amount of heat needed to raise 1 Gm water 1° Celsius.

The *degree* describes intensity of heat. This difference can be seen if we consider two pans of water. The first contains 1 liter of water at 80°C; the second 2 liters of water at 80°C. They are both equally hot, but the pan with 2 liters represents a greater quantity of heat. It would require twice as much fuel to heat 2 liters of water to 80°C.

Miss A weighs 110 lb. Her body temperature is 98.6°F, or 37°C. Mr. B weighs 190 lb. His body temperature is 98.6°F, or 37°C. Miss A and Mr. B represent the same *intensity* of heat, but 190 lb at 37°C represents a larger *amount* of heat than 110 lb at 37°C.

Fahrenheit and Celsius scales

The Fahrenheit and Celsius[1] scales measure intensity of heat in degrees. In scientific work, the Celsius scale is used. On the Fahrenheit scale the freezing point (f.p.) of water is 32° and the boiling point (b.p.) is 212°. There are some intriguing, if contradictory, stories as to why Fahrenheit chose these numbers. It is interesting to note that between 32° and 212° there are 180°, which is half a circle. On the Celsius scale there are 100° between the freezing and boiling points of water, the freezing point being 0° and the boiling point 100°. It is sometimes necessary to translate from one scale to another. We see that on the Fahrenheit scale there are 180° between the freezing and boiling points and on the Celsius scale, 100°. These two sets of numbers represent equal differences in intensity of heat. The relationship between them is therefore $100/180$, which is $5/9$, or $180/100$, which is $9/5$. To translate a value from Fahrenheit to Celsius, add 40, multiply by $5/9$, and subtract 40 from the result:

$$(°F + 40) \times 5/9 - 40 = °C$$
$$98.6° + 40 = 138.6°$$
$$138.6° \times 5/9 = 77°$$
$$77° - 40 = 37°C$$

To translate a value from Celsius to Fahrenheit, add 40, multiply by $9/5$, and subtract 40 from the result:

$$(°C + 40) \times 9/5 - 40 = °F$$
$$37° + 40 = 77°$$
$$77° \times 9/5 = 138.6°$$
$$138.6° - 40 = 98.6°F$$

[1] The Celsius scale, previously called the centigrade scale, is now named for Anders Celsius, the Swedish astronomer who developed it.

Energy

How can we remember when to use $5/9$ and when to use $9/5$? A moment's thought will answer the question. A change of 1° on the Celsius scale represents a greater change in the intensity of heat than a change of 1° on the Fahrenheit scale. Why? It takes 180° Fahrenheit to represent the same heat change represented by 100° Celsius. Fahrenheit readings will be numerically bigger. In changing to Fahrenheit, multiply by the improper fraction $9/5$ to obtain the larger numbers. In changing to Celsius, multiply by the fraction that will give you the smaller answer; namely, $5/9$. The rest of the calculation is the same for both conversions: add 40, multiply, and subtract 40 from the result.

Other equations can be used for these conversions; for example,

$$(°F - 32) \times 5/9 = °C$$

It is necessary to subtract 32 from the Fahrenheit reading to equate it to the freezing point of water, 0°, on the Celsius scale.

$$(98.6° - 32) \times 5/9 = °C$$
$$66.6° \times 5/9 = 37°C$$

To change to Fahrenheit, add 32 to account for the differences in freezing points.

$$(°C \times 9/5) + 32 = °F$$
$$37° \times 9/5 = 66.6° + 32 = 98.6°F$$

Which represents the higher temperature, 100°F or 39°C? Changing the Celsius value to Fahrenheit,

$$(39° + 40) \times 9/5 - 40 = °F$$
$$79° \times 9/5 = 142.2°$$
$$142.2° - 40 = 102.2°F$$

If we wish to work the other way, changing the Fahrenheit value to Celsius,

$$(100° + 40) \times 5/9 - 40 = °C$$
$$140° \times 5/9 = 77.7°$$
$$77° - 40 = 37.7°C$$

The higher temperature is 39°C.

heat transfer

Heat may be exchanged from one particle to another by several methods.

Conduction. In a solid, rapidly moving (hot) molecules activate adjacent molecules, causing them to move more quickly; they in turn, by their rapid movement, cause heat. In this way heat moves along a solid. This is called *conduction* of heat. Metals are good conductors of heat. If we heat one end of a metal rod, the other end quickly becomes warm. Such things as rubber, nylon, wool, flannel, asbestos, nonmoving (dead) air are poor conductors of heat. These are called *insulators.* To prevent loss of heat from a building, the walls may be built with a dead air space between. Clothing serves as an insulator to prevent loss or gain of heat by the body. A metal bed that has been in a room for several hours will be at the same temperature as the room. The room may be a comfortable 68°F (31.1°C); yet if you place your hand on the metal, it feels cold. The heat of your hand is conducted off by the metal, and your hand is actually colder while the bed becomes warmer. If an unclothed patient sits in a porcelain or metal tub, he may complain of the cold. Some pediatricians hold the metal disk of the stethoscope in their hand for a minute before placing it on a child's skin.

Convection. In a solid, atoms and molecules are limited to a small amount of space. This is not true in liquids and gases, where the atoms or molecules can move about and form currents. As the particles in a liquid or gas absorb heat, they move rapidly, and they also move away from each other, that is, they expand. As a substance expands, it becomes relatively lighter. Unheated or cooler liquids and gases are relatively heavier and tend to fall, pushing lighter fluids up. In this way, currents of moving gases and liquids are formed. These currents are called convection currents. If we open the window of a heated room on a cool day, the incoming cooler, denser air drops to the floor and moves along it, pushing the warmer, lighter air toward the ceiling. Air currents in confined spaces such as rooms or houses are referred to as drafts; in the great outdoors, they are breezes, winds, gales, etc.

Radiation. Radiation is a method of heat transfer by invisible rays. These rays are longer than visible light rays and are called *infrared rays.* They can pass through a vacuum. Heat from the sun reaches the earth as infrared radiation. If a person stands facing a fire in a fireplace, he may find that, although the part of his body facing the fire warms up, his back does not. He needs to turn around and expose his back to the radiant energy of the fire before his back also becomes warm.

Dark, rough objects absorb more radiant energy than do bright, smooth ones. For this reason Thermos jugs are made with two layers of a smooth, silvered material. Bags for transporting ice cream are silvered on the outside. People in the tropics wear light-colored clothing and live in white or light-colored houses.

body temperature

Heat is generated in the body by the oxidation of food and the liberation of the potential energy stored therein. To maintain a constant temperature the body must lose heat as rapidly as it is generated. The body's heat-regulating center is in the hypothalamus. This controls the rate of heat production and heat loss. In a cold environment, for example, the peripheral (surface) blood vessels constrict, reducing the amount of fluid brought to the surface of the body. In a warm environment the peripheral blood vessels dilate, permitting a larger amount of fluid to come to the surface for cooling by convection and radiation. The sweat glands become more active, secreting more perspiration, which, on evaporating, cools the skin. During muscular exercise, the rate of oxidation of food increases and more heat is produced.

The body adjusts by increasing the rate of heat loss.

Fever occurs when heat is produced more rapidly than it can be dissipated. The speed of chemical reactions increases with increase in temperature so that as the body temperature rises, more and more heat is produced. The heat-regulating center in a newborn infant does not immediately function efficiently. A new mother will be taught to keep the baby covered, usually with an insulating material such as wool, to prevent the loss of body heat by convection and radiation, to bathe the baby in a warm room free of drafts to prevent heat loss by convection, and to cover and dry the baby immediately after bathing to prevent heat loss by evaporation. It would hardly be necessary to admonish the mother not to place her naked baby on a metal or stonelike surface, for in this case there would be considerable heat loss by conduction.

cold

Cold is the absence of heat. If there were no molecular movement, there would be absolutely no heat. At this point we would have absolute zero. Theoretically, this point is $-273°C$, or $-459.4°F$. Although scientists have come very close experimentally to producing conditions of absolute zero, they have not as yet reached this point. It was mentioned that increasing the temperature increases the speed of chemical reactions. The opposite is also true: decreasing the temperature decreases the speed of chemical reactions. In some surgical procedures, for example, brain and heart surgery, it is desirable to slow down the metabolic processes of the body, reduce the body's need for oxygen, slow up the heart, reduce the rate at which body wastes are formed, and reduce the need to eliminate body wastes. This can be done by cooling the body. This is called *hypothermia.* Body temperatures of $48°F(8.8°C)$ have been successfully maintained during some kinds of open-heart surgery.

Several different methods are used to

Energy

reduce the body temperature. One consists in packing the patient in an ice-salt mixture. This is a time-consuming process. In a method that brings about rapid cooling the blood is routed through a tube that is submerged in an ice-salt or other cooling medium. A needle attached to the tube is inserted into a blood vessel. The tube passes through the cooling medium, and through the other end of the tube the cooled blood is returned to the circulatory system.

Hypothermic techniques permit the use of smaller amounts of anesthetics and reduce the undesirable aftereffects of certain surgical procedures.

heat and cold in therapy

Heat and lack of heat (cold) are used in many ways in treatment. Some of the more common ones are hot dry packs, hot moist packs, electric heating pads and blankets, hot-water bags and bottles, hot soaks, hot baths, paraffin baths, and heat from chemical reactions.

Cold compresses, cold soaks, ice packs, ice bags, and ice caps are some of the methods used to extract heat from the body.

Electricity

Every atom consists of a positively charged central core about which one or more negative charges spin. The atom itself is electrically neutral, but in the atoms of some substances the outer electrons are not held as tightly as in others. They can be brushed or rubbed off onto some other material. The material from which the electrons were removed is now short of electrons and has a positive charge. The material that picked up the electrons now has extra electrons and a negative charge.

static electricity

The accumulation of charges on the surface of an object or the accumulation of charges in an area in which they remain is called *static*, or *stationary*, *electricity*. Some substances permit electrons to pass through them with ease. These substances are conductors. For the early investigators in electricity, this was a very puzzling phenomenon. Electricity, whatever it was, appeared to pass right through solid material. Substances through which electrons can pass easily are the metals, solutions of acids, hydroxides, and salts, and melted salts. On the other end of the scale are the insulators—substances that resist the passage of electrons. Examples are rubber, mica, glass, silk, dry air, nylon, fur, hair, and wool. In between these extremes are substances with varying ability to conduct electrons: cotton, moist air, water, and carbon.

Insulators can hold an electric charge on their surfaces, whereas conductors permit the electrons to be dissipated through the material. The result is that conducting materials do not build up static charges.

We are all familiar with static electricity from everyday experience. What we experience is the discharge. On a dry day, particularly in the winter when the humidity is low, one may experience a shock upon touching a piece of metal. Combing clean hair, one can hear crackling sounds as the electrons are discharged. If the hair combing is done in the dark, one can see sparks of light. Woolen and nylon garments may cling together. A girl in a fur coat, seated in a car with plastic seat covers, slides across the seat to touch the metal handle of the car door. Result? Discharge.

This does not occur in the summer because then the temperature of the air is higher and more moisture can be held in the air. Moist air is a better conductor of electricity than dry air. The electrons, instead of accumulating on the surface, are conducted away by the moist air.

Potential Difference. In discussing energy, we mentioned a water wheel turned by the flow of water from a higher level down over the wheel to a lower level. The water at the

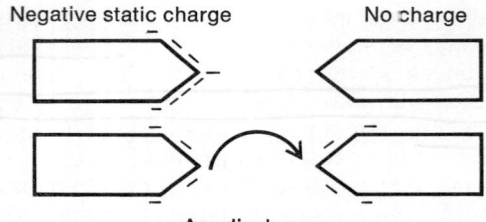

Fig. 6–1 Discharge of static electricity.

higher level had potential energy. Electrons can flow from an area in which there are a greater number of electrons to an area in which there are fewer electrons. In other words, they flow from negative to positive. Charged particles (*ions*) influence the area around themselves, and this area is called the *electric field*. Ions with unlike charges are attracted to each other, and those with like charges repel each other. To move a charge against an electric field requires work. A charge on which work has been done has higher potential energy, and the term used is *potential*. The unit of measurement is the *volt*. As a charge is built up, the potential increases; the difference in charge between two places is called the *potential difference*. Another term used to designate potential is *electromotive force* (emf).

Discharge. As two objects at different potentials approach each other, the insulating effect of the intervening air may be overcome; as a consequence, the electrons leap the gap between the two objects, and a spark passes from one to the other. This very rapid spark may be extremely hot. Static charges are dissipated more easily from points than from smooth, round shapes. (See Fig. 6–1.)

A nurse clad in nylon uniform and rubber-soled shoes is insulated from the floor. On a cold, dry day, the nurse, moving about, picks up a charge by friction: brushing against objects, brushing against the air, etc. The nurse approaches the patient lying quietly in bed and reaches forth her hand. Discharge occurs from the tips of the fingers (points).

The nurse may discharge herself by touching the metal of the bed before approaching the patient. Touching the metal with the palm of the hand rather than with the fingers will supply a larger area for discharge and result in a less unpleasant sensation.

Machinery with moving parts can also develop static charges. In this case, the machine is usually *grounded*. The earth (the ground) is a vast storehouse of electricity. It is described as *ground potential*. Objects, wires, pipes, etc., extending into the ground are at ground potential. By attaching a wire from the machine to a pipe that is eventually connected to the ground, the machine becomes grounded. If excess electrons accumulate on the machine, they are conducted off along the wire down into the ground. If the machine loses electrons and becomes positively charged, electrons move up the pipe, through the wire, and onto the machine, bringing it back to ground potential.

A flash of lightning is a discharge, essentially the same as the discharges of which we have been speaking. A potential difference builds up between the earth and a cloud, the difference becomes great enough to overcome the insulating effect of the air, and the spark jumps the gap. The charge is dissipated best at points. We are warned against seeking shelter under a tree during an electric storm. The tree, of course, is the point. Tall buildings are equipped with lightning rods, pointed conductors connected by heavy wires to the ground. As a positively charged cloud approaches the rod, electrons pass up the wire from the ground and "leak" off the points, reducing the potential difference.

In the operating room, where sparks in the presence of highly flammable anesthetics are to be avoided at all costs, the floors are made of special conducting materials, all machines with moving parts are grounded, personnel wear leather shoes with metal tips at heels and toes, nylon and other synthetic uniforms are taboo, and patients are wrapped in cotton rather than woolen blankets.

current electricity

Streams of electrons passing through matter or through a vacuum are electric currents. Electrons pass most easily through conducting materials. In these materials, the electrons are not bound tightly to their nuclei and can move from one atom to the next when electrons are introduced into the material. The incoming electrons take the places of the original electrons. In this way, the electrons move through the material, setting up a flow of electric current.

The unit of electric current is the *ampere*, which is a measure of the amount of current flowing in 1 sec. The instrument used to measure the amount of current is the ammeter. If very small amounts of current are to be measured, a milliammeter is used. A milliampere or milliamp is one-thousandth of an amp.

The unit of electric intensity or pressure is the *volt*. The instrument used to measure the intensity of electricity is the voltmeter. Here again very small differences in charge are measured in millivolts. The relationship between amperes and volts may be better understood by the analogy between water and electricity. The pressure of water in a pipe is the equivalent of voltage while the amount of water flowing in a given period of time is the amperage. A flashlight battery has a considerable amount of electricity available, but the voltage is so low that one can touch both poles without getting a shock.

resistance

All substances, even good conductors, offer some opposition to the passage of electrons. This opposition is the *resistance*. As current flows through a wire, the resistance offered by the wire causes it to heat up. Toasters, electric stoves, heat lamps, etc., operate on this principle. Certain types of operations use so-called electrocautery, in which heat generated by the passage of an electric current through a resistant wire coagulates tissue including blood vessels, reducing the amount of bleeding. The incandescent light bulb emits light because the filament in the bulb heats up to white heat.

When alternate pathways are available for the passage of an electric current, the electrons follow the path of least resistance; that is, they pass through the better conductor. If a live electric wire were to fall onto the top of an automobile, people in the car would probably be unharmed as long as they remained in the car, because the car is insulated from the ground by the rubber tires and there is no available pathway for the electrons to pass from the wire through the car to the ground. However, the human body can conduct an electric current. If a person in the car should attempt to leave the car so that part of his body touches the ground while another part of his body is still touching the car, the electricity would use his body as a pathway to the ground and, depending upon the voltage, he would get a shock or might even be electrocuted. Theoretically, if he could jump from the car in such a way that at no time did his body simultaneously touch both the car and the ground, he would be safe.

Ohm's law

Ohm's law gives us the relationship between current, resistance, and potential. The amount of current is equal to the potential (electromotive force), divided by the resistance.

$$I \text{ (current, in amperes)} = \frac{E \text{ (potential, in volts)}}{R \text{ (resistance, in ohms)}}$$ [1]

---

[1] The ampere is the amount of current that will plate out 0.001118 Gm silver from a silver solution such as silver nitrate in one second.

The volt is the amount of electromotive force needed to drive a current of one ampere through a resistance of one ohm.

The ohm is the resistance of a circuit in which a potential difference of one volt produces a current of one ampere. It is the resistance offered to a current by a column of mercury 106.3 cm long weighing 14.45 Gm at 0°C.

This law tells us that as the voltage increases, the current increases; as the resistance increases, the current decreases. Equipment designed for 110 volts (the usual house current in this part of the world) will burn out if used on the 220-volt lines found in some areas. The reason for this is that the increased voltage forces more electric current through the equipment than it is designed to handle. *Fuses* are designed to burn out if too much current passes through them. They are made of high-resistance, low-melting metals. They are designed to permit the passage of electric current up to specified amounts: 10 amps, 15 amps, etc; if the current exceeds this amount, the metal heats up, melts, and breaks the circuit, preventing the passage of any further current. Fuses will break or blow if the circuit is overloaded or if there is a short circuit as would occur if two wires in the line were to touch. The obvious answer to a blown fuse is to fix the difficulty and not to put in a larger fuse. Circuit breakers do essentially the same thing as fuses.

sources of electric current

Chemical energy can be transformed into electric energy in a cell. One or more cells comprise a battery. A voltaic cell consists of two bars of different metal connected by a wire and submerged in a solution of acid, hydroxide, or salt. Acids, hydroxides, and salts are electrolytes and dissociate into charged particles, called *ions*. When zinc and copper rods are submerged in a solution of hydrocholoric acid, the zinc atoms tend to dissolve. They dissolve as zinc ions, with a charge of +2 (Zn^{++}), leaving the 2 electrons behind on the rod. Hydrochloric acid exists in solution as hydrogen ions (H^+) and chloride ions (Cl^-). The positively charged hydrogen ions are repelled by the positively charged cloud of zinc ions in the vicinity of the zinc rod. The hydrogen ions collect near the copper rod. The electrons left on the zinc rod by the dissolved zinc ions travel through the wire to the copper rod. Here the positively charged hydrogen ions each pick off an electron to form neutral hydrogen atoms. Two atoms combine to form molecules, and hydrogen gas bubbles out of the solution at the copper rod. We have a circuit consisting of a flow of electrons through the wire and a flow of ions through the solution.

Some types of cells in common use are the dry cell and the storage cell. The dry cell is not really dry. The electrolyte, usually ammonium chloride, is absorbed in some inert material. In the storage cell, there is a plate of lead and one of lead oxide submerged in a solution of sulfuric acid. The action in the storage cell is reversible. The chemical reaction can liberate electrons as a source of electric current. This is called *discharging*. The cell can then be connected to a source of electric current, and the reaction, due to the supply of electrons from the electric current, can be forced in the other direction. This is called *charging*. Theoretically, a storage cell should last forever. In actual practice this is not the case. One or more storage cells make a storage battery. These batteries are used in automobiles to turn the motor and are frequently kept as stand-by equipment in the hospital in case of emergency.

Magnets can be used to generate electricity. A magnet has two poles: a north pole and a south pole. If two magnets are brought into close proximity in such a way that the north pole of one and the south pole of the other approach each other, the two magnets will clamp together. These opposite poles are mutually attractive. If the reverse is done, i.e., north pole to north pole or south pole to south pole, the magnets repel each other. This action is similar to the attraction of unlike charges and the repulsion of like charges.

The passage of an electric current through a wire coil imparts to the coil the properties of a magnet. Conversely, moving a magnet near a conducting wire, or moving the wire near the magnet, sets up a flow of electrons in the wire; in other words, an electric current. This is essentially the way electricity is generated in a dynamo or generator.

Mechanical energy is needed to do the moving. This may be supplied by the energy of falling water or by a turbine driven by expanding steam.

An electric motor changes electricity into motion and is the reverse of the dynamo. The electric current flowing through wire coils causes the coils (or armature) to move, thus producing mechanical energy, which can work for us.

the body as a source of electric current

We realize that the body can conduct an electric current. The body fluids are dilute solutions of electrolytes, primarily salts. Dry skin offers a relatively high resistance to the passage of an electric current, but moist or wet skin has a much lower resistance. Also, on the surface of the skin, there is sodium chloride from perspiration. Sodium chloride is an excellent conductor of electricity. Occasionally, we read in the paper of the tragic accidental electrocution of a person in the bath. Here we have a combination of faulty electric apparatus which supplied the current, plus wet skin, which makes the individual a good conductor.

We have discussed chemical energy as a source of electricity. Considering all the chemical changes that occur in the body and that body fluids are electrolytes, it should not be surprising that electric activity accompanies many body changes.

When the charges on either side of a membrane are unequal, a difference in potential exists across the membrane and the membrane is said to be polarized. This is the same as saying that an electromotive force (EMF) exists across the membrane. In the body, these forces are measured in millivolts (mv). As the heart contracts an electric current sweeps across the cardiac muscles and the muscles lose the electric potential; the organ is thus depolarized. During the resting stage or recovery phase, the heart again becomes charged or polarized.

The transmission of nerve impulses is also caused partly by electric depolarization of the nerve fiber (partly by chemical transmission at the nerve endings). Here again, the electric currents are measured in millivolts. The nerves cannot be restimulated until they become repolarized. Brain cells, being nerve tissue, generate electric waves even during sleep.

measuring instruments

Electrocardiograph. The heart muscle, in expanding and contracting, generates electric impulses. These currents can be measured by a device called the electrocardiograph. Usually three electrodes are attached to different parts of the body, and the potential differences between pairs of electrodes are measured. An electrolyte, such as sodium chloride paste, is placed between the electrodes and the skin to ensure good conduction. The record of the electric impulses is called the electrocardiogram (EKG). The EKG shows a pattern of waves. By studying these wave patterns, the physician can detect abnormalities in heart function.

Electroencephalograph. By attaching electrodes to different areas of the skull, measurements can be made between electrode pairs to give an indication of the electric currents generated by the brain cells. These electric currents are very small and are amplified and recorded as an electroencephalogram (EEG). Here again the pattern appears as waves. Different patterns appear depending upon the patient's state; variations occur in sleep, wakefulness, drowsiness, fright, mental activity, etc. The electroencephalograph is used to diagnose brain tumors, hemorrhages, epilepsy, etc.

Electron Microscope. The optical microscope consists of lenses and a focusing device; in addition, light is needed. The object under the microscope can be observed by the eye or recorded on a photographic plate. The extent of magnification is limited by the wavelength of light. In the electron microscope, electrons are used in place of light. Their very small size enables us to "see" much smaller particles. To focus the electron beam, a magnetic field (an electrostatic field) is used. The image is recorded on a photographic plate or thrown onto a fluorescent screen. Many recent advances in science are due to the use of this tool, which enables the scientist to "see" things heretofore invisible.

Electronic Devices. Many currents and sounds generated by the body are of very low intensity and must be amplified to be studied, analyzed, and interpreted. Until recently, amplification systems were large and bulky.

With the development of the transistor, it became possible to use extremely small amplifiers. This affords an opportunity for electronics engineers to develop new tools that can be used in medicine. In addition, digital computers can correlate and analyze experimental results with great rapidity. A 3-oz device that counts heartbeats can be worn all day while a patient carries on his activities. A tiny radio sending station can be swallowed and can send messages as it travels through the digestive tract. A device has been developed that can report changes in the volume of air in the lungs of an infant. These are but a few of what will probably soon be a long list of new tools.

Study Exercises

1. Define the following terms: energy, potential energy, kinetic energy, heat, cold, and calorie.
2. List seven forms of energy.
3. State the laws of conservation of matter and of energy.
4. How does Einstein's equation $E = mc^2$ explain the devastating effects of the explosion of an atomic bomb?
5. A patient's temperature is recorded as 38°C. Express this reading as degrees on the Fahrenheit scale. A temperature reading is 95°F. Change this to the corresponding Celsius value.
6. Explain how heat is transferred by conduction, convection, and radiation.
7. Name two good conductors of heat and three insulators.
8. Give a scientific explanation for each of the following observations: (a) on a breezy day, a slender eight-year-old child developed a chill while standing on the beach in a wet bathing suit, (b) a boy playing ball on a hot day had a highly flushed face and perspired profusely, (c) a patient in a room with an open window was given a tepid bed bath but complained that it was cold.
9. What is hypothermia?
10. What is the effect of an elevated body temperature on the rate of chemical change in the body? What is the effect of a reduced body temperature?
11. Define static electricity and current electricity.
12. Name three good conductors of electricity and five insulators. On which type of material does a static charge tend to accumulate?
13. What are electrolytes?
14. What is meant by ground potential?
15. Why are machines with moving parts grounded when they are used in the operating room?
16. What happens when a static charge is discharged?
17. A nurse in a nylon uniform and rubber-soled shoes received a shock when she touched the metal handle of a cabinet. Explain when and why this happened.
18. What is an electrocardiogram? What is an electroencephalogram?
19. Mention two ways in which the body can generate electricity.
20. Why does the electron microscope give so much greater magnification than the optical microscope?

chapter seven FLUIDS

Ideas for preview or review
  A gas will occupy all available space.
  The composition of a mixture may vary over a wide range.
  Atmospheric pressure is measured with the barometer and can be reported as weight per unit area or as the height of a column of mercury.
  Density is weight per unit volume.
  Water cannot freely flow into a bottle filled with air unless the air can escape.
  Conversely, water cannot freely flow out of a bottle unless air can flow in to take the place of the water.
  In a container with flexible walls, fluids can flow out without being replaced because the sides of the container collapse, e.g., a rubber hot-water bag.

GASES

Matter in the gaseous state is mostly empty space. The gas molecules are far apart, and except in highly compressed gases, there is very little attractive force between molecules. Gas molecules move about (diffuse) independently of each other and exert pressure by their bombardment or impact against the walls of the container.

Pressure

Pressure is force per unit area. As we mentioned in explaining the process of boiling, atmospheric pressure is approximately 15 lb per sq in., or 1 kg per sq cm. The external pressure on the surface of the body of an adult is more than 12 tons. We do not feel this pressure because it is counterbalanced by an equal internal pressure. If the internal pressure were not so great as the external pressure, we would be crushed. One can see how this human adaptation to living under pressure poses one of the problems of putting a man into space, where the pressure is much less. It also explains the need for pressure suits and pressurized cabins for high-altitude flying. In the autoclave, the pressure is usually built up to about 15 lb per sq in. above atmospheric pressure, or to twice the atmospheric pressure.

Laws Governing Behavior of Gases

Boyle's law

It seems obvious that as we increase the pressure on a gas, the volume should decrease. Imagine a small rubber balloon filled with air. As pressure is exerted from the outside, we expect the size of the balloon to decrease. Robert Boyle experimented accurately with the volume of a gas as the pressure was increased and found that the volume was halved as the pressure doubled. He tried the experiment with many gases and decided that *for any gas at a constant temperature, the volume varies inversely as the pressure.* The law may be expressed as

$$P_1V_1 = P_2V_2$$

or

$$\frac{P_1}{P_2} = \frac{V_2}{V_1}$$

This is known as Boyle's law. In other words, if the pressure is doubled, the volume is one-half its former value; if the pressure is halved, the volume is doubled.

Think of respiration in terms of Boyle's law. On expiration, the volume of the chest cavity decreases, the pressure of air in the lungs increases, and air is exhaled. On inspiration, the volume of the chest cavity increases, the pressure in the lungs decreases, and air is inhaled.

Charles' law

Charles' law describes the variation in the volume of gas with changes in temperature if the pressure is not changed. The essence of this law we know from everyday experience. If we heat a gas, it expands; if we cool it, it contracts. Charles' law states that *the volume of a gas is proportional to the absolute temperature.* Expressed mathematically,

$$\frac{V_1}{V_2} = \frac{T_1}{T_2}$$

(Absolute temperature is a scale on which zero is $-273°C$.) Since air contracts on cooling, there are more molecules of gas per unit area in cold air than in warm air and the density of cold air is greater. Cold air tends to fall. In a room the air temperature may be comfortable for adults but cold for a child creeping on the floor.

The behavior of gases as described by Boyle's and Charles's laws can be expressed in one equation called the general gas equation or the universal gas law, which states

$$PV = RT$$

where P is the pressure, V is the volume, T is the absolute temperature, and R is a constant.

An examination of this equation shows that if the absolute temperature is increased, either the pressure or the volume must increase.

If we prevent a gas from expanding or contracting by having a fixed volume, variations in temperature will cause direct variations in pressure. If we heat the gas, the pressure increases; if we cool it, the pressure decreases. Spray-type cans containing materials under pressure, such as aerosol-bomb insect sprays, shaving creams, and hair sprays, usually warn on the label: "Store in a cool place."

One should never leave such a container near a stove or radiator or in the direct rays of the sun because the increased temperature will increase the speed of the molecules, thereby increasing the number of times the molecules strike the walls of the container (increasing the pressure), and the container may explode.

Dalton's law of partial pressures

An understanding of the *law of partial pressures* is essential if we hope to understand gas exchange in the lungs and the tissues. If we have gases in a mixture that do not combine chemically, the total pressure of the gas mixture is the result of all the pressures exerted by all the molecules in the mixture. If the mixture was composed of 20 per cent oxygen and 80 per cent nitrogen, 20 per cent of the total pressure would be due to oxygen molecules and 80 per cent due to nitrogen molecules. If the total pressure of the mixture was 760 mm of mercury, the *partial pressure* due to oxygen would be 152 mm, and that due to nitrogen, 608 mm.

Dalton's law states that *each gas in a mixture exerts its own pressure (partial pressure), and the total pressure of the mixture is equal to the sum of the partial pressures.*

Each gas in the mixture behaves as if it were there alone, its contribution to the total pressure being dependent upon the number of molecules of that particular gas present in the mixture. If a 1-liter container of oxygen gas at 760 mm or 1 atm pressure is connected to another 1-liter container in which there is nitrogen gas at 1,520 mm or 2 atm pressure and the gases are allowed to mix, both gases will diffuse to occupy the 2 liters. The pressure of oxygen in the resulting mixture will be 380 mm or $1/2$ atm. We obtain this result by applying Boyle's law:

$$P_1V_1 = P_2V_2$$

Original pressure 760 mm, original volume, 1 liter; final pressure x, final volume, 2 liters. Solving for x, the final pressure is one-half the original pressure. Performing the same calculations for nitrogen, we find that its final pressure is 760 mm or 1 atm. The total pressure of the mixture is the sum of the partial pressures or 1,140 mm or $1 1/2$ atm.

Gases move from areas of greater pressure to areas of lower pressure; they move from areas in which there are a greater number of the gas molecules to areas in which there are fewer of those molecules. Let us assume that we have two mixtures of gases separated by a membrane; the total pressure of each mixture is the same but the partial pressures are different. The gases will diffuse from the area of greater partial pressure to the area of lesser partial pressure. When the partial pressure of oxygen in alveolar air is higher than the partial pressure of oxygen in the pulmonary capillaries, oxygen will diffuse into the capillaries. If the partial pressure of carbon dioxide in the lungs is higher than the partial pressure of carbon dioxide in the capillaries, carbon dioxide will also diffuse from the lungs into the capillaries. In certain cases of hyperventilation, the concentration of carbon dioxide in the blood drops below normal. The patient is directed to breathe into a paper bag and to rebreathe the expired air. In this way, the concentration of carbon dioxide in the lungs is increased and because of the higher partial pressure of carbon dioxide in the lungs, carbon dioxide diffuses back into the blood.

Graham's law

We know from the kinetic molecular theory and from experience that if we increase the temperature, we increase the speed of molecules, or their rate of diffusion. In addition to this fact, the rate of diffusion of gases is related to the weight of the molecules. As we might suspect, lighter molecules move more rapidly than heavier ones.

The rate of diffusion of molecules is inversely proportional to the square root of their densities. This observation is known as Graham's law. Air is approximately 20 per cent oxygen and 80 per cent nitrogen. For certain respiratory difficulties, patients are given a mixture of 20 per cent oxygen and 80 per cent helium. The weight of nitrogen is 28 and that of helium is 4. This gives us the relationship 28:4, and the square-root proportion is approximately 5.3:2. We see that, in addition to being a lighter mixture, helium will diffuse $2 1/2$ times faster than nitrogen.

Henry's law

We all know that unopened bottles of soda water contain carbon dioxide gas dissolved under pressure. There are no visible bubbles of gas in the unopened bottle. When the bottle is opened, the pressure is released and the carbon dioxide comes out of solution as bubbles of gas. Henry's law states that *the weight of a gas dissolved by a fixed volume of a liquid at a constant temperature is proportional to the pressure.* Thus it is that divers and others who work under high pressures have an increased amount of air, principally nitrogen, dissolved in the blood. Sudden decrease in the pressure releases air from solution and causes it to form bubbles in the blood stream. These bubbles travel though the blood vessels until they reach the capillaries. A bubble of gas is bigger than the diameter of a capillary, and the bubble becomes wedged in the capillary, cutting off the blood supply to the cells supplied by that capillary. These cells, deprived of food and waste removal, die. This causes the very painful condition known as the "bends" and explains why an attack may occur hours after the victim has left the high-pressure situation. The remedy is, of course, to put the man back under high pressure and dissolve the bubbles. The proper name for the "bends" is *caisson disease,* and it can be avoided by slow decompression in passing from an area of high pressure to one of lower pressure.

Increasing the pressure increases the solubility of a gas in all the body fluids, not only the blood. The pressure of fluid in the cells, as well as that of fluid bathing the cells, rises. Under normal pressure, very little oxygen is dissolved in the body fluids, but increasing the pressure on the body will increase the amount of oxygen dissolved. This knowledge is applied in *hyperbaric techniques.* Gangrene is caused by an organism that flourishes in the absence of gaseous oxygen. A patient with this disease can be placed in an atmosphere of oxygen under high pressure. The body fluids and cells become saturated with oxygen. This environment is unfavorable to the growth of the microorganism and the patient has a chance to recover.

Let us remember that the solubility of a gas varies with the temperature. Gases are more soluble at lower temperatures than at higher temperatures. A combination of reduced temperature (hypothermic) and increased pressure (hyperbaric) would cause the greatest amount of gas to dissolve.

Archimedes' principle

We usually think of buoyancy in connection with liquids, but gases have buoyant effects also. Archimedes' principle states that *an object submerged in a fluid is buoyed up by a force equal to the weight of the amount of fluid displaced by the object.* If a balloon displaces 100 ml air, it is buoyed up by a force equal to the weight of 100 ml air. Since water is denser, its buoyant effect is greater.

One cc of water weighs 1 Gm. Will an object float in water if its total volume is 10 cc and it weighs 5 Gm? The answer is of course yes, because the object weighs less than does 10 cc of water. The object would displace 5 cc of water and the rest of the object would be above the water.

If an object has a volume of 5 cc and weighs 10 Gm it will sink in water. If this object is weighed while it is suspended, submerged beneath the surface of the water, the object will weigh less than it does in air. By how much? The volume of the object is 5 cc; 5 cc of water weigh 5 Gm; the buoyant effect of the water is 5 Gm, so the object will weigh less by 5 Gm.

Several devices used to measure the density or specific gravity of liquids are based upon Archimedes' principle. The general name for these floating instruments is the hydrometer, but when they are devised for specific measurements, the name reflects the use to which the instrument is applied. The lactometer measures the specific gravity of milk; the urinometer, the specific gravity of urine. In a dilute sample of urine, the urinometer sinks lower than it does in a more concentrated sample. The weight of 1 ml of dilute urine is less than the weight of 1 ml of more concentrated urine. Fewer milliliters of the more concentrated urine will equal the weight of the urinometer, less urine will be displaced, and the urinometer will float higher in the fluid. (See Fig. 7-1.)

In hydrotherapy, when underwater exercises are prescribed, the buoyant effect of water is utilized to give some support to body parts.

Pascal's principle

Years ago, automobiles had mechanical brakes. One of the disadvantages to these brakes was that if you stepped on the brakes hard, not all four wheels grabbed evenly and you could go off the road. Today, cars have hydraulic brakes which use brake fluid. You step on the brakes and all four wheels grab

Fluids

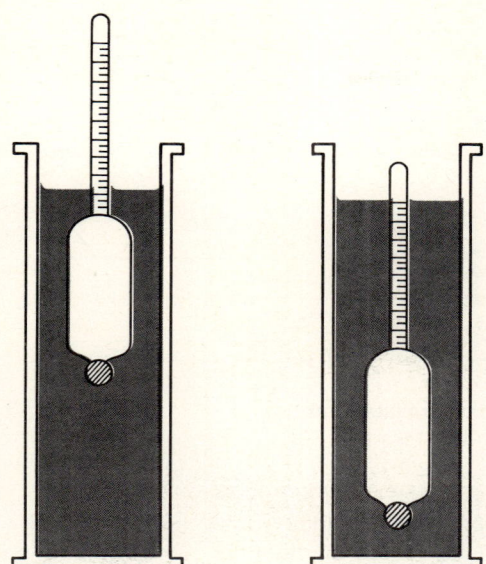

Fig. 7–1 Urinometer: urine, sp. g. 1.030., water, sp. g. 1.000.

evenly. This is an application of Pascal's principle which states that *pressure applied to a confined fluid is distributed equally, uniformly, and undiminished throughout all parts of the fluid.*

Eye irrigations are always given at very low pressures. The eye is filled with fluid so that pressure on the anterior aspect of the eye will be distributed throughout the eye to the sensitive retina in the back of the eye.

One problem with patients confined to bed for long periods of time is the development of pressure sores, particularly in those areas of the body that bear the patient's weight. A solution to this problem is a water blanket or water mattress. This solution is another application of Pascal's principle. Now the patient's weight is distributed over his entire body instead of being primarily on the pressure points. The same principle applies to an air mattress.

In the usual method of measuring blood pressure using the sphygmomanometer, an inflatable cuff is wrapped around the patients arm and air is pumped into the cuff. The increased amount of air in the cuff increases the pressure which is transmitted to the blood vessels. When the pressure is sufficiently high, the flow of blood in the vessels stops.

It is interesting to note that, since the circulatory system is closed, an occlusion or partial obstruction that increases the pressure in one part of the system causes the pressure in the entire system to rise. Again, during pregnancy, women wear girdles that will give support but are not too tight. The developing baby grows in a fluid encased in a membrane, where pressure exerted on any part of the system can be distributed throughout.

The hydraulic press and the hydraulic lift are two other applications of Pascal's principle. A piston of small area is connected by a tube filled with fluid to a piston of large area. Pressure applied to the small piston is transferred undiminished to the large piston. But here again, we get nothing for nothing; if the small piston is 10 sq cm and the large piston, 100 sq cm, moving the small piston 10 cm will lift the large piston 1 cm. This is assuming, of course, that there are no friction losses.

Medical Applications

artificial pneumothorax

Between the lungs and the walls of the chest is a small space called the *intrapleural space*, or *pleural cavity*. The pressure in this space is less than that of the atmosphere. If the chest wall or the lungs were punctured, air would enter; the pressure would increase, and the increased pressure would cause the lung to collapse.

simple underwater drainage (closed chest drainage)

Since the pressure in the pleural cavity is normally subatmospheric, when the chest is

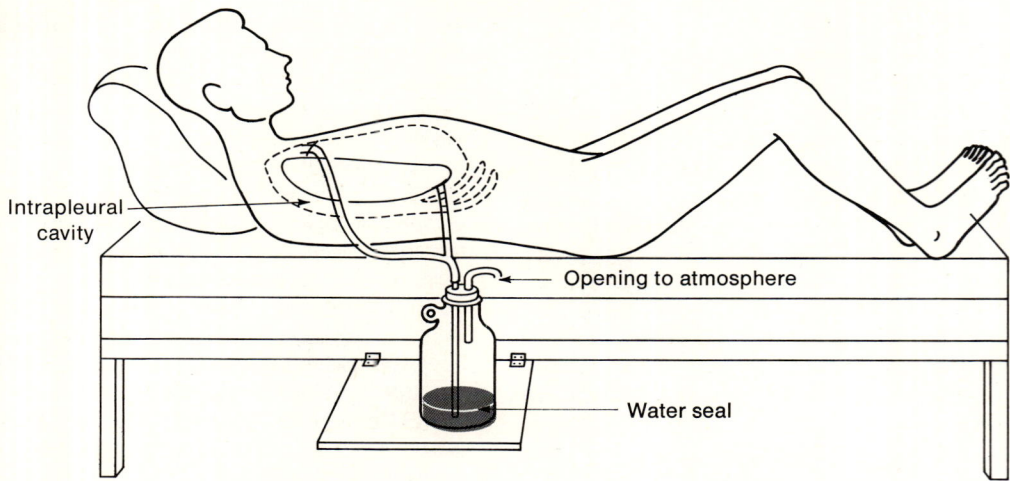

Fig. 7-2 Simple underwater drainage (closed chest drainage).

opened in surgery, air passes from an area of greater pressure to one of lower pressure and enters the pleural space, causing a pneumothorax. At the end of the operation, before the chest is completely closed, tubes are inserted anteriorly and posteriorly into the pleural space. The other ends of the tubes reach beneath the water surface in bottles partially filled with water and open to the atmosphere. Because the ends of the tubes are below the surface of the water, air cannot enter through them. This is called a *water seal*. Air and liquid can escape into the bottles, but the water seal prevents the back flow of air. The bottles are kept below the level of the patient's bed. (See Fig. 7-2.)

On expiration the chest cavity becomes smaller and the pressure of air in the pleural cavity increases (Boyle's law). Some of this air then bubbles out through the bottle. On inspiration the chest volume increases and the pressure of the air in the pleural cavity decreases, but because the tubes are submerged in water, air cannot go back. However, since the atmospheric pressure is greater than the pressure in the cavity, the pressure on the surface of the water causes the water to rise a bit in the tube. What you see is air bubbling out, then water rising in the tube, then air bubbles, etc. As the air is removed from the space, the pressure decreases and the lungs expand. When all the bubbling activity has ceased, the lungs have expanded to the point where they have stopped off the tubes in the chest.

the respirator

One type of respirator is called the *iron lung*. It is used when the patient is unable to expand and contract his chest. We usually associate this condition with paralysis, as in poliomyelitis. The respirator consists essentially of an airtight chamber the volume of which can be increased and decreased. In the ordinary type, the patient's body is inside the chamber and his head is outside. A rubber collar fits around the patient's neck and prevents the escape of air from the chamber. A diaphragm or bellows attached to the chamber can move down and up or in and out, thus changing the volume of the chamber. Since the patient's head is outside the chamber, his lungs are open to the atmosphere. As the bellows moves out, the volume of the chamber increases, the pressure in the chamber decreases, and the patient's chest expands. As the patient's chest expands, the chest volume increases and the pressure of the air in the lungs decreases. Since the pressure in the lungs is less than the pressure of the atmosphere, air enters the patient's lungs through his nasal passages. When the bellows moves in, the volume in the chamber decreases, the pressure increases, and the chest contracts. Contraction of the chest increases the pressure in the lungs. Since the pressure in the lungs is now greater than the atmospheric pressure, air is exhaled. Do you recognize Boyle's law here?

LIQUIDS

In the liquid state, molecules are very much closer together than in gases. Although liquids flow, their molecular behavior resembles that of solids more than that of gases because the molecules do not have a tendency to separate from one another as do those in the gaseous state.

Cohesion and Adhesion

The attraction between like molecules is called *cohesion*, and the magnitude of the attraction is called the *cohesive force*. *Adhesion* is the attraction between unlike or different molecules. Because of their cohesive forces molecules of liquids tend to cling together and form a *sphere*, which provides the *smallest surface area for the greatest volume*. A freely hanging sphere is attracted by gravity and assumes a pear shape, which we call a drop. The cohesive forces also explain the phenomenon of surface tension.

A molecule in the body of a liquid is attracted to and by other like molecules around it, above it, and on all sides. A molecule on the surface of a liquid has no like molecules above it; therefore, the downward and sideward forces are more effective, and the molecules on the surface are packed closer together, forming a sort of taut molecular skin. This force at the surface can be demonstrated by floating a steel needle on the surface of water. If the needle is placed flat on the surface, it will float; if it is dropped on the water in a slanted position, it will sink. Surface tension can be determined experimentally by measuring the force required to pull a floating steel ring from the surface. Surface tension exists at the boundaries between two liquids as well as between liquid and air. In several laboratory tests one liquid is floated upon the surface of another. Such a test is Heller's test to detect albumin in the urine.

Detergent action is due to a lowering of surface tension. Bile salts function like soaps and lower the surface tension of intestinal fluids to permit an intimate mixture of fat and intestinal enzymes. Increasing the temperature lowers the surface tension. We know from experience that a combination of hot water and detergent gives the best cleaning action.

Capillary Action

If the adhesive forces between the molecules of a liquid and the molecules of the container are high, the liquid will wet the walls of the container. This is the case with water and glass and water and many other kinds of material. In wetting glass, the water molecules creep a bit along the glass surface. If the container is a tube, water creeps up the sides; since water has a high surface tension, which tends to keep the surface as small as possible, other water molecules follow along and the liquid rises in the tube. This process goes on until the adhesive forces are balanced by the weight of the liquid in the tube. It follows that liquids rise higher in tubes of narrow bore. A test for the clotting time of blood makes use of capillary action. The finger is pricked, and one end of a very narrow tube (capillary tube) is placed in the drop of blood. The blood rises in the tube. At specified times pieces of the tube are broken off, and the examiner observes the time it takes for the blood to clot in the tube.

Sterile setups covered with a dry sterile cloth are kept free from contamination, but if water is splashed on the covering or if the setup is placed on a damp surface, one can no longer expect the material to be sterile. Many cloth fibers serve as miniature tubes through which water creeps by capillary action, thus explosing the sterile setup to possible contamination. A setup with a wet or damp covering must be replaced. Why?

A terry-cloth towel left on the side of a tub or basin with one end submerged permits the water to creep along the towel fibers and form a puddle on the outside of the basin.

The curved surface of water in a cylinder (meniscus) is also explained by adhesion and surface tension.

Fluidity and Viscosity

These two terms represent different viewpoints of the same phenomenon: fluidity means *the tendency to flow;* viscosity refers to *the resistance to flow.* A viscous liquid is sluggish. Whipping cream is described as heavy or thick, but, in actuality, so-called heavy cream is less dense (relatively lighter) than light cream. There is a higher proportion of fat in heavy cream. Whipping cream is more viscous than light cream. The tendency to flow (fluidity) increases with increase in temperature, and the viscosity increases with decrease in temperature. The expression "slow as molasses in January" describes this physical fact. An elevation of body temperature increases the fluidity of body fluids. Lowering the body temperature, as in hypothermic procedures, causes body fluids to become more viscous.

Hydrostatic Pressure

The name itself tells us that this is pressure exerted by water at rest. The hydrostatic pressure of a fluid is determined by its density and height. The height of a column of water is measured from its surface to the outlet.

The word "hydrostatic" refers to water, but all liquids exert hydrostatic pressure. It is easy to see that a liquid in a container exerts a sideward pressure against the sides of the container. If this were not so, a hole in the side of a container below the water line would not leak and we know that it does. In addition to sideward and downward pressure, liquids exert upward pressure. Must not an upward force support a ship on the surface of the sea? When we consider that the liquid is at rest, we must conclude that any part of the liquid exerts a force in all directions and that in any part of the liquid these forces must all be equal. This is what we mean by "at rest" or *equilibrium:* the forces counterbalance each other, with the result that there is no apparent change. If the forces in any part of the liquid were not counterbalanced by equivalent forces, the liquid would flow in the direction that would tend to balance the uneven forces.

The resting liquid exerts force in all directions, and when the force is considered in terms of the area against which it acts, it is described as *pressure.*

In considering the pressure on the bottom of a container of fluid, or, let us say, at the outlet nozzle attached to an enema bag, the total amount of water in the container is not the crucial point. The important point is: What is the pressure on the bottom or at the outlet? This is determined by the amount of liquid directly above the bottom or the outlet. We could restate the question and say: What is the downward force exerted on 1 sq cm of the bottom of this container, or what is the force on 1 sq cm at the outlet nozzle? The force on 1 sq cm is the weight of the column of fluid directly above it. The weight of the fluid directly above is determined by the height of the fluid and by its density. Since the term "pressure" includes the concept of unit area, the second, more elaborate statement is not necessary once one has grasped the meaning of the term.

Containers of different shapes, although they may be filled to equal depths with water, need not contain the same amount. The pressure on the bottom of each will, however, be the same, because the distances between the bottom of each container and the surface of the water is the same in each case. If we take a few containers of various shapes—for example, an Erlenmeyer flask (which is wider at the bottom than it is at the top), an ordinary straight-sided glass tumbler, and a funnel, into the bottom of which we have inserted a cork plug—and if we fill each container to a depth of 10 cm, we may discover that the Erlenmeyer flask requires about 500 ml, or 500 Gm; the tumbler, about 350 ml, or 350 Gm; and the stoppered funnel about 50 ml, or 50 Gm water. The question is: is the pressure on the stopper in the bottom of the funnel, which contains only 50 Gm water, the same as the pressure on the bottom of the Erlenmeyer flask, which contains 500 Gm water? To answer this question, we check to see if the depth of the water is the same. If it is, the answer is that the pressure (per unit area) is the same in each case.

This principle of hydrostatic pressure can be expressed by the equation

$$p = h \times d$$

where p is pressure per square area, written $p/\text{sq cm}$ or p/cm^2, h is the height of the column of liquid, expressed in centimeters, and d is the density of the fluid, expressed in grams per cubic centimeter, written Gm/cc or Gm/cm^3.

Why is it important that health workers understand this principle? When administering an enema or douche, the pressure at the nozzle is dependent upon the height

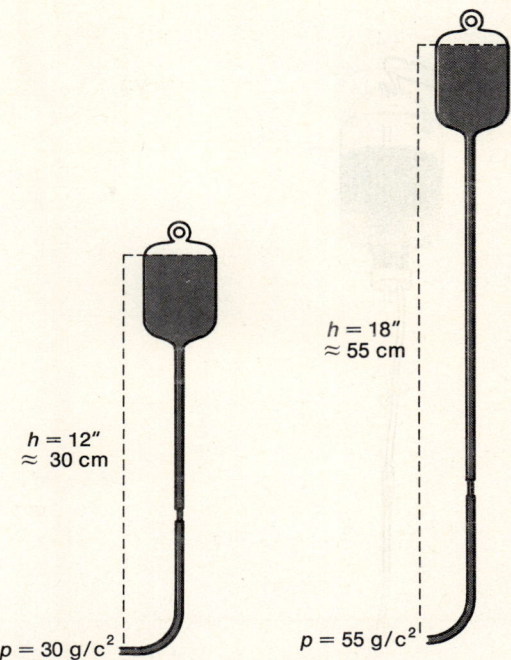

Fig. 7-3 Hydrostatic pressure. The volume of water in each bag is the same.

of the column of fluid, not on the total volume. In other words, the pressure *is* related to the vertical distance between the surface of the liquid and the outlet (Fig. 7-3).

Intravenous feedings, blood transfusions, or other infusions also depend upon the principle (Fig. 7-4). Raising the bottle will increase the pressure at the outlet of the needle or catheter and will also increase the rate of flow of the liquid. Lowering the container will decrease the pressure and decrease the rate of flow.

siphons

A siphon may be a tube or a hose bent so that one arm is longer than the other. The siphon is used to transfer liquid from a higher to a lower level over an intermediary elevation. The tube is filled with liquid so that there are

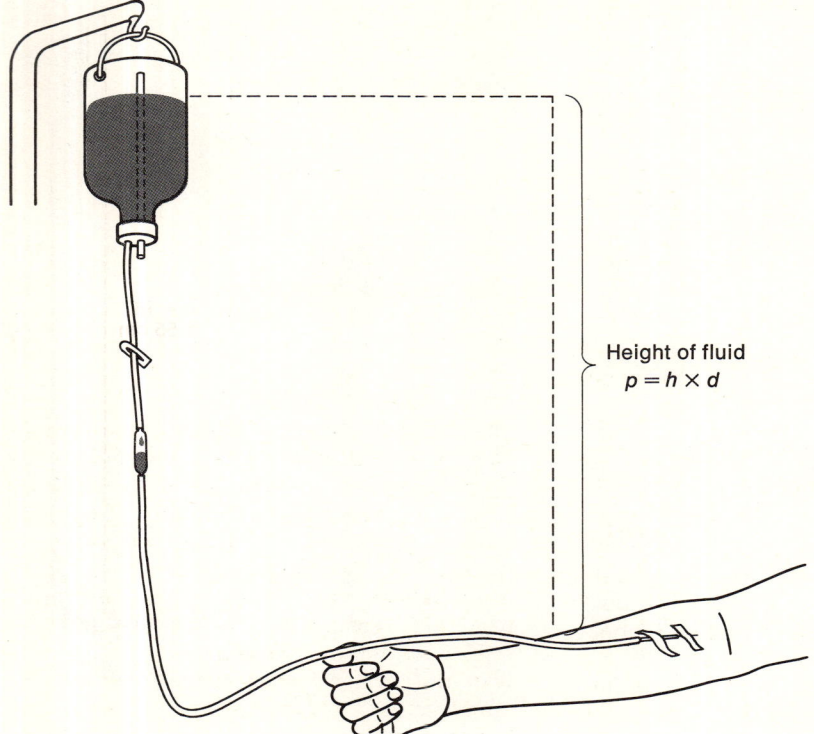

Fig. 7–4 Hydrostatic pressure.

no air bubbles in the line, and the short arm is submerged below the surface of the liquid on the higher level.

The weight of the fluid in the longer arm is greater than that in the shorter arm, and the fluid tends to flow out through the long arm. Atmospheric pressure bearing down on the surface of the liquid in which the short arm is submerged forces fluid up into the short arm. The flow is then up the short arm and out the long arm. The length of a siphon arm is measured from the surface of the liquid into which a tube is submerged to the top of the hump or bend. As long as the arms are of uneven length and as long as no air gets into the line, the siphon will continue to work. If air enters the line, the tubes empty and the operation stops.

Gastric lavage operates on the principle of the siphon. The patient swallows one end of a flexible rubber tube. Through a funnel on the other end, liquid is poured into the stomach. Before all the fluid has been poured into the tube, the funnel end of the tube is pinched to prevent the entry of air. The funnel end is then lowered over a receptacle to catch the siphoned material. The funnel end of the tube must be bent down below the level of the patient's stomach so that the funnel end becomes the long arm. Fluid then flows out through the long arm. A siphon could also be used to empty the bladder (Fig. 7–5).

Tidal Drainage. The Munro tidal drainage apparatus is used to alternately fill and empty the bladder. The apparatus consists of a bottle suspended higher than the patient. From this bottle one arm of a T tube connects to the patient's bladder, and the other arm runs into a bottle placed below the level of the patient. An air vent keeps the system at atmospheric pressure. This much of the apparatus serves to fill the bladder. Fluid flows from the top bottle down the T tube by gravity and fills the bladder. Now to empty the bladder the procedure is as follows: inserted in the arm of the T tube connected to the bladder is a siphon. As the fluid flows through the T tube, it fills not only the bladder but also the siphon arm. When the siphon has been filled, the whole system empties by siphonage into a drainage bottle. When the fluid drops below the level of the air vent, air enters the siphon system and stops the siphonage.

More fluid then fills the system until the top of the siphon is reached again; then everything is emptied again by siphonage until air enters and stops the siphon, etc. We can see that the amount of fluid that enters the bladder is determined by the height of the siphon arm, because as soon as the siphon arm is filled, not only does no more fluid enter the bladder but the whole system empties. By increasing the height of the siphon arm, more fluid will enter the bladder; by shortening the siphon arm, less fluid can enter.

This apparatus is sometimes used to re-educate bladder muscles (Figs. 7–6 and 7–7).

suction

Most persons think of suction as a pulling force. We speak of sucking a soda through a straw. Do we pull the liquid up through the straw? If one stands on the station platform while the fast express train goes by, one observes that papers, leaves, etc., on the platform are "sucked" under the train. The propeller-driven airplane rises in the air. Does some force reach down from the sky and pull it up, or is it somehow pushed up?

Suction depends on a pressure difference between two areas. Fluids flow from the high-pressure area to the low-pressure area. If we remove some air from a container with rigid walls, the pressure in the container will fall. Outside the container the air is at atmospheric pressure; inside it is below atmospheric pressure. If we open the container, air from the outside will rush in. The outside air at the higher pressure moves into the container until the pressure in the container reaches atmospheric pressure. Any container open to the atmosphere is at atmospheric pressure. It is easy to see from this illustration that air is not pulled into the bottle but that the higher pressure outside the bottle forces the air in. This is the principle by which all suction machines operate: *fluid flows from higher-pressure areas to lower-pressure areas.*

When we suck the soda through the straw, what do we do? With the lips closed about the straw, we inhale through the mouth, thus removing some of the air in the straw. The pressure in the straw is then lower than atmospheric pressure. Atmospheric pressure on the surface of the liquid in the glass forces the soda up into the straw. If

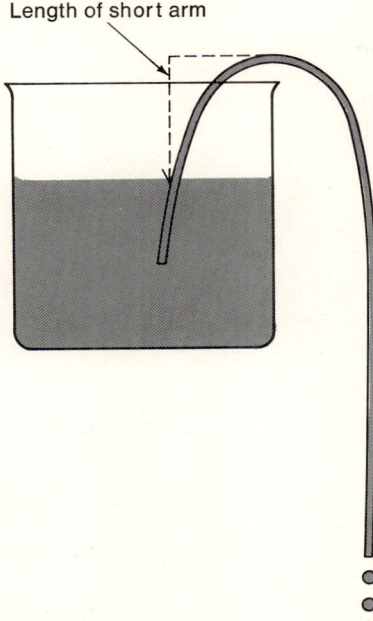

Fig. 7–5 A siphon.

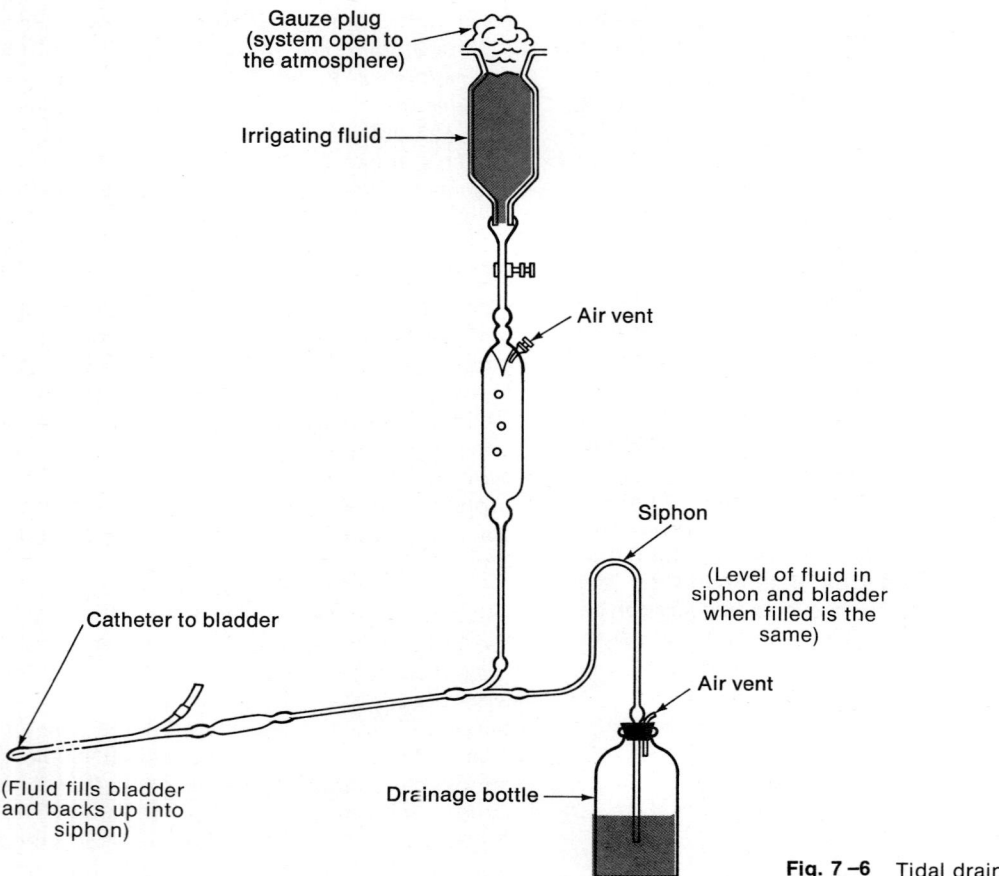

Fig. 7-6 Tidal drainage.

we permit air to enter the top of the straw, the liquid will fall. If, without letting air get into the straw, we place a finger over the top of the straw, we can lift it out of the glass with the liquid still in it. What keeps the liquid in the straw? The answer is atmospheric pressure. The medicine dropper works the same way. By squeezing the bulb on the dropper, we force out some of the air. When the dropper is dipped into a liquid and the bulb released, the greater pressure on the surface of the liquid forces the liquid into the dropper. Syringes function the same way.

The fast express train races past the station, moving air with it as it goes. As the air speed increases, its lateral pressure decreases. (This is an example of the working of the Bernoulli principle.[1]) The air by the platform, being at a higher pressure than the disturbed air by the tracks, moves into the lower-pressure area and, in this process, moves leaves and papers along.

The wings of the propeller-driven airplane are designed with the upper surfaces curved and the undersurfaces flat. Moving air passing over the curved upper surface of the wing has to travel farther than air passing

---

[1] The Bernoulli principle states *the greater the forward velocity of a fluid, the lower the lateral pressure.*

Fluids

over the straight undersurface. Since it has to travel farther, it must move faster than the air moving under the wing. When the propeller moves rapidly, it forces the air across the wings. Since the air passing over the wing moves more rapidly, its lateral pressure becomes lower. The air pressure under the wing is greater than the pressure above the wing, and the plane is pushed up into the air.

Suction apparatus used to drain body cavities consist of a drainage jar connected to the cavity and some method of removing air from the jar. If the pressure in the drainage jar is lower than the pressure in the body cavity, fluid will flow from the cavity into the jar. A device in common use is the Wangensteen suction apparatus. The recent models are powered by electricity, but the old-style ones used a flow of water from a higher level to a lower level to reduce the pressure in the drainage jar.

Wangensteen Apparatus. A drainage jar, placed below the patient's bed, has a tight-fitting cap through which run two tubes. One tube, which reaches to the bottom of the

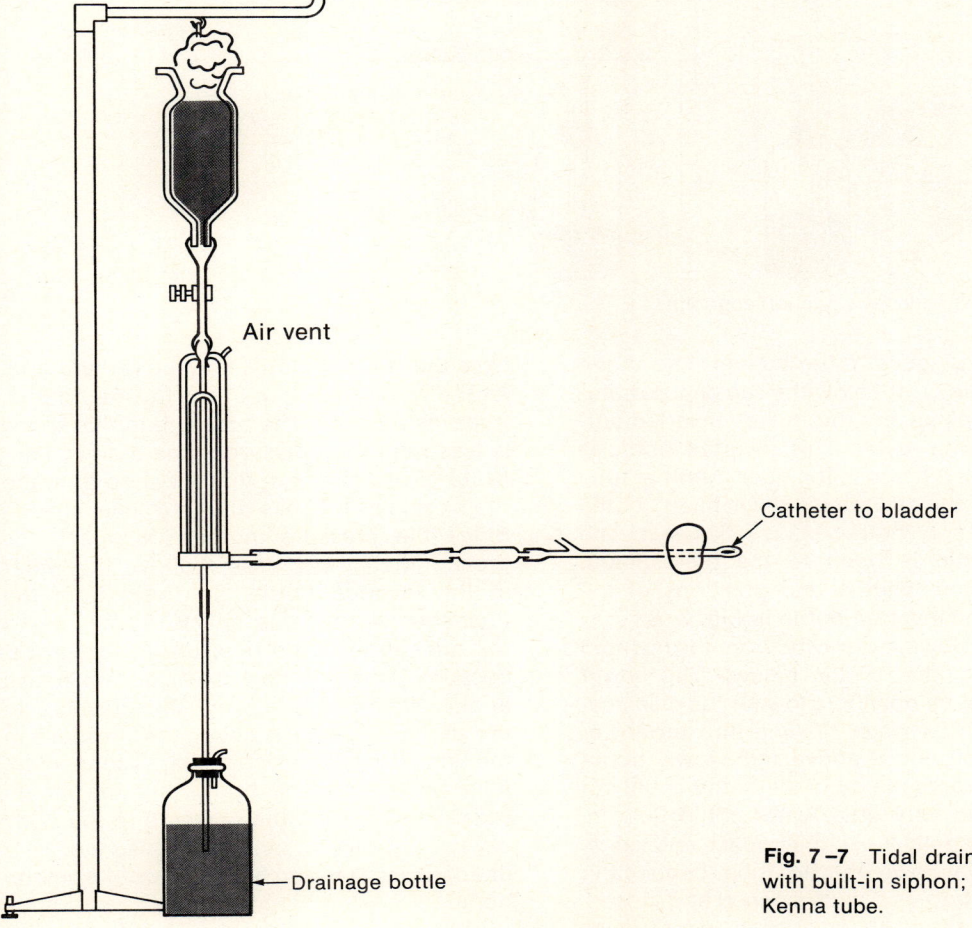

Fig. 7–7 Tidal drainage with built-in siphon; McKenna tube.

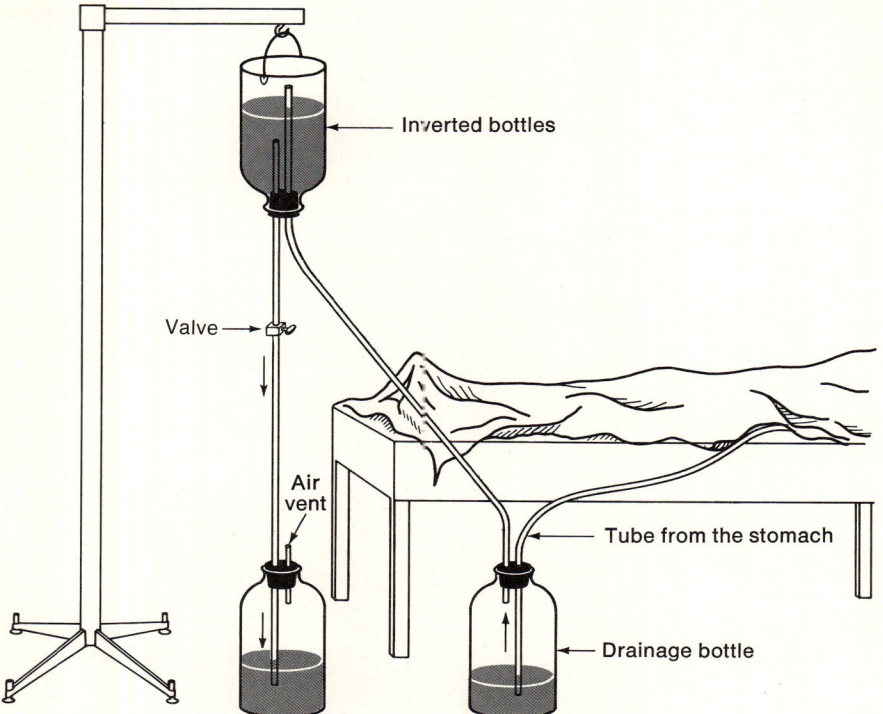

Fig. 7–8 Wangensteen suction apparatus.

drainage jar, goes to the patient. The other, which begins just below the cap on the drainage jar, runs up and through an inverted bottle filled with water. This inverted bottle is hung about 3 ft above the floor. Another tube connects the inverted water-filled bottle with an empty bottle, placed below it. The empty bottle is open to the atmosphere. (Observe the lengths and positions of the tubes in the inverted bottle in Fig. 7-8.)

Water flows, by gravity, from the inverted bottle into the bottle below. The lower bottle must be open to allow for the displacement of air by water flowing into the lower bottle from the one above. If the lower bottle were not open, the air in the bottle could not escape and very little water could flow in. Remember that air occupies space and exerts pressure. A valve in the tube connecting these two bottles can be opened and closed.

When the valve is opened, water flows from the inverted bottle into the lower bottle. As the water flows out of the inverted bottle, it leaves a space. The pressure in this space is less than the pressure in the drainage jar. (Note on the diagram that the tube from the drainage jar reaches up into this space.) Since the pressure in the drainage jar is greater than the pressure above the water in the inverted bottle, air flows from the drainage jar into this space. As air leaves the drainage jar, the pressure inside drops, because there are now fewer air molecules in the drainage jar. When the pressure in the drainage jar falls below the pressure in the body cavity to which the jar is connected, fluid flows from the cavity into the jar. In other words, when the pressure in the body cavity is greater than the pressure in the drainage jar, gas and liquid flow out from the body.

In actual practice, the two bottles involved in the flow of water (the inverted bottle and the lower bottle) are mounted on a stand equipped with a swivel and are furnished with duplicate sets of tubes. By this arrangement, it is possible to switch the bottles. When the inverted bottle empties and the lower bottle fills, the nurse can turn the swivel and swing the lower bottle into the top position and the inverted bottle into the lower position. In this setup, there are three tubes, which enable the nurse, by adjusting the valves, to connect the drainage bottle to whichever water bottle is in the top position. This procedure enables the apparatus to function indefinitely without the necessity of emptying or refilling the water bottles. Since this apparatus works without electric power, it is often kept as emergency standby equipment.

More recent models are electrically powered. A fan, a pump, or a heating element is used to reduce the pressure in a small bottle attached to the drainage jar. The explanation, however, is exactly the same. When the apparatus is connected to a source of electric current, a motor drives the fan or pump or the heating element heats up. In the case of a fan, the blades are so constructed that, as they turn, air is scooped away from the small bottle. The removal of air causes the pressure to drop. A tube runs from a drainage jar beneath the patient's bed to this small bottle in the "suction machine." As the pressure in the small bottle drops, the pressure in the drainage jar becomes proportionally greater and air leaves the drainage jar via the tube to the small bottle. As the air leaves the drainage jar, the pressure drops. When the pressure is less than that in the body cavity being drained, fluid flows out from the body into the drainage jar (Fig. 7–9).

Sometimes an air pump is used to reduce the pressure in the small bottle. This little motor-driven pump is essentially the same as the hand-powered pump used to inflate bicycle tires. It consists of a cylinder with intake and outlet valves and a movable piston. The valves are so constructed that only one valve is open at a time. The intake valve opens into the cylinder, and the outlet valve opens to the outside. Both valves in the bicycle pump are at the bottom and on opposite sides of the cylinder. When the bicycle pump is not in use, the piston rests on the bottom of the cylinder so that there is practically no space between the bottom of the piston and the bottom of the cylinder. As the piston is pulled up, the space between the bottom of the piston and the bottom of the cylinder increases. The air in this space expands, and the pressure decreases (Boyle's law). The atmospheric pressure outside the pump is greater than the pressure in the cylinder. The greater pressure pushes open the intake valve, which opens inward, and air from the

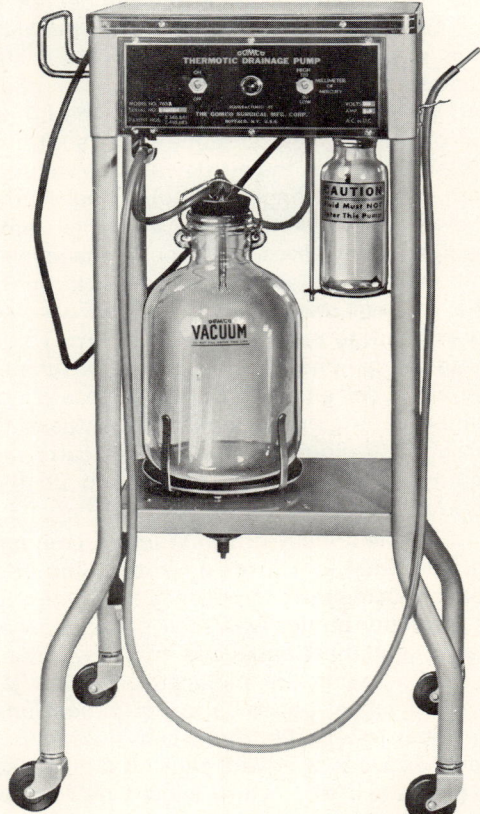

Fig. 7–9 Electrically powered suction apparatus. *(Courtesy of Comco Surgical Manufacturing Corp.)*

outside enters the cylinder. When the piston is pushed down, the volume decreases, the air pressure in the cylinder increases and becomes greater than the atmospheric pressure outside; the greater pressure forces open the outlet valve, which opens outward, and air flows out of the cylinder—from greater pressure to lower pressure.

Observe that this pump can be used two ways. By attaching a tube to the outlet valve, the pump can be used to force air into a container. We do this when we inflate the bicycle tire or fill an air ring or an air mattress. But if we attach a tube to the intake valve and run the tube to a bottle, what will happen to the air in the bottle? In this case air will flow into the cylinder on the upstroke of the piston, and the pressure in the bottle will decrease.

Most of us have observed that the vacuum sweeper accomplishes essentially the same thing as the air pump; air flows in one part of the machine, the vacuum part, and out of the machine at some other spot, the pressure part. As the air flows in the vacuum part, it carries along with it dust and debris. The dust and debris behave in the same fashion as the leaves and paper at the railway station when the fast express passes through. The vacuum sweeper can also be used to spray paint, to mothproof clothes, etc. To do these things, one attaches a hose or tube to the air outlet and runs the hose to a jar of paint or mothproofing liquid. The air passing through the liquid in the jar, will carry the liquid along through a nozzle and do the spray job.

The electrically driven Wangensteen apparatus, whether it uses a fan or pump or a heating element, accomplishes the same end as the water bottles with their gravity flow of water. In all these situations, there is a drainage jar beneath the patient's bed. This jar has two tubes; one runs to the patient and the other to a bottle from which fluid (either water or air) is removed, causing a decrease in pressure. This in turn causes air to flow from the drainage jar into the bottle, dropping the pressure in the drainage jar. When the pressure in the drainage jar drops below the pressure in the body cavity to which it is connected, e.g., the stomach or intestines, gas and liquid flow out of the body and into the drainage jar.

(Let us note in passing that prolonged suctioning may cause the patient to lose fluid and electrolytes to a point where there is danger of dehydration or electrolyte imbalance. It is vitally important that the nurse keep an accurate record of the amount and kind of fluid lost in these cases.)

other drainage methods

In addition to the methods described above, other devices can be used to reduce the pressure in the drainage jar. One of these is a venturi tube. This is a short tube that can be screwed on to the water outlet or faucet in a sink or basin. The top of the tube, the part that screws on to the faucet, is wider than the bottom. A small open side arm is attached to the constricted part of the tube. As water flows through the narrow part of this tube, its rate of flow must increase over the rate of flow through the wider part. This same phenomenon can be observed in the behavior of rivers. Where the river bed is wide, the river flows calmly, but in the narrow parts are found the rapids. In both situations, all the water must flow past. To do this, it must flow more rapidly through the narrow parts.

In flowing fluids, both liquids and gases, as the rate of flow increases, the sidewise pressure decreases. This observation is named for the man who discovered it, Daniel Bernoulli, and is called the Bernoulli principle. As the water flows rapidly through the narrow part of the venturi tube, the lateral pressure drops. The venturi tube is connected to the atmosphere by the open side arm. When the pressure in the tube drops below atmospheric pressure, air rushes into the tube through the side arm.

If the side arm is connected to a drainage jar, air will flow from the drainage jar through the side arm, and as air leaves the drainage bottle the pressure inside will drop.

Still another method is the use of the siphon. This is called the *siphon suction apparatus*. It consists of three bottles: a drainage bottle and two bottles for the siphon. The drainage bottle and the bottle containing the short arm of the siphon are placed on a stand or table, and the bottle with the long siphon arm is placed on the floor. A tube runs from the patient to the drainage bottle, and another tube runs from the drainage bottle to the bottle containing the short siphon arm. This latter tube starts and ends just beneath the cap on both bottles. A tube that reaches from the bottom of the bottle that we are calling the short-siphon-arm bottle to the bottle on the floor comprises the siphon. The bottle on the floor must be open to the atmosphere. At the start of the procedure, the short-siphon-arm bottle is filled with water, and the long-siphon-arm bottle is empty. With the exception of the openings for the tubes, the drainage bottle and the short-siphon-arm bottle are sealed off from the atmosphere but are connected to each other and to the patient by the tubes. When the siphon arms have been filled with water, water flows from the short-siphon-arm bottle into the long-siphon-arm bottle below it. As the water flows out of the short-siphon-arm bottle, it leaves a space in which the pressure is lower than it is in the drainage bottle. Air then flows from the drainage bottle into the short-siphon-arm bottle. As air leaves the drainage bottle, the pressure falls below the pressure in the body cavity to which it is connected and fluid flows from the patient into the drainage bottle.

Observe that the physical principle involved in all these setups is identical. In those involving flowing water, the nurse can observe the flow rate and can check the tubes to be certain that they are open. The electrically powered machines are housed in a case on a portable stand. Dials or buttons enable the operator to regulate the rate of the suction. They also have a light signal to indicate that all is working properly. Since these machines are plugged into an electric outlet, there is always the possibility that, if the machine is not carefully placed, a person passing may inadvertently loosen or dislodge the connection. If the light on the machine goes out, the electric plug is the first thing to check.

"bubble drainage"

The devices described are used principally to drain or aspirate the gastro-intestinal tract. Frequently, in thoracic or urologic surgery or where mucous membranes are involved, it is desirable to have a very small pressure difference between the drainage bottle and the body cavity. It may also be necessary to eliminate fluctuations in pressure difference. This is accomplished by inserting into the line between the pump and the drainage bottle a bottle partially filled with water and containing a tube that extends below the surface of the water and above the bottle cap. The tube above the bottle cap is open to the atmosphere. It is in this bottle that bubbles of air form.

Sometimes to ensure sterility or as a safety precaution other bottles are also used. Essentially, the apparatus consists of the drainage jar, which is attached by one tube to the patient and by another to the bubble bottle; the bubble bottle, which is connected to the drainage jar and the pump; and the pump.

To understand how this apparatus works, let us realize that the pump, the bubble bottle, the drainage bottle, and the body cavity are all connected by tubes so that a change of pressure in any part of this system will affect all the other parts. Let us also recall that pressure can be expressed in several ways: pounds per square inch; grams per square centimeter; centimeters or millimeters or inches of mercury; and centimeters or millimeters of water. When describing small pressures or pressure differences, centimeters or millimeters of water are used in

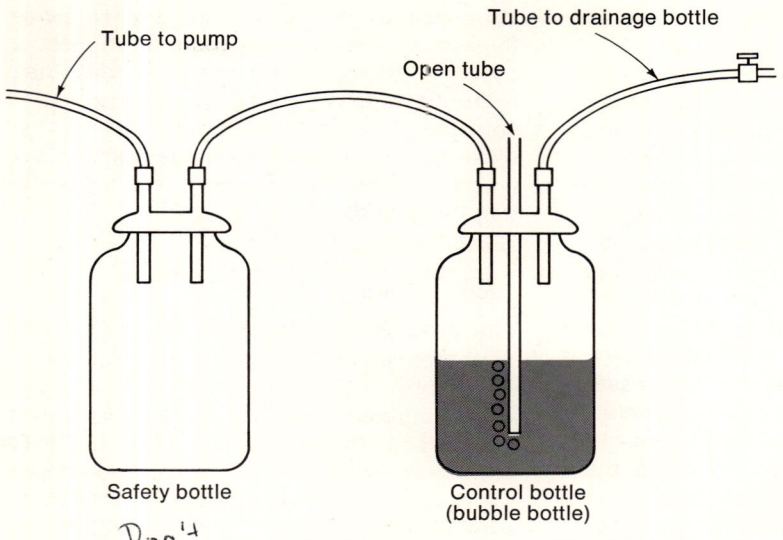

Fig. 7–10 Part of "bubble drainage" system.

preference to measurements with mercury. If the procedure calls for a pressure difference of 5 cm, or, what is the same thing, 50 mm, water, that is the same as 0.367 cm, or 3.67 mm, mercury. It is obviously easier, in this case, to work with water rather than mercury measurements.

As we have mentioned, in postthoracic surgery not only are small pressure differences needed, but these pressures should be steady. The bubble bottle solves both these problems. It is the regulator for the system.

Let us look at the diagram and figure out what would happen if we disconnected the drainage jar and worked with only the pump and the bubble bottle (Fig. 7–10). The pump, by its piston action, allows air to flow from the bubble bottle out through the pump. We have the regulating tube submerged 5 cm below the surface of the water, but since we have left open the connection that should go to the drainage jar, as fast as the air flows out through the pump it flows in through this open connection. It does not flow in through the regulating tube because fluids, like many other things, always take the path of least resistance, and the regulating tube is closed off from the atmosphere by 5 cm water. Now if we stopper the connection to the drainage jar, what happens? The electrically powered pump continues to work, and in a short while we see bubbles of air coming into the bottle from the atmosphere, down through the 5 cm water in the tube and joining the air above the water in the bottle. This air will eventually leave the bottle through the pump. The question now arises: At what point did the air bubbles form in the bottle?

We have concluded that before we stoppered the connection that should go to the drainage jar, air flowed into the bottle through this open connection. We know that as air flows out through the pump, the pressure in the bottle drops and that if the atmospheric pressure is greater than the pressure in the bottle, air will flow into the bottle. With the bottle sealed off from the atmosphere with 5 cm water, air cannot enter the bottle until enough air leaves the bottle through the pump that the pressure of the air above the water plus the pressure of 5 cm water adds up to a total pressure that is less than that of the atmosphere. In other

Fluids

words, when the pressure of the air above the water in the bottle drops to the point where the difference between the atmospheric pressure and the pressure in the bottle is greater than the pressure exerted by 5 cm water, the water can no longer seal out the air and it bubbles into the bottle.

When the regulating tube is submerged 5 cm below the suface of the water, the air pressure in the bottle can fall to 756.3 mm mercury and then air will enter in bubbles through the water. We can figure this out when we recall that 1 atmosphere of pressure is equal to the pressure of 76 cm mercury, or 1,034 cm (34 ft) water. This gives us the following relationship: 1 cm water is equal to 0.0735 cm, or 0.735 mm, mercury. The 5 cm water offered a resistance to the entrance of air equal to 5 × 0.735 mm mercury, or approximately 3.68 mm mercury. The atmospheric pressure is 760 mm mercury; if we subtract 3.68 from 760, we realize that when sufficient air has left through the pump that the air pressure is 756.3 mm, a 5-cm water seal is no longer sufficient to shut out air from the outside at 760 mm pressure.

It should also be evident from this that the pressure in the bottle cannot drop below 756 mm no matter how fast the pump works because air keeps bubbling into the bottle at this point. If the pump should slow up for any reason, the rate of flow into the bottle will slow up also. If the pump speed increases, the rate of flow of air into the bottle will increase accordingly.

If the procedure calls for a pressure difference of 10 cm water, the pressure in the bottle can drop to 760 mm − (10 × 0.735), or 752.6 mm mercury and air will bubble into the bottle and keep the pressure at this point.

Note that we have carried through this explanation with the drainage jar closed off from the bottle. This should impress us with the idea that the appearance of bubbles in the regulator bottle does not indicate that all is well. The tube from the patient to the drainage jar or the tube from the drainage jar to the regulator bottle could be clogged, and the regulator bottle would be bubbling more rapidly than would be the case if all the tubes were working properly. Why would this be so?

Let us make all the proper connections (Fig. 7–11). We have a tube from the pleural cavity to the drainage jar. Usually one tube

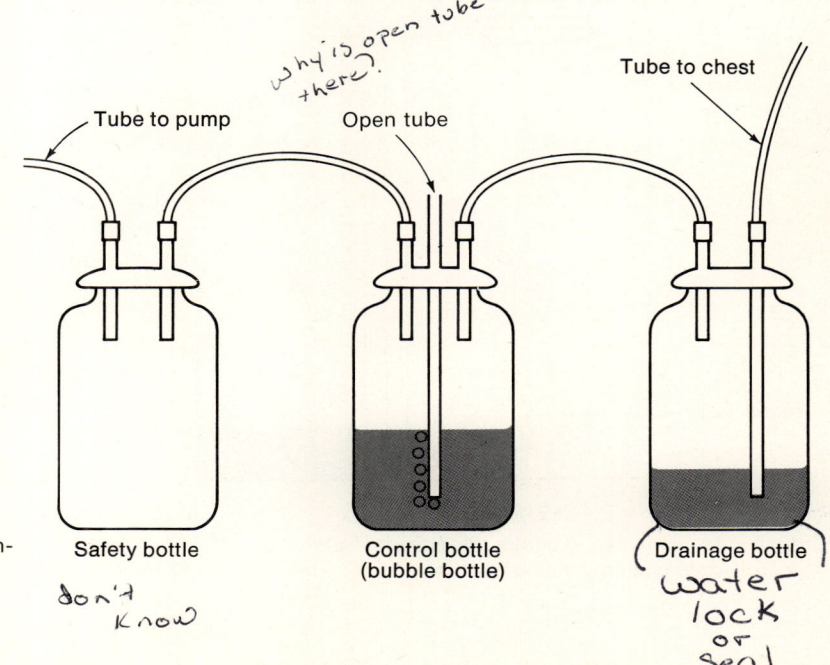

Fig. 7–11 "Bubble drainage" system.

is inserted anteriorly through which gas and some liquid can flow, and another tube is inserted posteriorly through which liquid drains. These two tubes are joined into one with a Y connection before the tube reaches the drainage jar. We have a tube from the drainage jar to the regulator bottle and a tube from the regulator bottle to the pump. The regulator bottle has a glass tube submerged 6 cm below the surface of the water and extending out through the bottle cap into the air. The pump is turned on. As air flows from the regulator bottle into the pump, the pressure in the regulator bottle drops; but now if the pressure in the regulator bottle is lower than that of the drainage jar to which it is connected, air will flow from the drainage jar into the regulator bottle; when the pressure in the drainage jar drops below the pressure in the cavity, fluid flows from the cavity into the drainage jar. This process continues until the pressure in the drainage bottle drops to $760 - (6 \times 0.735)$, or about 755.5 mm mercury, at which point air will bubble into the bottle.

Although the pressure in the regulator bottle will stay at 755.5 mm, this pressure is lower than the pressure in the cavity and fluid will continue to flow into the drainage jar and from the drainage jar into the regulator bottle. Therefore, when all the tubes are open and working properly, less air enters through the regulator tube because some air enters the bottle from the drainage jar.

The nurse must inspect the tubes to see that they are patent and must make certain that the regulator tube remains at the proper depth beneath the surface of the water. If some of the water in the regulator bottle should evaporate or be spilled, the difference in pressure between the drainage jar and the body cavity would be less than that desired by the doctor.

The deeper the tubes are submerged below the surface of the water, the more effective is the water seal and the larger is the pressure difference that can be maintained between the drainage jar and the body cavity.

A single disposable plastic container

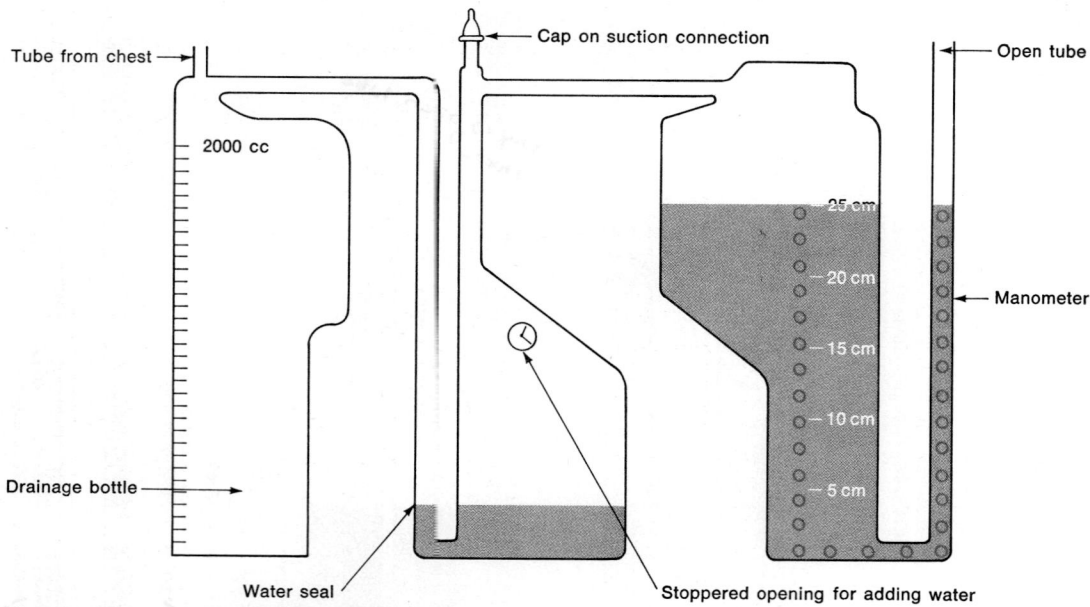

Fig. 7-12 Simplified diagram of thoraseal disposable triple-chamber suction bottle.

with three chambers has been devised which can be used as either simple underwater drainage (closed chest) or as bubble suction (Fig. 7–12). When used as closed-chest drainage, only the first two chambers are used. The rest of the device serves as an opening to the atmosphere. The stopper on the middle chamber is removed and sterile water is added through the hole to the desired level to form a water seal. The stopper is then replaced tightly. When the tube from the drainage bottle is attached to the chest, the water seal prevents air from entering the pleural space.

When used for bubble suction, sterile water is added to the third chamber through the manometer. The third chamber acts as the regulator or bubble bottle. The cap on the suction connection is removed and the tube leading to the pump or other source of suction is attached. Thus, as air is removed by the pump, when the desired pressure drop has been achieved (as determined by the depth of the water in the regulator bottle), air will bubble in through the manometer.

The first chamber is marked in cubic centimeters so that the volume of drained fluid can be determined. The second and third chambers are marked in centimeters so that the depth of water can be ascertained.

Study Exercises

1. State Boyle's law. How does Boyle's law help to explain external respiration (inhalation, exhalation)?
2. State Charles' law. Why does cold air collect at the floor and warm air at the ceiling?
3. State Dalton's law. What would be the effect of inhaling a gas mixture in which the partial pressure of carbon dioxide was greater than the partial pressure of carbon dioxide in the blood?
4. State Graham's law. Why is it easier to breathe a mixture of 20 per cent oxygen and 80 per cent helium than a mixture of 20 per cent oxygen and 80 per cent nitrogen?
5. State Henry's law. Caisson disease is an occupational hazard for persons who work under high pressures. Explain why this is the case.
6. State Archimedes' principle. Why is it easier to stay afloat in salt water than in fresh water?
7. Define cohesion, adhesion, surface tension, capillarity, fluidity, and viscosity.
8. In an automobile, why is it necessary to replace summer oil with winter oil?
9. A patient is receiving an intravenous infusion. What would be the effect of raising the infusion bottle? What would be the effect of lowering the bottle?
10. Explain how a siphon can be used to wash out the stomach.
11. In tidal drainage, what determines the amount of fluid entering the bladder?
12. How does Boyle's law explain simple underwater drainage and the operation of the respirator?
13. What is meant by subatmospheric pressure? Describe three ways that a container can be brought to subatmospheric pressure.
14. What is the function of the control bottle in the "bubble drainage" system?
15. If some of the water were spilled from the control bottle (not enough to expose the open tube to the atmosphere), what effect would this have on the pressure in the drainage bottle?

chapter eight IMPORTANT GASES

Ideas for preview or review

In a mixture the individual components retain their properties. Air is a mixture.

Decomposition is a chemical change in which a complex substance is broken down into simpler substances.

Density is defined as mass per unit volume.

A calorie (cal) is the amount of heat needed to raise 1 Gm water 1°C.

Pressure is defined as force per unit area.

The pressure of the atmosphere accepted as standard pressure is the pressure at sea level under certain chosen conditions. It is a pressure that will support a column of mercury 760 mm high and is called simply 760 mm.

OXYGEN

In a short course such as this, many substances cannot be studied with any degree of completeness. In the interest of science, however, at least one substance should be studied thoroughly: oxygen, one of the commonest of the gases and one that concerns us vitally. It will be worth while to show its occurrence, discovery, preparation, properties, and uses, since we shall discuss many subjects of a general nature using oxygen as the specific example.

Occurrence

Oxygen occurs *free* in nature. The word "free" here means that it is uncombined and free in the economic sense—it is ours for the taking. We find it, in this free state, chiefly in the atmosphere, where it is mixed with nitrogen roughly in the ratio of 1:4.

The relatively inert nitrogen serves to dilute the more active oxygen and thereby keep oxidation processes moderate. The percentage of oxygen in the atmosphere remains remarkably constant. Although it is true that oxygen is continually consumed by the respiratory process in all animals, by combustion, and by slow oxidation of metals, at the same time oxygen is being produced by all the growing plants through the photosynthesis of carbohydrates. This would mean, perhaps, that the atmosphere in the country might temporarily have a higher percentage of oxygen than that in the slums of a large city. However, the continual changing and mixing of the atmosphere by the winds accounts for the constant composition. In addition to the free oxygen in the atmosphere, a small amount is dissolved in the waters of the earth.

Oxygen occurs *chemically combined* in many compounds. Eight-ninths of water by weight is oxygen, the other ninth being hydrogen. Oxygen is contained in the silicates and carbonates of rocks and in sand, or silica, SiO_2. Nitrates, which occur in large beds such as the deposit of saltpeter, $NaNO_3$, in Chile, contain oxygen. Finally, the great bulk of the organic compounds—notably carbohydrates, fats, and proteins—contain oxygen.

Discovery

Oxygen was discovered by Joseph Priestley, an English minister, in 1774, while he was heating mercuric oxide. It was left to the Frenchman Lavoisier to work out the properties of this new gas and give it a name. He called it *oxygen,* meaning *acid former,* for he believed it to be a constituent of all acids. We know today that not all acids contain oxygen, although many, such as sulfuric acid, nitric acid, phosphoric acid, and all the organic acids, do.

Preparation

When we discuss the preparation of a substance, it is implied that we wish to obtain it in sufficient quantity to collect it easily, study it, and use it. We are not interested in theoretical methods of formation for their own sake. To produce a substance in quantity we naturally seek a cheap, abundant source and a method that is easily carried out. There is nothing cheaper than air, and we naturally turn to this abundant source for our oxygen.

fractional distillation of liquid air

It has already been stated that the atmosphere is a mixture of oxygen and nitrogen. If we could remove the nitrogen, the residual gas would be oxygen. A moment's consideration, however, shows that any chemical sufficiently active to react with the relatively inert nitrogen certainly would react with the more active oxygen. It is therefore impossible to separate oxygen from nitrogen by any *chemical* method. Instead we must use a *physical* process.

If we cool the atmosphere sufficiently and then greatly increase the pressure on it, we may liquefy it. Liquid air is a solution of liquid nitrogen and liquid oxygen that may be easily separated into its components because of the difference in their boiling points.

In liquid air nitrogen boils at −195.5°C and oxygen at −182.9°C. Accordingly as the temperature of the mixture is allowed to rise, the nitrogen boils off first and is collected in tanks; the oxygen then vaporizes and is collected. The mixture has been separated into two fractions by distillation; hence, the process is called *fractional distillation.*

electrolysis of water

The next most abundant source of oxygen is water, in which it is chemically combined with hydrogen. Pure water is a poor conductor of electricity. In order to carry out this

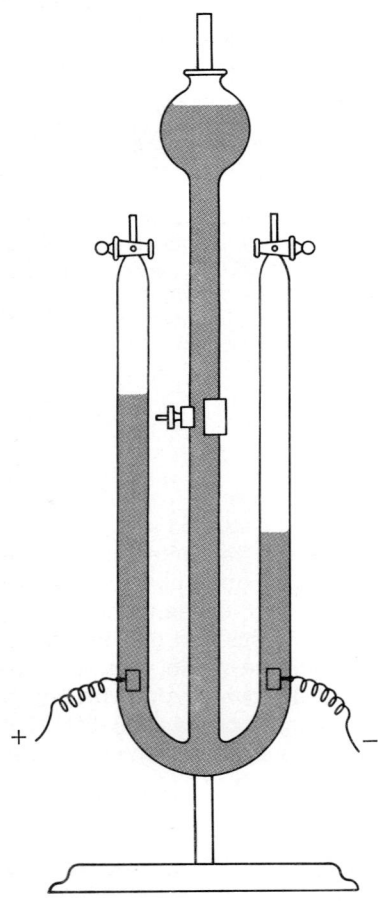

Fig. 8-1 The Hofmann apparatus for the electrolysis of water.

Important Gases

decomposition, therefore, a solution of sulfuric acid is used in the Hofmann apparatus shown in Fig. 8–1. Hydrogen forms at the cathode (negative electrode) and oxygen at the anode (positive electrode) in the ratio of 2 volumes of hydrogen to 1 volume of oxygen. The sulfuric acid is not used up in this reaction and might be considered as a catalyst.

This method is relatively slow, but it produces both gases in a very pure state. It would not serve as a means of preparing the gases for general student use in the laboratory and can be used commercially only when a cheap source of direct current is available.

decomposition of compounds of oxygen

Theoretically, any compound of oxygen should be a source of oxygen. Practically, however, many of the compounds are so stable that under ordinary conditions it is impossible to obtain sufficient energy to decompose them. For instance, calcium oxide, even under the intense heat of the oxyhydrogen flame, does not decompose.

A satisfactory laboratory method involves the heating of potassium chlorate, $KClO_3$, a white solid which melts at 357°C and decomposes to form oxygen and potassium chloride at 400°C according to the equation[1]

$$2KClO_3 \rightarrow 2KCl + 3O_2 \uparrow$$

This reaction can be greatly accelerated by mixing a small amount of manganese dioxide, MnO_2, with the potassium chlorate. When this is done, oxygen begins to form at 200°C and the evolution is smooth and easily controlled. Here the reaction is being catalyzed by the manganese dioxide, and the reaction should therefore be written

$$2KClO_3 \xrightarrow{MnO_2} 2KCl + 3O_2 \uparrow$$

Potassium chlorate solution has been used as a gargle and a mouthwash, presumably because of its oxidizing ability. However, its effect on bacteria depends upon osmosis, and a solution of this salt is no better than a solution of any other salt of the same concentration.

Properties

physical

Oxygen is colorless, odorless, and tasteless. For all practical purposes the gas may be considered insoluble in water. For instance, we collect it over water without any noticeable loss. However, oxygen is sufficiently soluble in water to supply the needs of fish, and a small amount is dissolved in the blood under normal atmospheric pressure.

The question of the density of oxygen gives us an opportunity to discuss a method of determining the relative density of gases. In comparing gases we may consider the weights of equal volumes; a liter of oxygen, for example, under standard conditions (0°C and 760 mm pressure) weighs 1.429 Gm, and an equal volume of hydrogen weighs 0.0896 Gm. Oxygen, therefore, is 16 times heavier than hydrogen. However, it is not necessary to remember such figures, for we may make use of the fact that a mole[1] of any gas is contained in 22.4 liters under standard conditions of temperature and pressure, a volume known as the *molar volume*. Accordingly 32 Gm (2 × 16) of oxygen, or 2 Gm (2 × 1) of hydrogen, or 44 Gm (12 + 32) of carbon dioxide each occupy the same volume, 22.4 liters, under standard conditions. We now see that oxygen is 16 times heavier than hydrogen and that carbon dioxide is 22 times heavier. But we are not so much interested in that fact as we are in their weights compared to air, which, being a mixture, does not have a molecular weight. We may calculate the weight of 22.4 liters air, though, and we find it to be 28.8 Gm. To

---

[1] This reaction is typical of the decomposition reaction of *all* chlorates. Such a generalization eases the burden of remembering chemical facts.

[1] One gram-molecular weight is equal to 1 mole.

determine the relative density of any gas, we merely calculate its molecular weight (atomic weights are always available) and compare it with 28.8; if the weight is greater than this figure, the gas is heavier than air, and vice versa. Oxygen, with a molecular weight of 32, is heavier than air. This being the case, bottles filled with oxygen and loosely covered should stand upright on the bench to reduce losses. Oxygen should be fed into the top of an oxygen tent, as its greater density causes it to fall toward the patient's head.

chemical

To determine the chemical properties of oxygen we must make it take part in a series of experiments.

First, oxygen does not burn. If it did, since the air is about 20 per cent oxygen, if one lit a match, the air would ignite and then where would he be?

If a piece of sulfur in a deflagrating spoon is placed in a jar of oxygen, nothing happens. If, however, the sulfur is heated until it begins to burn and then replaced in the oxygen, it burns with a bright blue flame. A colorless gas with a choking odor, sulfur dioxide, is formed by the reaction

$$S + O_2 \rightarrow SO_2 \uparrow$$

At the same time a finely divided white solid, sulfur trioxide, is formed in small quantities.

$$2S + 3O_2 \rightarrow 2SO_3$$

If the experiment is repeated with phosphorus, a dazzling white light is observed and a white solid, phosphorus pentoxide, is formed according to the reaction

$$4P + 5O_2 \rightarrow 2P_2O_5$$

Carbon in the form of charcoal reacts in a less spectacular fashion. *A flame is always caused by burning gases or vapors.* The charcoal was made by heating wood out of contact with the air, thus driving off the volatile substances. This process is called *destructive distillation.* Since there is nothing left in charcoal to produce a flame, it merely glows brightly when heated and placed in the oxygen. Carbon dioxide, a colorless gas, is formed by the reaction.

$$C + O_2 \rightarrow CO_2 \uparrow$$

Iron in the form of steel wool reacts vigorously when heated and thrust into the oxygen. The product here formed is magnetic iron oxide.

$$3Fe + 2O_2 \rightarrow Fe_3O_4$$

These few reactions were chosen as fairly typical ones, and from them we may draw the conclusion that oxygen is a very active substance, reacting with both nonmetals and metals to form the corresponding oxides. There are a few exceptions to this general statement. Oxygen does not react with the halogens—fluorine, chlorine, bromine, and iodine—nor with the metals silver, gold, or platinum. Of course, there would be no reaction with the inert gases.

Our discussion so far has involved only the reaction of oxygen with elementary substances. Compounds composed of combustible elements unite with oxygen to form the oxides of those elements. For instance, methane, a hydrocarbon (composed of carbon and hydrogen), burns to form carbon dioxide and water.

$$CH_4 + 2O_2 \rightarrow CO_2 \uparrow + 2H_2O \uparrow$$

The formal *test for oxygen* is made by inserting a glowing splint into a container of the gas; if the splint bursts into flame, the gas is proved to be oxygen.

In summary, it has been shown that oxygen (*a*) is a very active elementary substance, (*b*) supports combustion, and (*c*) is noncombustible.

Uses

From the point of view of a nurse the most important use of oxygen is therapeutic. We find it used whenever a concentration of oxygen is required higher than that in normal air; for instance, in reviving an asphyxiated person, in anesthesia, or for the relief of a patient suffering from pneumonia or any condition in which the lung capacity is reduced by fluid. Further, for infants whose first moments of life are precarious because of difficulty in breathing or slow expansion of the lungs, the resounding slap that introduced most of us to this troubled world has now been replaced by more scientific methods. They are now given a mixture of oxygen and carbon dioxide, the latter acting as a respiratory stimulant. An obstructed passageway to the lungs, reducing the availability of oxygen to the lungs, is another cause for the use of oxygen therapy. Finally, there must be sufficient blood reaching the lungs to absorb the necessary amount of oxygen, and the hemoglobin content of the blood must be sufficiently high to provide the means of transporting the oxygen needed. Oxygen therapy gives marked relief to patients with cardiac insufficiency. Hemoglobin carries the oxygen in the blood by the formation of oxyhemoglobin, an easily decomposed compound. Obviously, any condition that reduces the hemoglobin concentration of the blood reduces the amount of oxygen made available within the body. Such conditions are found following a hemorrhage or following the inhalation of a gas, such as carbon monoxide, that forms a fairly stable compound with the hemoglobin. In summary, any patient who shows signs of *anoxemia* (oxygen want) is benefited or made more comfortable through oxygen therapy.

As was mentioned in connection with Henry's law of the solubility of gases, the amount of oxygen transported by the blood can be increased by increasing the pressure and thus increasing the amount of oxygen dissolved in the plasma or liquid part of the blood. Special chambers have been constructed called *hyperbaric chambers*. In these chambers, the pressure can be increased to 3 times that of the atmosphere, greatly increasing the oxygen content of the cells.

Persons who expect to attain high altitudes carry with them apparatus for increasing the oxygen content of inspired air, especially if the change in altitude is to take place rapidly. In a gradual ascent the body is able to compensate to a great extent for the reduction in the partial pressure[1] of the oxygen. In the sudden ascent of a flier, however, there is insufficient time for the body to adjust itself to the new conditions. Oxygen masks and pressurized cabins are the answer to these problems.

Industrially, oxygen is used in huge quantities in the cutting and welding of steel to produce the high-temperature flame of the oxyhydrogen or oxyacetylene blowtorch. The construction of steel skyscrapers, the building of steel bridges, and the assembling of prefabricated sections of ships are just a few examples of what can be done with these torches.

Combustion

Combustion is a chemical reaction between two substances that evolves heat and light. Notice that in this definition nothing has been mentioned about oxygen. Oxygen is not required by definition for combustion. A jet of hydrogen burns in oxygen to form water according to the equation

$$2H_2 + O_2 \rightarrow 2H_2O \uparrow$$

If, now, the jet is allowed to burn in a jar of chlorine, combustion takes place with the formation of hydrogen chloride, a gas that fumes in moist air.

$$H_2 + Cl_2 \rightarrow 2HCl \uparrow$$

Combustion involving oxygen, then, is only one kind of combustion. Since that type is the most common and the one that interests us most, we shall devote our attention to the

---

[1] Each gas in a mixture exerts its own pressure (partial pressure), and the total pressure of the mixture is equal to the sum of the partial pressures. (Dalton's law.)

conditions that influence combustions of that type.

Three conditions are necessary for a combustion or fire to take place.

a. There must be a combustible[1] substance present, i.e., one that will react with oxygen, in our case.
b. A supporter of the combustion must be present—in our present discussion, oxygen.
c. The temperature of the combustible substance must be raised to the point at which the reaction will continue without any further application of heat from an outside source. This temperature is called the *kindling temperature*. These three conditions must exist simultaneously to produce a fire; if any one of the three conditions is removed, a fire must cease to exist. Here, then, we have the cause of all fires and at the same time suggestions for extinguishing fires.

The first condition is obvious. Some substances will not react with oxygen; they are noncombustible. It is just as important to us that we have noncombustible substances as that we have combustibles, since without them we would not be able to confine a fire. Materials such as brick, stone, or clay are noncombustible and so find application as linings for fireboxes or as fire walls and building materials.

Kindling temperature has been defined as the temperature to which a substance must be heated in order for it to continue to burn of its own accord. It is common knowledge that different substances burn at different temperatures, i.e., have different kindling temperatures. It is easier to start a wood fire than to start a coal fire. White phosphorus ignites at such a low temperature that it must be stored and cut under water.

[1] Synonyms for combustible are flammable and inflammable.

Now let us see if a given substance has a fixed kindling temperature. A steel file cannot be ignited even if held in the bunsen flame. Steel wool, on the other hand, can be ignited with a match. Iron, then, does not have a fixed kindling temperature. Again, in building a log fire we do not hold a lighted match under a log and expect any action. Instead we use small chips, twigs, or perhaps excelsior, substances whose low kindling temperatures are attainable with a match flame. We then use the heat produced by these materials to raise the temperature of larger pieces to their kindling temperature, and so on, until the kindling temperature of the log is attained. Wood also does not have a fixed kindling temperature.

Why should the kindling temperature of a given substance vary? The answer involves the fundamental principles concerning the rate of a reaction. If a reaction is to take place at all, the molecules of the reactants must come together, or collide. Any condition that increases the frequency of these collisions should increase the rate of the reaction. The more finely we divide the substance, the greater will be the surface area and the greater will be the number of molecules subject to collision. The rate of reaction is increased by increasing the surface area. Consequently, a finely divided material has a lower kindling temperature than a large chunk of the same material.

While we are discussing the conditions affecting the rate of reaction, we may carry the discussion a little further to explain why different substances have different kindling temperatures. According to the kinetic molecular theory, the molecules of a substance are in continual random motion. The extent of this motion depends upon the kinetic energy of the molecules, which in turn is a function of the temperature. As the temperature of a substance is increased, the velocity of the molecules increases and in turn the number of collisions increases. Further, the increased temperature makes the collisions more effective in producing the reaction between the molecules. We now have the explanation of the fact that the kindling temperature varies with different substances.

Another factor that affects the rate of reaction and consequently varies the kindling temperature of a substance is the concentration of the reactants. The concentration of a solid or a liquid cannot be readily changed, but we may vary the concentration of a gas. Thus we find the kindling temperature of a given substance is lower in pure oxygen than in air, or stating it the other way, substances burn better in oxygen than they do in air. Because of the increased concentration of the gas there are more collisions, and therefore there is a higher rate of reaction. A variation of this idea explains the function of drafts in a stove or the use of a bellows on an open fire. The concentration of the oxygen in the air is not changed, but because of the increase in the rate of flow of the air, more molecules of oxygen come in contact with the burning substance in unit time, which is equivalent to an increase in the concentration of the oxygen.

Quiet Burning versus Explosion

When a jet of pure hydorgen is ignited, the gas burns quietly with a colorless flame, giving off a large amount of heat. As the gas issues from a tip attached to a source of hydrogen, it unites with the oxygen of the air and the process is a slow one. In this manner it would take some time to burn the hydrogen. On the other hand, if hydrogen were previously mixed with oxygen in the ratio of 2:1, the proportions of the gases in water, we would be very conscious of the amount of energy evolved for there would be a loud explosion. We must now consider the cause of the difference in the two experiments.

The same *amount* of heat is always evolved in a given chemical reaction so long as the same amount of reactants is used in each instance, regardless of the length of time required for the action. In the case of the explosion, all the heat was evolved in a split second; the heat was *intense* for a short time. We measure quantity of heat in calories, but we express intensity of heat in degrees Celsius and measure it with a thermometer. Because of the high temperature there was a tremendous expansion of water vapor formed in the reaction. The air was disrupted, setting up sound waves that reached our ears, and we were conscious of a loud noise. No matter whether we think of an explosion as noise or as the bursting and rending of something, it always results from the same cause, a sudden expansion of gases. A solid explosive *forms* gases suddenly upon detonation. The difference between an explosion and quiet burning is the difference in the time element of the exothermic reaction, which depends upon whether the gas was *previously* mixed with oxygen or allowed to mix slowly and gradually as it burned. Accidental explosions are not uncommon. Extensive damage has been caused by explosions when suicides have turned on the gas in a room containing an open flame, such as a pilot light.

This discussion should enable the student to explain why extreme precautions to prevent fires or explosions must be observed in the operating room and in rooms in which patients are confined in oxygen tents. In the operating room, vapors of volatile, highly flammable anesthetics such as diethyl ether, cyclopropane, and ethylene may pervade the air in the room. Concentrated oxygen is mixed with inhalation anesthetics to supply the patient with oxygen for respiration. Only a spark is needed to start a fire or even to cause an explosion. In the oxygen tent we have combustible material, e.g., bed coverings, mattress, pillows, and patient. The kindling temperature of these combustible materials is lower in concentrated oxygen than in air. The nurse, therefore, avoids using volatile, easily combustible materials on the patient while he is in an oxygen tent.

Alcohol rubs and volatile oils are not used. To avoid the possibility of reaching the kindling temperature of the combustible material present, smoking is prohibited. Particularly when the air is dry, the nurse avoids wearing nylon or silk uniforms and rubber-soled shoes. Nylon, silk, rubber, and wool are poor conductors of electricity. In the winter, when the air is dry, it is not uncommon for a nurse, moving about the ward, to experience a shock upon touching a patient or a metal bed. The shock is a tiny spark caused by discharging electrons. This spark though of very short duration, is extremely hot, hot enough to ignite highly combustible material in the presence of concentrated oxygen.

Slow Oxidation

So far all the reactions discussed in this chapter have been combustions, i.e., rapid oxidations. Slow oxidation of many substances also occurs. It has been shown that the same *amount* of heat is evolved in either type of oxidation. During the corrosion of a metal we are not conscious of the evolution of heat, for it is dissipated to the surroundings as it is evolved so that the metal never gets hot. However, we are conscious of the heat evolved in our bodies as a result of the slow oxidation of our food materials. Oxygen from our lungs is taken to the cells by hemoglobin of the blood in the compound oxyhemoglobin, which is crimson. In the presence of the enzyme oxidase in the cells the following reactions take place:

Carbohydrate + oxygen → carbon dioxide + water + energy
Fat + oxygen → carbon dioxide + water + energy
Protein + oxygen → carbon dioxide + water + urea + energy

The above reactions are overall reactions; no attempt is made at this time to discuss what happens to the food materials step by step. Also, the story of the energy factor is rather vague, for the energy undergoes transformations within the cells to emerge finally as heat at the surface of the skin and lungs. On the skin surface the heat is dissipated by radiation or by the evaporation of water. The material products are eliminated as follows: carbon dioxide in a combined form is carried by the venous blood, which is now a darker hue, the color of reduced oxyhemoglobin, back to the lungs, where it is finally exhaled; the water is chiefly eliminated through the kidneys and excreted in the urine, although some of it is vaporized in the lungs and from the skin; the urea is chiefly eliminated through the kidneys and excreted in the urine.

Decay is accomplished through slow oxidation. By this process undesirable waste proteins are converted into the harmless, even desirable, products nitrogen, carbon dioxide, and water. The process is too complicated to be discussed in detail here; it will suffice to say that it makes this world a pleasanter place in which to live. The natural purification of water is accomplished by the oxidation of objectionable or harmful substances by the atmospheric oxygen. The action is speeded up by spraying the water into the air. This process is called *aeration*. One form of disinfection depends upon slow oxidation.

Spontaneous Combustion

We have already discussed the formation of heat in a slow oxidation process, and it has been pointed out that an increase in the temperature increases the rate of reaction. A convenient rule in this respect states that the speed of a reaction is doubled for each 10°C rise in temperature. This rule really needs qualification, but it will serve our general purposes.

Finely divided phosphorus may be deposited on a piece of cloth by pouring a carbon disulfide solution of it on the cloth. After the solvent has evaporated, the phosphorus will undergo slow oxidation. The heat evolved is not dissipated for the cloth is a poor heat conductor. As a result each particle becomes a little hotter. Because of this increase in temperature the reaction proceeds faster, with a greater evolution of heat. The action increases in intensity until finally the temperature of the material becomes greater than the kindling temperature and the whole thing bursts into flame. This is *spontaneous combustion,* which may be defined as a *rapid* oxidation resulting from the accumulation of heat produced by a *slow* oxidation. Obviously, to start this train of events a substance must be present that will undergo slow oxidation. The traditional example of spontaneous combustion has been oily rags. It should be pointed out, however, that the oil involved must be a drying oil such as linseed oil, which is used in paint and which reacts with oxygen to form a solid. Wet hay decaying in a haymow causes many barn fires, and the decay of organic material in wet soft coal often starts fires in the coal bunkers of ships. Since it is the accumulation of heat that aids and abets the slow oxidation, such fires may be prevented by a free circulation of air to remove the heat. A painter's cloth hung loosely on a nail is fairly safe, but the same cloth in a pile or stuffed into a drawer is a distinct fire hazard. The safest thing to do with the cloth is to burn it; the next best thing is to hang it up as described; if neither of these procedures is desirable, place it in a covered metal container.

NITROGEN

Approximately 80 per cent of air is composed of this colorless, odorless, tasteless gas. Its solubility is practically the same as that of oxygen, but the gas is slightly lighter. It neither burns nor supports combustion. Liquid nitrogen, N_2 (boiling point, $-195.5°C$), is used to remove warts; cotton dipped in liquid nitrogen is applied to the wart, and the tissue freezes solid and soon falls off.

Nitrogen in the form of its compounds is essential for all living things, from the simplest viruses to the most complex organisms. Plants are constantly using up the available nitrogenous compounds in the soil, which must then be replaced. Electric storms in the atmosphere cause a chemical reaction between the nitrogen and the oxygen in the air, forming soluble nitrates which precipitate with the rain. Electric storms have been called the "poor man's fertilizer." Certain nitrogen-fixing bacteria (microscopic plants) in the soil also change nitrogen gas into usable nitrates. The nitrogenous compounds synthesized by plants make nitrogen available for animals. Both plants and animals die and decay, and the nitrogen compounds are returned to the earth. Some other kinds of bacteria in the soil convert some of these compounds to nitrogen gas. This is the *nitrogen cycle;* relatively inert nitrogen from the air to the soil is made available to plants, from plants to animals, from both of these living forms back to the soil on death, and then back to nitrogen gas in the air.

NITROUS OXIDE

Nitrous oxide, N_2O, is the general anesthetic that is sometimes called "laughing gas." A colorless gas that is fairly soluble in cold water, it has a faint, sweetish odor and taste that makes it pleasant to inhale. These properties, plus the fact that there is no sense of suffocation, make the gas a desirable preliminary anesthetic in the gas-ether sequence. Nitrous oxide is also used in dentistry and obstetrics.

CARBON MONOXIDE

This is a colorless, odorless, and tasteless gas that is lighter than air. It is one of the most insidious poisons in everyday life, and the only way to combat this menace is through education of the public. If even one life is saved as a result of the knowledge gained here, the time will be well spent. Carbon monoxide, CO, is formed during the incomplete combustion of hydrocarbons; for instance, using methane, CH_4, to represent all such compounds,

$$2CH_4 + 3O_2 \rightarrow 2CO + 4H_2O$$

In a coal fire, the carbon dioxide formed at the bottom of the fire comes in contact with hot carbon to form carbon monoxide.

$$CO_2 + C \rightarrow 2CO$$

Carbon dioxide coming in contact with hot metal may be reduced to carbon monoxide.

$$CO_2 + Zn \rightarrow CO + ZnO$$

Now let us consider each of these methods and see how accidental poisoning occurs and how it may be averted. It is common practice in many homes to warm a room, particularly a bathroom, by means of a portable kerosene heater. In such a small closed room it does not take long to deplete the oxygen sufficiently to start the production of carbon monoxide. Such rooms should be ventilated to prevent the formation of the deadly gas.

In another common instance, the hydrocarbon is gasoline. The engine of *every* automobile is giving off carbon monoxide in its exhaust whether it is running in a garage or out in the open. The headache often experienced in a long line of traffic is undoubtedly caused by this gas. Tunnels and tubes must be equipped with adequate ventilating apparatus to sweep out the gas that accumulates in them. It is possible for carbon monoxide to seep inside closed automobiles, particularly those heated by hot air. For this reason, closed cars should always be ventilated in cold weather. Many accidents on the highways, especially those of heavy trucks and trailers, can be traced to the driver's becoming affected by carbon monoxide in the closed cab of the truck. Recently, road checks on all types of closed vehicles have revealed a remarkably high carbon monoxide content in the air in the cars. Finally, an automobile engine should *never* be allowed to run in a *closed* garage. Be sure that the garage doors are open before the engine is started. Not only that—be sure that the doors will *stay* open and will not blow shut. An automobile engine will produce enough carbon monoxide in a single-car garage to render a person unconscious in less than 3 min. The open doors will provide the necessary ventilation.

Coal fires will probably always be used unless the use of atomic energy becomes practical. Everyone is familiar with the light-blue flame of the burning carbon monoxide on the top of a coal fire. There is no danger provided the stove is used intelligently and provided further that adequate means of removing the gases are operating effectively. Faulty smoke pipes and chimneys, leaky furnaces that should be resealed, and improper adjustment of dampers produce the hazards of carbon monoxide. It is a general rule that if coal gas can be detected in the home after the heater has been stoked, carbon monoxide probably is also present.

Overheated gas heaters provide the third hazard. Such heaters should always be connected to the chimney to remove any carbon monoxide formed in the oxidation of the hot metal.

The poisonous action of carbon monoxide is due to the fact that the gas combines with the hemoglobin of the blood to form carbonyl-hemoglobin, also called *carboxyhemoglobin*. Therefore, insufficient oxygen is carried in the blood, and the result is tissue anoxia. Carboxyhemoglobin is bright red in color, giving the victim a flushed appearance. This compound is formed about 200 times as readily as oxyhemoglobin and can be only slowly decomposed by air. High concentrations of oxygen are more efficient, and the presence of 5 per cent carbon dioxide in the inspired air greatly assists. The carbon dioxide acts as a respiratory stimulant and reduces the pH of the blood, which aids the decomposition of carboxyhemoglobin. Mechanical respirators usually supply 95 per cent oxygen and 5 per cent carbon dioxide. Prolonged exposure to small amounts of carbon monoxide, as could be the case with garage workers, can lead to chronic carbon monoxide poisoning, characterized by a general feeling of malaise and lassitude.

Some figures on the toxicity of carbon monoxide are timely. Unconsciousness comes with 1 part in 2,000 parts of air, and 1 part in 800 is fatal. Stated differently, 10 ml carbon monoxide per kg body weight is fatal. To make this more real, we may bear in mind that a person weighing 154 lb weighs 70 kg.

It is hoped that this extensive discussion of carbon monoxide and its dangers will make no one apprehensive but rather will produce a sense of awareness of the danger sufficient to reduce the fatalities—at least by one.

CARBON DIOXIDE

Carbon dioxide, CO_2, is a colorless, odorless, tasteless gas, heavier than air. It is only slightly soluble in water under room conditions, but if the pressure is increased, the solubility is increased. Soda water and carbonated beverages contain carbon dioxide under a pressure of about 4 atmospheres. Such solutions must be tightly capped to retain the gas under normal pressures. When the bottle is opened, the excess gas escapes, giving the beverage its effervescent characteristics. Since the solubility of gases decreases with an increase of temperature, such beverages should be opened when cold to retain as much gas as possible.

Everyone today is familiar with solid carbon dioxide under the name of Dry Ice.[1] It can easily be prepared by allowing the liquid CO_2 to flow into a cloth bag. In vaporizing, heat is required and is supplied by the liquid itself, since the cloth insulates the liquid from the warmth of the air. The loss of heat by the liquid lowers its temperature below its freezing point ($-78.5°C$), and the bag will be found to contain carbon dioxide snow.

It has been pointed out that carbon dioxide is neither a combustible gas nor a supporter of combustion. These facts, coupled with its density and lack of odor, make it an efficient fire extinguisher.

Not only does carbon dioxide dissolve slightly in water, but it also reacts with water, forming carbonic acid.

$$CO_2 + H_2O \rightleftharpoons H_2CO_3$$

In carbonated beverages the acid imparts a pleasant sour taste.

Carbon dioxide reacts with metal hydroxides to form first the carbonate and then the bicarbonate.

$$Ca(OH)_2 + CO_2 \rightarrow CaCO_3 \downarrow + H_2O$$
$$CaCO_3 + H_2O + CO_2 \rightarrow Ca(HCO_3)_2$$

The first of these reactions is used to test for carbon dioxide. The second reaction indicates the erosion of limestone by natural waters. The soluble bicarbonate is also responsible for the calcium ions that produce the hardness in temporary hard water.

Carbon dioxide is used in its solid form to destroy warts. Since Dry Ice is not as cold as liquid nitrogen, this process takes much longer. The gas is used as a respiratory stimulant postoperatively, in pneumonia, in asphyxia due to electric shock, in drowning, and in poisoning, such as carbon monoxide poisoning. The gas is also used to stimulate respiration in newborn infants. It is used also to control hiccups.[2]

HELIUM

Helium, He, is one of the inert gases. It is the second lightest substance (hydrogen being the lightest) and has a very high diffusibility. It is colorless and nonflammable. Air is approximately 80 per cent nitrogen and 20 per cent oxygen. Helium is used in place of nitrogen to form a mixture of 80 per cent helium and 20 per cent oxygen. This mixture is about one-third as heavy as air and much easier to breathe. It is used in some cardiac conditions with respiratory involvement, for asthmatic

---

[1] Dry Ice is the registered trade-mark of Pure Carbonic, Inc.

[2] While small amounts of carbon dioxide act as a respiratory stimulant, in large amounts the gas has a narcotic or depressant effect on the respiratory center in the brain and lowers the rate of respiration.

patients, in respiratory depression and in obstruction of the air passages.

CHLORINE

Chlorine, Cl_2, is a greenish-yellow gas that has a choking odor. It is extremely toxic. It is not only soluble in water, but it reacts with water to form hydrocloric and hypochlorus acids.

$$Cl_2 + H_2O \rightleftharpoons HCl + HOCl$$

Diluted chlorine water has a marked bacteriocidal action. Hypochlorous acid and its salts are used as bleaches and germicides. These salts can be formed by bubbling chlorine gas into a solution of a metal hydroxide. Such solutions have long been used for bleaching purposes in the home under the names of Javelle water and Clorox. Solutions used in the treatment of suppurating wounds, where not only a germicidal effect but the dissolving of necrotic tissue is desired, are Dakin's, Carrel-Dakin's, and Labarraque's solutions. These substances are effective because they act as oxidizing agents.

Study Exercises

1. Explain briefly how oxygen can be obtained from the atmosphere. Is this a chemical or a physical method of preparation?
2. What are three different chemical methods for preparing oxygen?
3. Why does each of the following increase the rate of reaction? (a) reduction in particle size; (b) increase in temperature; (c) increase in the concentration of the reactants.
4. Define combustible, flammable, inflammable, noninflammable, nonflammable, kindling temperature, and spontaneous combustion.
5. How may the kindling temperature of a substance be reduced?
6. Why do the following extinguish a fire? (a) liquid carbon dioxide, (b) a blanket, (c) water.
7. What is the difference between the quiet burning and the explosion of a combustible gas?
8. List five situations in which the use of concentrated oxygen is to be preferred to atmospheric oxygen.
9. Why is oxygen introduced into the top of an oxygen tent rather than the bottom?
10. Give a chemical or physical reason for each of the following situations: (a) rubber-soled shoes are not worn in the operating room; (b) smoking is prohibited in areas in which concentrated oxygen is being used; (c) grease is never used on the cocks of oxygen tanks; (d) the air in the operating room is analyzed before electrocautery procedures are employed; (e) electric signal devices, such as buzzers, are not permitted in oxygen tents.
11. Why is nitrous oxide a desirable anesthetic in dentistry?
12. Account for the great toxicity of carbon monoxide.
13. What are three ways in which carbon monoxide may be formed in everyday life and so provide a poison hazard?
14. What precautions would remove the danger of carbon monoxide poisoning in these cases?
15. What is the test for carbon dioxide?
16. What is the reaction of carbon dioxide with water?
17. Give two uses of carbon dioxide in the hospital.
18. Why is a helium-oxygen mixture preferred to a nitrogen-oxygen mixture for certain respiratory involvements?
19. Name three chlorine disinfectants used in the hospital.

chapter nine IMPORTANT LIQUIDS

Ideas for preview or review

Matter in the solid state has definite shape and definite volume. Liquids have definite volume but assume the shape of the container, and they flow. Gases have neither definite shape nor volume, they fill a container, and they also flow.

Density is mass per unit volume.

A calorie is the quantity of heat needed to raise 1 Gm water 1°C.

Hyperacidity indicates a greater than normal concentration of acid in gastric juice.

WATER

Water is by far the most common chemical compound in our daily lives. While it is possible for nearly all substances to exist in the three physical states if the conditions are sufficiently varied, we are not familiar with all these substances in all three states within our normal living conditions. We think of iron as a solid, and though some of us have seen molten iron, relatively few of us have seen iron vapor in an electric arc produced by iron electrodes. Again, while we are familiar with oxygen as a gas, to become a liquid it must be cooled below $-118°C$ and compressed; only below $-218°C$ is it a solid. Yet in one day we may easily come in contact with ice, liquid water, and water vapor. Three-quarters of the earth is covered with water, either as ice in the Arctic and Antarctic zones or as liquid water in the others. Water vapor is always present in the atmosphere. About 70 per cent of the body weight is due to water, and one way of reducing (not recommended) is to lose some of this water by one method or another. Water is essential to life; we may exist many days without food, but a few days without water is the limit.

Physical Properties

Obviously, the physical properties of water will vary depending upon its physical state. As everyone knows, the liquid is colorless, odorless, and tasteless. Water vapor is colorless. The common conception that steam is visible is erroneous, for what we see is not water vapor but a cloud of minute droplets of liquid water formed by the condensation of the water vapor upon cooling.

density

As the temperature of water is lowered, its density increases until maximum density is attained at 4°C. At this temperature the density is 1 Gm per ml, i.e., 1 ml water weighs 1 Gm. Below this temperature the density decreases, which means that ice is lighter than water. It also means that water expands as it freezes. This expansion has both advantages and disadvantages. The harvesting of ice is simple because ice floats. However, the great expansive force produced in the freezing of water bursts water pipes and car radiators. Water pipes have to be protected, and antifreeze must be used in the automobile cooling system. Again, in the freezing of the toes and fingers the water in the tissues freezes; as it expands, the cells are ruptured. If this destruction is too extensive, amputation is necessary. Finally, the water at the bottom of deep ponds is at 4°C. It is easy to imagine the situation if the density of water continued to increase as the temperature decreased below 4°C. At 0°C the water would freeze, but at the bottom of the pond. This procedure would continue until the pond was frozen solidly, and no fish could survive. As it is, the coldest water is at the top in the winter; the ice forms on the surface, and fish continue to live in the warmer water below the ice.

specific heat

The specific heat of a substance—sometimes referred to as its *heat capacity,* which is perhaps a more descriptive name than the other—is *the number of calories of heat required to raise the temperature of 1 Gm of that substance 1°C.* Let us see what this means from a practical point of view. It shows that in choosing a bed warmer, for instance, we should choose that substance which has the highest heat capacity, for in coming back to room temperature it must give out the amount of heat that has been absorbed.

Water has an abnormally high specific heat, as can be seen by noting the specific heats of a few common substances: water, 1.0; alcohol, 0.5; and iron, 0.1. These figures mean that, whereas water requires 1 cal to raise 1 Gm of it 1°C, iron requires only 0.1 cal. In other words, water has a heat capacity of 1 cal per Gm and iron has a capacity of only 0.1 cal per Gm. A given weight of water will absorb and give out 10 times as much heat as the same weight of iron, if both are heated to the same temperature. The efficiency of a hot-water system for house heating lies in this high specific heat of water. In the body, we depend upon the high heat capacity of the water in the system to absorb the heat formed in the cells and transfer it to the skin where it can be lost.

Purification

Natural water contains many chemical impurities. Consider a lake or a mountain stream. Salts may be dissolved out of the soil of the bed; gases of the atmosphere may be trapped and then dissolved; or gaseous compounds in the air may react with rain to form acids that erode the rocks to form soluble salts. There is also a certain amount of solid material of colloidal size in suspension. Chemical impurities may be classified as solid, soluble, and gaseous or volatile impurities.

filtration

Solid impurities may be removed by filtration. The type of filter medium naturally depends upon the size of the particles. Filter beds of sand are efficient but slow. The process can be speeded up by adding a coagulant, aluminum sulfate, which reacts with water to form aluminum hydroxide, a gelatinous precipitate. The hydroxide as it settles carries with it much of the suspended matter. With this treatment, water may be filtered through a fairly coarse medium at a greater speed.

distillation

Soluble impurities, obviously, pass through any filter, since they are of molecular or ionic size. These dissolved solids can be removed by distillation, the process in which a liquid is vaporized at its boiling point and then passed through a cooling chamber where it is condensed. By this process the liquid is purified of all substances that are not volatile at its boiling point. Large-scale distillation is usually prohibitive in cost, although under special conditions, when necessary, it may be carried out practically. For instance, it is common practice on large naval vessels and ocean liners to distill sea water for freshwater needs.

ion exchange

So-called ion-exchange resins[1] are now available for the purification of water. Water purified in this way approaches distilled water in its purity. During the Second World War apparatus using these resins was standard equipment on life rafts, and many lives were saved by the ability of the resins to convert sea water into potable water.

boiling

Since the process of distillation depends upon the vaporization of a liquid, it should be obvious that any impurity that is a gas or substance volatile below the boiling point of the liquid will pass over with the vapor of the liquid and the distillate will not be pure. Distillation is not a method for purifying water containing such an impurity. Since all gases are insoluble in boiling water, a volatile impurity may be removed by boiling the water. Distilled or boiled water tastes flat because of the absence of dissolved air. The taste may be improved by shaking the water in a bottle with some air in it or by pouring the water back and forth from one glass to another several times.

chlorination

Destruction of bacteria and the organic matter that supports their growth is accomplished through chlorination. Free chlorine is the most efficient agent for this purpose. Ordinarily 0.5 parts per million (ppm) are sufficient for bacterial action in relatively pure water, but 20 ppm are required in heavily polluted water.

[1] There are two types of ion-exchange resins. The first one exchanges hydrogen for the metal ions in the water to form acids. The second absorbs these acids.

Hard Water

Distilled water shaken with soap solution produces a froth or suds that persists for many minutes. Hard water, on the other hand, produces no suds but forms a white, sticky precipitate that adheres to everything it touches and is difficult to remove by rinsing. This precipitate arises through the union of Ca^{++}, Mg^{++}, and Fe^{+++} ions in the water with, for instance, stearate ions, $C_{17}H_{35}CO_2^-$, coming from soap, $NaC_{17}H_{35}CO_2$.

$$Ca^{++} + 2C_{17}H_{35}CO_2^- \rightarrow Ca(C_{17}H_{35}CO_2)_2 \downarrow$$

Throughout this discussion, Ca^{++} ion will be used as the representative ion in hard water; similarly, sodium stearate will be used to represent all soluble soaps. The Ca^{++} ions form an insoluble soap that cannot produce suds. Thus the softening of hard water depends upon the removal of these offending ions.

temporary hard water

This kind of hard water contains the bicarbonates of calcium, magnesium, and iron. It results from the erosion of carbonate rocks by natural water containing carbon dioxide according to the equation

$$\underset{\text{Insoluble}}{CaCO_3} + H_2O + CO_2 \rightarrow \underset{\text{Soluble}}{Ca(HCO_3)_2}$$

The following are three methods of softening temporary hard water.

Boiling. If temporary hard water is boiled, insoluble calcium carbonate will be formed, removing the Ca^{++} ions by precipitation.

$$Ca(HCO_3)_2 \rightarrow CaCO_3 \downarrow + H_2O + CO_2 \uparrow$$

Soap added to this boiled water should produce suds. It is because the water may be softened by boiling that it is called *temporary* hard water. This reaction takes place spontaneously in nature when temporary hard water slowly drips from the roof of a cave. The calcium carbonate very slowly accumulates, forming stalactites on the roof and stalagmites on the floor.

Adding a Base. The soluble bicarbonate is converted into the insoluble carbonate by the addition of a base. The fundamental reaction is

$$HCO_3^- + OH^- \rightleftharpoons H_2O + CO_3^=$$

In the household, ammonia water is used according to the equation

$$HCO_3^- + NH_3 \rightleftharpoons NH_4^+ + CO_3^=$$

or $Ca(HCO_3)_2 + 2NH_3 \rightarrow (NH_4)_2CO_3 + CaCO_3 \downarrow$

Commercially, quicklime, which forms calcium hydroxide in the water, is used.

$$Ca(HCO_3)_2 + Ca(OH)_2 \rightarrow 2CaCO_3 \downarrow + 2H_2O$$

Adding Washing Soda.

$$Ca(HCO_3)_2 + Na_2CO_3 \rightarrow CaCO_3 \downarrow + 2NaHCO_3$$

permanent hard water

This type of hard water contains soluble calcium, magnesium, and iron salts that cannot be decomposed by boiling, e.g., $CaCl_2$ and $CaSO_4$. Such salts leach out of the soil and eventually appear in water supplies and in sea water. Permanent hard water *cannot* be softened by boiling or by the addition of a base. It can be softened by adding a salt containing an ion that will form an insoluble calcium salt, e.g., washing soda, Na_2CO_3, or borax, $Na_2B_4O_7$.

The Permutit Process. This process is convenient for softening water, because the apparatus is installed in the water system and all water used is reduced to "zero" hardness. This process employs a natural or artificial mineral called *zeolite* that we may represent by the formula Na_2Ze. In operation, hard water flows through a tank containing the mineral and the following exchange of ions takes place:

$$Na_2Ze + CaCl_2 \rightarrow 2NaCl + CaZe \downarrow$$

When all the zeolite has been exhausted by conversion to the calcium form, it is regenerated by allowing a salt solution to flood up into the tank. The reverse of the above equation represents the action of regeneration.

The use of ion-exchange resins has already been discussed.

some practical aspects of hard water

Not only is the softening of water important in household uses, but it is particularly important that water used in steam boilers be softened. Some idea of the extent to which minerals deposit when hard water is boiled and evaporated may be gained by examining the inside of a teakettle, if one lives in a hard-water district. In a steam boiler, this deposit of minerals not only insulates the tubes but by corroding the iron makes the boiler dangerous to use.

Chemical Properties

The great stability of water has already been discussed. Because of this stability water was once believed to be an element. But if water is dropped onto a plate heated to about 2,000°C, a momentary decomposition into hydrogen and oxygen takes place. It is evident that it must be a compound of these two elements.

electrolysis

Water may be decomposed by a direct current of electricity provided a substance such as sulfuric acid is present. Hydrogen and oxygen are formed in the ratio of 2:1 by volume as shown in the equation

$$2H_2O \rightarrow 2H_2 \uparrow + O_2 \uparrow$$

action with active metals

If a small piece of metallic sodium, wrapped in copper gauze to weigh it down, is dropped into water in a dish, the hydrogen formed may be collected in an inverted test tube by the downward displacement of water. The equation for the reaction is

$$2Na + 2HOH \rightarrow H_2 \uparrow + 2NaOH$$

Important Liquids

The gas may be tested as follows: a *lighted* splint is applied to the gas; if the gas burns with a slight "pop," it is hydrogen. Of course, any combustible gas mixed with the proper amount of air will explode, but hydrogen is almost the only odorless gas that acts this way. Further, hydrogen is the only gas that burns to form water as the only product of the combustion.

$$2H_2 + O_2 \rightarrow 2H_2O$$

The burning-splinter test, however, is sufficient for our purposes.

action with metal oxides

Water reacts with certain of the oxides to form hydroxides.

$$H_2O + K_2O \rightarrow 2KOH$$
$$H_2O + CaO \rightarrow Ca(OH)_2$$

The hydroxides formed in these reactions have a common property that will be discussed in detail later: they all turn pink litmus[1] paper blue. A solution which shows this characteristic reaction to litmus is known as a *base*. Since these oxides react with water to form bases, they are called *basic oxides*. Further, since these oxides are the bases *without water,* they are called *basic anhydrides*.

action with nonmetal oxides

Oxides of nonmetals react with water to form acids containing oxygen in the group.

$$CO_2 + H_2O \rightarrow H_2CO_3$$
$$SO_3 + H_2O \rightarrow H_2SO_4$$
$$P_2O_5 + 3H_2O \rightarrow 2H_3PO_4$$

It is important to note in this connection that the nonmetal in the acid has the same valence number that it had in the oxide. The formation of the acid can be shown by testing the solution with blue litmus paper, which turns pink. These nonmetal oxides, since they form acids with water, are called *acidic oxides* or *acid anhydrides*.

Hydrates

Water reacts with many substances, particularly salts, to form *hydrates*. Some common hydrates are $CuSO_4 \cdot 5HO$, $Na_2CO_3 \cdot 10H_2O$, and $MgSO_4 \cdot 7H_2O$. The water of hydration or water of crystallization is a definite part of the compound, and the number of molecules of water follows the law of definite composition.

Hydrated cupric sulfate is blue, but the anhydrous cupric sulfate is a greyish white. If water is added to the anhydrous variety, the hydrated form is produced again and the color turns blue. The reverse action will also take place if the anhydrous variety stands in an atmosphere containing water vapor. Since this is a reversible reaction, the equation is written with a double arrow.

$$CuSO_4 \cdot 5H_2O \rightleftharpoons 5H_2O + CuSO_4$$

It should be obvious, then, that a solution of cupric sulfate will always contain the same substance, hydrated cupric ion $Cu(H_2O)_4^{++}$, whether it is prepared from the hydrated or the anhydrous salt.

Plaster of paris is a hydrate which finds great use in the making of casts and molds. It is prepared commercially from gypsum, $CaSO_4 \cdot 2H_2O$, a naturally occurring rock. The rock is very carefully heated to drive off three-quarters of the water. Expressing this chemical fact we would write

$$\underset{\text{Gypsum}}{2CaSO_4 \cdot 2H_2O} \rightarrow 3H_2O + \underset{\text{Plaster of paris}}{(CaSO_4)_2 \cdot H_2O}$$

When plaster of paris is treated with water, the reverse action, called *setting*, takes place.

$$(CaSO_4)_2 \cdot H_2O + 3H_2O \rightarrow 2CaSO_4 \cdot 2H_2O$$

Since the hydrate is insoluble, it crystallizes and the crystals form in an interlocking

---

[1] Litmus is one of the many organic substances the color of which is changed by acids and bases. They are called *indicators* for they indicate whether a substance in solution has acidic or basic characteristics. Acids turn blue litmus paper pink, and bases turn pink litmus paper blue.

manner to produce a hard, firm mass. The actual procedure for making a cast need not be discussed here, but one point is important: plaster of paris expands upon setting. It is this property that makes it desirable in making a mold, for it forces itself into every crack and crevice. Thus it forms an exact negative, from which an exact reproduction of the original can be prepared. In applying a plaster cast to some part of the body, however, allowance must be made for expansion of the plaster. Otherwise, when the setting is complete, the cast may impair circulation. The feeling of warmth that the patient experiences during the setting is due to the heat of crystallization that is given off.

Efflorescence

Some crystalline hydrates lose their water of crystallization spontaneously upon exposure to the air and fall to a powder. Hydrated sodium carbonate (washing soda), for instance, reacts in this way.

$$\underset{\text{Crystals}}{Na_2CO_3 \cdot 10H_2O} \rightarrow 10H_2O + \underset{\text{Powder}}{Na_2CO_3}$$

This process is called *efflorescence* and may be defined as the *spontaneous loss of water by a hydrate*. Under ordinary conditions hydrated cupric sulfate does not effloresce. There must be a reason why one hydrate is efflorescent and another is not. When the vapor pressure of a hydrate at room temperature is greater than the partial pressure of the water vapor in the atmosphere, the salt will lose its water of hydration on exposure, or effloresce; if the opposite conditions obtain, the hydrate will be stable.

Deliquescence

Deliquescence is the opposite of efflorescence in effect only; i.e., deliquescent substances absorb moisture from the atmosphere, and as this moisture accumulates, they are dissolved in it. We need not be concerned with the cause of deliquescence. It will be sufficient to realize that in order for a substance to be deliquescent it must be extremely soluble at room temperature. A lump of sodium hydroxide exposed to the air soon appears wet, and eventually it will form a solution of itself. The fumes formed by gaseous hydrogen chloride in moist air are caused by the absorption of water and the formation of tiny droplets of hydrochloric acid solution; on the other hand, hydrogen, which is insoluble, does not fume.

some practical aspects of efflorescence and deliquescence

Substances that are efflorescent or deliquescent must be stored in tightly stoppered bottles in order to keep them in their original conditions.

The property of deliquescence is utilized when granular calcium chloride is used to "lay" the dust on roads. Sprinkled on the road, it soon forms small puddles that not only dampen the dirt but cause it to pack solidly. Deliquescent substances are also used extensively as drying agents. Anhydrous calcium chloride, for instance, is used in *desiccators,* containers which provide a perfectly dry atmosphere for the storage of materials that require such an atmosphere. Wet gases passed through granules of calcium chloride are freed of their moisture. Anhydrous cupric sulfate is added to *absolute alcohol* to act in a dual capacity. First, it removes any water that the alcohol has absorbed from the air. Second, as it reacts with the water, its color slowly changes to blue. Thus it serves as an indicator of the condition of the absolute alcohol. If it is light blue it shows that the alcohol has absorbed some water from the air but that the water has been removed from the alcohol; if it is dark blue, it indicates the same fact but, in addition, raises a doubt as to whether all the water has been removed from the alcohol.

Hydrolysis

Hydrolysis is *a double decomposition reaction in which one of the reactants is water.* Literally, the term means "decomposition by water." Hydrolysis will be discussed later, but it is introduced here as a reaction of water. We shall have many examples of hydrolysis in the future.

The Physiologic Importance of Water

By way of summarizing the properties of water, let us consider briefly the role of water in life.

a. Living tissues are mainly water.
b. All body fluids are dilute water solutions. Digested food as well as body wastes are transported dissolved in water.
c. Water enters into chemical reactions in the body. Chemical digestion consists of a series of reactions with water (hydrolysis).

   Sucrose + water → glucose + fructose
   Proteins + water → amino acids

d. The high specific heat and the high heat of vaporization of water help to control body temperature. In addition, these properties account for the use of hot and cold baths, sponges, and wet packs. Heat of fusion makes possible the use of ice caps and ice packs. The heat of water is used in these ways for its relaxing effect and frequently to relieve pain, and cold may be used either for stimulation or for its analgesic, anesthetic, or antipyretic effects.
e. Water, either alone or mixed with other substances, is used to clean the body externally and internally. Internally it is used in lavages, douches, and enemas. Its effectiveness here is due to its great solvent action and in some cases to water pressure.

OTHER IMPORTANT LIQUIDS

Hydrogen Peroxide

Hydrogen peroxide, H_2O_2, is an unstable water-clear liquid. It decomposes spontaneously to form water and oxygen. To reduce the rate of decomposition, an inhibitor or negative catalyst, acetanilid, is added to the commercial product and the peroxide is stored in dark or opaque containers away from light. Cells contain an organic catalyst (enzyme), catalase. When hydrogen peroxide comes in contact with tissue, it liberates oxygen very rapidly. Since oxygen, particularly atomic oxygen, is an active oxidizing agent, hydrogen peroxide is useful as an antiseptic.

Ethyl Alcohol

Ethyl alcohol, C_2H_5OH, is a colorless volatile liquid with a slight fruity odor. It is widely used as a solvent. Many substances insoluble in water are soluble in ethyl alcohol. The terms *extract,* *tincture,* and *elixir* refer to mixtures of substances with ethyl alcohol. In nursing, one meets tincture of iodine, tincture of benzoin, tincture of belladonna, tincture of digitalis, cascara sagrada extract (a laxative), elixir phenobarbital, and terpin hydrate and codeine elixir.

Ethyl alcohol is used also, depending upon the concentration, as an antiseptic and a disinfectant. Applied as a rub, it has an astringent effect; as a bath or sponge, it is cleansing and cooling and, used in this way, tends to reduce body temperature.

Isopropyl Alcohol

Isopropyl alcohol, C_3H_7OH, is sometimes used externally in place of ethyl alcohol. It is more toxic than ethyl alcohol but has a slightly greater antiseptic effect. It can be used for rubs, sponges, and baths as well.

Ether

Ether, $(C_2H_5)_2O$, a clear colorless liquid, has a sharp but not unpleasant odor. It is highly volatile, and the vapors are extremely flammable. It is the most widely used general anesthetic for long operations. In addition, it is an excellent solvent and can be used to clean the skin of oils, adhesive, etc.

Study Exercises

1. Why does the land area in close proximity to a large body of water experience a late spring and fall?
2. How may water containing the following impurities be purified for drinking? (*a*) a volatile substance; (*b*) sodium chloride; (*c*) clay; (*d*) bacteria.
3. What is hard water? Differentate between permanent and temporary hard water. How may permanent hard water be softened?
4. What is a hydrate? Write the equations for the following reactions: (*a*) the formation of plaster of paris from gypsum; (*b*) the setting of a plaster cast.
5. In applying a plaster cast, more water was used than was required for the hydration process. To speed up the evaporation of the excess water, the heat from a gooseneck lamp was applied. The nurse was cautioned to be careful not to overheat the cast. Why?
6. What is meant by efflorescence?
7. Give a physical or chemical explanation for each of the following situations: (*a*) magnesium oxide and water may be used to treat gastric hyperacidity; (*b*) an alcohol rub is used to prevent bed sores (decubitus ulcers); (*c*) hydrogen peroxide is used as an antiseptic for minor wounds; (*d*) in a preoperative preparation, the skin area was washed with ether. (*e*) water that is to be used in intravenous infusions is triple distilled. (*f*) the outer surface of a hot-water bottle or bag is always dried thoroughly before it is given to a patient.

chapter ten LIQUID MIXTURES

Ideas for preview or review

The composition of mixtures may vary.

Mixtures can be separated into their component parts by physical means.

Gram-molecular weight equals the sum of the gram-atomic weights as represented in the formula of the compound. Gram-equivalent weight equals the gram-atomic weight divided by the valence.

Valence is the combining ability of elements.

Valence number is the number of electrons lost, gained, or shared by an atom in a chemical reaction.

Filtration is the passage of a fluid (liquid or gas) through porous material to separate suspended material.

Osmosis is the passage of solvent molecules through a semipermeable membrane.

Dialysis is the passage of dissolved particles (solute) through a permeable membrane.

The anode is the positively charged electrode in a cell.

All the body fluids are liquid mixtures. The blood is an example of all three types of liquid mixtures: *solution, colloid,* and *suspension.* Glucose and amino acids in the blood are in *solution.* Serum albumin, serum globulin, and fibrinogen form *colloids.* Red blood cells and fat droplets are *suspended* in the fluid. Each of these three types of mixture has its own set of properties. What are the fundamental differences among them?

GENERAL PROPERTIES OF SOLUTIONS, SUSPENSIONS, AND COLLOIDS

A mixture is composed of at least two different substances. In the case of solutions these are called the *solvent* and the *solute.* If we had a mixture of 10 ml ethyl alcohol and 90 ml water, we might say the alcohol was dissolved in the water since the water is present in the larger amount. If, however, the situation were reversed and we had a mixture of 90 ml ethyl alcohol and 10 ml water, would we say that the water was dissolved in the alcohol? Actually solution is a mutual process between the two substances and involves the breaking down, if necessary, of the substances involved to the size of average molecules. Traditionally, we speak of the liquid as the solvent and a gas, a solid, or another liquid as the solute.

What is the size of average molecules? Average molecules are considered to be of the order of 0.00000001 cm in size. Writing this number as a power of 10, it would be 1×10^{-8} cm, which means 1 in the eighth decimal place, or 100 millionths of a centimeter. Particles this size are invisible not only to the naked eye but under the most powerful electron microscope. Physiologic salt solution looks like pure water. Glucose solution looks like pure water. Any colorless solution looks like pure water. The dissolved particles are invisible. The dissolved material does not settle out on standing, and all parts of the solution remain the same. A sample taken from the lower part of a solution is the same as one taken from the top. We say the mixture is *homogeneous.* Both the solvent and solute particles can pass through a filter and a permeable[1] membrane.

---

[1] A permeable membrane permits passage of both solvent and dissolved solute.

At the other end of the scale we may consider a <u>suspension</u>. In milk of magnesia, a common laxative and antacid, and in calamine lotion, used for skin conditions, the particles are so large that they may be seen with the naked eye. In fact, the mixture when shaken or stirred appears opaque. When the mixtures are allowed to stand, the particles settle out. Containers for such mixtures bear the legend "shake well before using." These particles are about one ten-thousandth of a centimeter, 0.0001 cm, or larger. Suspended particles can be held back by a filter, thus separating the mixture into its parts. It would also follow that these particles are so large that they cannot pass through a membrane.

A special kind of suspension is composed of droplets of one liquid distributed throughout another. This is called an *emulsion*. The droplets may quickly separate, in which case the mixture is called a temporary emulsion, or they may remain suspended for some time, in which case they are called permanent emulsions. Usually some substance must be added to an emulsion to make it permanent. Like other suspensions, emulsions are cloudy or opaque. French dressing (olive oil and vinegar) must be shaken before use because the liquids quickly separate. French dressing is a temporary emulsion. Mayonnaise is made of the same ingredients plus an emulsifying agent from egg. Mayonnaise is a permanent emulsion. Bile salts emulsify fat so that it can be distributed as fat droplets in a water solution containing the lipase enzymes.

Between these extremes are colloidal mixtures. Those in which the particles are only slightly larger than in solutions are called *colloidal solutions,* and those in which the particles are slightly smaller than suspensions are called *colloidal suspensions.* The whole class may be called *colloidal dispersions.* Colloidal particles will pass through an ordinary filter but are retained by a membrane. These particles are so small that they do not settle out on standing and are invisible to the eye. Some of the larger ones, however, reflect an iridescent light, causing the mixture to have an opalescent appearance. The opal, a gem stone, is made up of particles of colloidal size (Table 10–1).

Ways of Expressing Concentration

The word *concentration* serves two purposes. In the question: What is the concentration of

Table 10–1 PROPERTIES OF LIQUID MIXTURES

Solutions	Colloidal Dispersions		Suspensions
	Colloidal Solutions	Colloidal Suspensions	
Clear	Clear	May be opalescent	Cloudy or opaque
Do not settle	Do not settle	Do not settle	Settle out
Run through filters	Run through filters	Run through filters	Retained by filters
Pass through membranes	Retained by membranes	Retained by membranes	Retained by membranes
Size of average molecules, angstrom units	Giant molecules or clumps of molecules	Giant molecules or clumps of molecules	Visible particles
	Size, millim cra	Size, 10 to 100 micra	Size, 1μ and larger
Dissolved particles contribute to osmotic pressure	Colloidal particles contribute to osmotic pressure		Suspended particles exert no osmotic pressure
	Colloidal particles carry electric charges		
	Colloidal particles have great adsorptive ability		

Liquid Mixtures

physiologic salt solution?, it refers to the relative amount of the dissolved material, whether large or small. If we say that a concentrated solution of nitric acid is corrosive to tissue, we refer to a mixture in which the amount of solute, nitric acid, is large in comparison with the amount of solvent water. A dilute solution has a relatively small amount of solute for a relatively large amount of solvent. To say that all body fluids are dilute solutions gives us the idea that they are mostly water, but it does not give an accurate idea of the amounts of material present.

per cent solution

"Per cent" means parts in 100. A 1 per cent solution is 1 part of solute for every 100 parts of solution. A 5 per cent solution is 5 parts of solute for every 100 parts of solution.

A 1 per cent sodium chloride solution is made by adding 99 Gm water to 1 Gm sodium chloride; total, 100 Gm solution, 1 part of which is sodium chloride. A 5 per cent glucose solution is 5 Gm glucose and 95 Gm water; total, 100 Gm solution, 5 parts of which are glucose. In 200 Gm 5 per cent glucose solution there is 10 Gm glucose and 190 Gm water. Thus, for every 100 Gm solution, there is 5 Gm glucose. These solutions are per cent by weight. Since water has a density of 1 Gm per ml, one can measure the amount of water needed in a graduated cylinder and it need not be weighed out on a balance.

Method. Specific directions for the preparation of 1,000 Gm physiologic salt solution, 0.9 per cent sodium chloride, are as follows:

Changing 0.9 per cent to a decimal = 0.009,

```
      1,000 Gm
    × 0.009
      9.000 Gm NaCl

    1,000.0 Gm solution
  −    9.0 Gm NaCl
      991.0 Gm water
```

Weigh on the balance 9.0 Gm sodium chloride.

Measure in a graduated cylinder 991.0 ml water. Since 1 ml water weighs 1 Gm, 991 ml water weighs 991 Gm.

Dissolve the 9.0 Gm sodium chloride in the 991.0 ml water, making a total of 1,000 Gm solution containing 9 Gm solute.

Look at this another way. What is the per cent concentration of 1,000 Gm solution containing 9 Gm solute? In other words, what weight of solute is dissolved in 100 Gm solution?

Let x = the weight of solute in 100 Gm solution.

Setting up the proportion,

$$\frac{9 \text{ Gm solute}}{1{,}000 \text{ Gm solution}} = \frac{x}{100 \text{ Gm solution}}$$

Cross-multiplying,

$$1{,}000x = 900$$
$$x = \frac{900}{1{,}000}$$
$$= 0.9 \text{ Gm NaCl in 100 Gm solution}$$

The answer is 0.9 per cent concentration.

When a solution is to be introduced directly into the blood stream, its concentration must be exact. Physiologic salt solution must be 0.9 per cent, not 1 per cent. An error of this magnitude might be serious. For solutions used as baths or soaks, the concentration is not critical. To prepare a 2 per cent boric acid solution to be used as a wash, add 2 Gm boric acid to 100 ml water. This is roughly a 2 per cent solution. Here we have 2 parts of boric acid in 102 parts of solution rather than 100 parts, which is a 1.96 per cent and not a 2 per cent solution.

We have been describing per cent by weight solutions; per cent by volume solutions are made when both solute and solvent are liquids. Ethyl alcohol 70 per cent by volume is composed of 70 ml ethyl alcohol and 30 ml water. This does not total 100 Gm.

milligrams per cent

In medical literature, a common method of expressing the concentrations of dilute solutions is milligrams per cent, written mg %. If the concentration of glucose in the blood was 100 mg in 100 ml, this would be called 100 mg %. Since 100 mg is 0.1 gm, in terms of real percentage this concentration is 0.1 per cent. Milligrams per cent obviates the use of decimals in very dilute solutions, e.g., concentration of phosphate in blood, 3 mg %, concentration of uric acid in blood, 2 mg %. In terms of percentage, these figures would be 0.003 per cent and 0.002 per cent, respectively.

molar solution

In molar solution and the following concentration, normal solution, the concentrations refer to amount of solute in 1 liter of solution.

A solution is 1 molar ($1M$) when 1 gram-molecular weight of solute is dissolved in 1 liter of solution. One liter of $1M$ sodium chloride solution would contain 1 gram-molecular weight of sodium chloride. We obtain the gram-molecular weight by adding the gram-atomic weights.

$$\text{Na} \quad \text{Cl}$$
$$23 + 35 = 58 \text{ Gm-mol. wt.}$$

How much water do we need? We don't know. We weigh 58 Gm NaCl, put it into a graduated or standard flask, and add water to the liter mark. The total volume of solution is now 1 liter. Each milliliter contains one-thousandth of a molecular weight of NaCl, or 0.058 Gm.

If we pour 500 ml of this solution into another container, we now have in each container half a gram-molecular weight of NaCl, or 29 Gm NaCl in 500 ml solution. Have we changed the concentration? Obviously not. It is still $1M$ because we express molar concentration in terms of the amount in a liter of solution. If we have 29 Gm in 500 ml, we would have 58 Gm in 1,000 ml.

Suppose we take 500 ml $1M$ sodium chloride solution and add 500 ml water. What is its concentration now? How much sodium chloride is dissolved in 1 liter of solution? The answer is 29 Gm. What is the molecular weight of sodium chloride? The answer to that is 58 Gm. The solution contains one-half the molecular weight in a liter and is therefore $\frac{1}{2}$ molar, or $0.5M$.

Method. To prepare 1 liter of $1M$ magnesium chloride, we compute the gram-molecular weight.

$$\text{Mg} \quad \text{Cl}_2$$
$$24 + 2(35) = 94 \text{ Gm-mol. wt.}$$

Weigh 94 Gm magnesium chloride and add enough water to make 1 liter of solution. A $2M$ solution contains 2 gram-molecular weights in 1 liter. A $0.1M$ solution contains 0.1 gram-molecular weight in 1 liter.

normal solution

A solution is 1 normal ($1N$) when 1 gram-equivalent weight of solute is dissolved in 1 liter of solution.

The equivalent weight is found by dividing the atomic weight by the valence. For elements whose valence is 1, the atomic weight and the equivalent weight are the same (Table 10–2). In the blood, we have the cations Na^+, K^+, Ca^{++}, and Mg^{++}. An atomic weight of Ca^{++} or Mg^{++} will react with twice as much Cl^- as will Na^+ or K^+. If we change these to equivalent weights, their chemical reactivity will be equal.

There are two easy ways to figure the weight of solute in normal solutions.

Method 1. One liter of $1N$ magnesium chloride solution contains 1 equivalent weight of magnesium $\left(\frac{\text{atomic weight 24}}{\text{valence 2}}\right)$, or 12 Gm, and 1 equivalent weight of chlorine $\left(\frac{\text{atomic weight 35}}{\text{valence 1}}\right)$, or 35 Gm. The total weight of solute is $12 + 35 = 47$ Gm.

Table 10-2 COMPUTATION OF GRAM-MOLECULAR AND GRAM-EQUIVALENT WEIGHTS

Substance	Gram-molecular Weight	Gram-equivalent Weight
NaCl	Na^+ Cl^- $23 + 35 = 58$	Na^+ Cl^- $23 + 35 = 58$
$MgCl_2$	Mg^{++} Cl^- $24 + 2(35) = 94$	Mg^{++} Cl^- $12 + 35 = 47$
$CaCl_2$	Ca^{++} Cl^- $40 + 2(35) = 110$	Ca^+ Cl^- $20 + 35 = 55$

A $2N$ solution of magnesium chloride would contain per liter 2 equivalent weights of magnesium, 12, × 2 = 24 Gm, and the equivalent weight of chlorine, 35, × 2 = 70 Gm; the total weight of solute is 94 Gm.

A $0.1N$ solution of magnesium chloride would have 0.1 of an equivalent weight of magnesium and 0.1 of an equivalent weight of chlorine, or 12 × 0.1 + 35 × 0.1, or 1.2 + 3.5 Gm; the total weight of solute is 4.7 Gm.

Method 2. The molecule is always electrically neutral. The total positive charge and the total negative charge cancel out, giving no charge. In the formula $MgCl_2$, we have magnesium, valence +2, and 2 chlorine atoms, each with a valence of −1 (+2 − 2). The result is zero.

In this second method, instead of working with the individual components of a solution, we may start with the gram-molecular weight and divide by the positive valence. (The negative valence is the same.)

$$Mg\ Cl_2$$
$$24 + 2(35) = 94 \text{ Gm-mol. wt.}$$
$$\frac{94 \text{ (gm-mol. wt.)}}{2 \text{ (valence of } Mg^{++})} = 47 \text{ gm}$$

The result is naturally the same by either method.

Let us stress that equal volumes of solutions of the same normality contain equal amounts of chemically active material. Compare the amount of replaceable hydrogen in the solutions in Table 10-3.

Do not confuse this method of expressing concentration with physiologic salt solution, which is sometimes called *normal* saline solution. This solution is 0.9 per cent NaCl. What is its normality?

If a solution is 0.9 per cent, it would contain 0.9 Gm/100 = xGm/1,000 = 9 Gm per liter. The method of describing concentration called normal always refers to the amount of solute in a liter of solution. A $1N$ NaCl solution contains the equivalent weights in 1 liter.

$$Na\ Cl$$
$$23 + 35 = 58 \text{ Gm}$$

Setting up the proportion,

$$\frac{58 \text{ Gm}}{1N} = \frac{9 \text{ Gm}}{xN}$$
$$x = 0.155N$$

A modification of the normal solution is used to describe concentrations of parenteral fluids (fluids administered intravenously, intramuscularly, subcutaneously, etc.). Since the body fluids are extremely dilute, small

Table 10-3 AMOUNT OF REPLACEABLE HYDROGEN IN VARIOUS SOLUTIONS

Solution	Hydrogen, Gm	
	In 1 Liter of 1M Solution	In 1 Liter of 1N Solution
HCl	1	1
HNO_3	1	1
H_2SO_4	2	1
H_2CO_3	2	1
H_3PO_4	3	1

Table 10-4 EQUIVALENT WEIGHTS

Element	Ionic Weight, Gm	Valence	Equivalent Weight, Gm	mEq Weight, Gm	mEq Weight, mg
Na^+	23	1	23	0.023	23
K^+	39	1	39	0.039	39
Mg^{++}	24	2	12	0.012	12
Ca^{++}	40	2	20	0.020	20
Cl^-	35	1	35	0.035	35
HCO_3^-	1+12+3(16)61	1	61	0.061	61

units of weight are needed to describe their concentrations. These units are called milliequivalent weight (mEq). The milliequivalent weight is the equivalent weight divided by 1,000. Thus, the equivalent weight of sodium is 23 Gm and the mEq is 0.023 Gm or 23 mg. (Table 10-4).

The weight of sodium 0.023 Gm is called 1 mEq sodium; the weight 0.046 Gm would be 2 mEq sodium, etc. The label on the flask may read: "500 ml of this solution contains 154 mEq sodium and 154 mEq chloride." The solution contains

and
$154 \times 0.023 = 3.542$ Gm Na
$154 \times 0.035 = 5.390$ Gm Cl

A solution could contain 154 mEq sodium, 100 mEq chloride, and 54 mEq bicarbonate or any combination in which the sum of the equivalents of the positive ions equals the sum of the equivalents of the negative ions.

saturated solutions

The term *saturated solution* expresses a definite concentration because at any given temperature (unless the substances are completely miscible) there is a limit to how much solute will dissolve in a given amount of solvent when in contact with some undissolved solute. Many of these concentrations have been determined experimentally. At room temperature a saturated boric acid solution contains 4.8 Gm boric acid dissolved in every 100 Gm solution. We say it is roughly a 5 per cent solution. Since boric acid solution is used as an external wash or soak, we need not be as precise about the concentrations as we would be if a solution were to be used as an injection. If we added 6 Gm boric acid to 100 Gm water at room temperature, what would be the concentration of the solution? Approximately 5 Gm would dissolve, and the remaining 1 Gm boric acid would settle on the bottom of the flask. The solution would be saturated at 5 per cent concentration. Some of the boric acid on the bottom dissolves, but some of the dissolved boric acid crystallizes out as fast as the other material dissolves. We have two processes going on at the same rate: crystallization and solution. If both proceed at the same rate, there is no change in concentration and we have a system in equilibrium. Thus a *saturated solution* may be defined as *one in which the dissolved molecules are in equilibrium with the undissolved*. Note that, in order to have a saturated solution, some of the solute must still be present undissolved.

Any concentration less than that of saturation is called unsaturated at that temperature. A 4 per cent boric acid solution or a 2 per cent or a 3 per cent at room temperature is unsaturated.

supersaturation

If a solution is saturated at a high temperature and then cooled, the rate of crystallization usually exceeds that of solution and crystals continue to form until equilibrium is reestablished at the lower temperature. In some cases, however, if the saturated solution is filtered while hot to remove all crystals of the solute and any dust particles, it is possible to cool it to room temperature without crystallization. The solution at this lower temperature is more than saturated; it is *supersaturated*. Now, if the smallest crystal of the solute is introduced into the supersaturated solution or if particles of dust, which may act as nuclei, get into it, all the excess solute crystallizes out rapidly until equilibrium is reestablished at this lower temperature.

Heat of Solution versus Heat of Crystallization

When substances dissolve in water, heat is involved. This is called the *heat of solution*. If heat is absorbed in the process of solution, it will be given off in the same amount during crystallization. This is called the *heat of crystallization.* When a supersaturated solution crystallizes, considerable heat is evolved. This action has been used in a type of "chemical hot-water bottle." Let us say that a person habitually wakes up at 3 A.M. with cold feet or some other discomfort. If he took a hot-water bottle to bed with him, it would be cold by the time he needed it; but he can cause the chemical hot-water bottle to heat up when he needs it. It contains a supersaturated solution of sodium acetate. A twist of the stopper releases a crystal, crystallization takes place, and the heat of crystallization is evolved. To redissolve the crystals, it is merely necessary to let the bottle stand in hot water for a short time after use. If it is then set aside to cool, the supersaturated solution will form once more as the solution slowly returns to room temperature.

Solubility

We think of cane sugar as soluble and chalk as insoluble in water. Nothing is completely insoluble, but so little may dissolve that we can call some substances insoluble.

Fat is insoluble in water but soluble in ether and carbon tetrachloride. Generally speaking, most compounds of carbon are soluble in organic solvents, such as ether, chloroform, alcohol, and acetone, while inorganic compounds, if they do dissolve, are more soluble in water. Water has been called the *universal* solvent, but that would be impossible, wouldn't it, for in what kind of container could it be kept?

rate of solution

Finely divided solids dissolve more readily than large lumps, for this, like evaporation, is a surface process. In general, the *rate of solution increases with an increase of temperature*. A solution that is saturated at one temperature will not be saturated at a higher temperature until a new equilibrium between solution and crystallization is established.

Gases are an exception to this generalization. Gases are more soluble at lower temperatures. If we allow an open bottle of carbonated beverage to become warm, it goes flat. The carbon dioxide comes out of solution. There is more oxygen dissolved in cold ocean waters than in tropical seas. Would an elevated body temperature have an effect on the amounts of dissolved gas in the blood?

Agitating or stirring a mixture increases the speed of solution because it brings fresh solvent in contact with the solute.

Osmosis

Osmosis refers to the selective passage of solvent particles from one compartment to another. In biological systems, the solvent is always water. A membrane that permits passage of solvent particles but retains dissolved or solute particles is a semipermeable membrane. If such a membrane separates two compartments, each containing only pure water, water will flow back and forth across the membrane. The rate of flow will be the same and the volume of water in each compartment will remain the same. If the membrane separates pure water from a solution, again, water will pass back and forth across the membrane, but in this case there are fewer water molecules per unit volume in the solution than in the pure water, because of the presence of dissolved particles in the solution. As usual, particles pass from greater concentration to lesser concentration and more water will pass into the solution than passes out. The result is that the volume of pure water decreases and the volume of the solution increases while becoming more dilute. In the original experiments devised to measure this phenomenon, pressure was applied to the compartment containing the solution and the amount of pressure needed to prevent an increase in volume was measured. This was called the osmotic pressure of the solution. The osmotic pressure of a solution is therefore defined as *the pressure that must be applied to that solution to prevent the flow of solvent through a semipermeable membrane.*

The more particles there are dissolved in the solution, the greater will be its osmotic pressure. Osmotic pressure depends upon the number of dissolved particles, not the concentration. Small, light particles have an effect equal to the effect of larger, heavier particles. If the osmotic pressures of two solutions are the same, they are said to be isotonic. If the osmotic pressures are different, the solution with the higher osmotic pressure is hypertonic and the solution with the lower pressure is hypotonic. The solvent will pass *from hypotonic to hypertonic* solution, resulting in an equalization of the osmotic pressures of the two solutions.

Let us suppose that pure water is injected into the blood. Normally the fluid inside the red blood cells is isotonic with the fluid part of the blood, the plasma. With the addition of water, the fluid outside the cell becomes hypotonic and the fluid inside the cell is now hypertonic with reference to the plasma. Under these conditions, water will flow into the cell more rapidly than it will flow out. The volume of the cell increases, the cell swells, and if the pressure difference is great enough, the cell will rupture, liberating its hemoglobin contents into the plasma. This effect is an example of *hemolysis* and the blood is said to be *laked*. On the other hand, if a solution of higher osmotic pressure (hypertonic) is injected, water will flow out of the cell causing its contents to shrink. This shrinkage is called *plasmolysis or crenation.* In either case, the cell is destroyed. It must follow that solutions added to the blood should be isotonic with the blood.

Solutions isotonic with blood are sodium cloride, 0.88 to 0.92 per cent, called normal saline or physiologic salt solution. Ringer's solution more nearly reflects the composition of plasma, i.e., sodium chloride 0.86 per cent, potassium chloride 0.03 per cent, calcium chloride 0.033 per cent, and glucose 5 per cent. These figures indicate that it is the number of particles in solution that determine the osmotic pressure. Per unit volume, there are the same number of dissolved particles in 0.9 per cent sodium chloride as there are in 5 per cent glucose.

Sometimes, a pathologic condition, such as the accumulation of fluid in brain tissue, requires the use of a hypertonic solution. One would hope that the excess fluid would pass out of the tissue into the hypertonic solution.

Dialysis

A permeable membrane permits the passage of both solvent and dissolved particles. Here again, the particles pass from the area of greater concentration to the area of lesser concentration. This direction of passage is referred to as the concentration gradient. Substances can pass in either direction across a permeable membrane, depending upon their respective concentrations. This is the principle used in *hemodialysis,* the so-called artificial kidney. Catheters are inserted into the patient's vein and the blood flows through cellophanelike tubes through a vat or bath of solution and is returned to the patient's body. The solution in the vat is carefully prepared so that a substance to be removed from the blood is absent or in very low concentration and anything to be added to the blood is in higher concentration. The osmotic pressure of the solution also must be regulated to prevent either dehydration or overhydration of the patient. To illustrate, if urea and sodium are to be removed from the blood and potassium added, the concentration of potassium in the bath would be higher than in the blood, and urea and sodium would be lower. Potassium would diffuse in, and urea and sodium diffuse out.

The same principle is used in *peritoneal dialysis* where fluid of definite composition and osmotic pressure is introduced into the peritoneal cavity,[2] left for a specified period of time, and then drawn off. The greatest exchange of material occurs at the beginning of this procedure, because then the concentrations of the fluids are most different. As the dialysis proceeds the material being removed builds up in the dialyzing solution, the difference in concentration becomes less, and the rate of diffusion decreases.

Many substances in the body cross membranes by this same method. Higher concentrations of amino acids and glucose in the small intestines facilitate their passage from the intestines into the blood. In other cases, however, living membranes can selectively permit the diffusion of some substances and prevent the diffusion of others. In addition, living membranes can cause substances to diffuse *against a concentration gradient*. This means that substances can pass from an area of lesser concentration to an area of greater concentration. (This is the opposite of what we have been preaching all along.) In pregnancy, iron passes from the maternal to the fetal blood even when the concentration of iron in the fetal blood is higher than the concentration of iron in the maternal blood. This is fortunate for the baby as he must make his own blood and he needs iron to do so. This type of diffusion is called *active transport* and applies only to living membranes.

saline catharsis

The mechanism of the action of saline cathartics involves dialysis. A number of ions are slowly absorbed from the intestinal tract. These include magnesium, sulfate, phosphate, tartrate, and citrate ions. Solutions of salts containing these ions are retained for a comparatively long time in the intestines. Consequently, water from the body will pass freely into the intestinal tract until the solution is isotonic with the body fluids. This additional water will produce a watery stool that is easily evacuated. The action of saline cathartics, then, is similar to that of an enema. Since this water is lost, an individual may become dehydrated through continued use of hypertonic solutions of such cathartics. In fact, these compounds are often prescribed for the purpose of ridding the body of fluid. Obviously, if the patient cannot afford to lose body fluids and a cathartic action is still desired, sufficient liquid must be administered to offset the loss of body water.

Since the intestinal membrane is only relatively impervious to saline cathartics, some absorption of the salts takes place. These partially absorbed salts then act as a diuretic and increase the flow of urine. On the other hand, if the salt is given in a strongly hypertonic solution and an appreciable amount of body water is lost, the flow of urine may be reduced because of the dehydration.

The relative efficiency of saline cathartics depends upon the rate at which they are absorbed and on the osmotic pressure of the solution. Assuming the same rate of absorption, the efficiency will depend upon the amount of water required to make the

---

[1] The peritoneal cavity is a small space between two membranes, one covering the organs in the abdomen and the other adhering to the inner aspect of the body wall.

solution isotonic with the body fluids. Sodium sulfate, glauber's salt, removes more of the body fluid than does magnesium sulfate, epsom salts, when administered in equal dosage and is therefore more effective. Incidentally, glauber's salt is one of the cheapest saline cathartics. It was once thought that a salt such as magnesium sulfate, which has 2 ions that are slowly absorbed, was twice as effective as sodium sulfate, which has only 1. This is not true, for if 1 ion is not absorbed an equivalent amount of the oppositely charged ion must also remain unabsorbed.

THE COLLOIDAL STATE

Dispersed particles intermediate in size between those in solutions and those in suspensions are in the colloidal state. At one time colloids were thought to be gluelike substances that did not form crystals, and solutions were thought to be formed from crystalloids. Today it is possible to grind large particles in a mill until they are of colloidal size. Conversely, one can cause smaller particles to clump together. The particles resulting from both processes, when dispersed, exhibit typical properties of matter in the colloidal state. Since protein molecules fall into the range of colloidal particles in size and since protoplasm is protein in nature, the properties of the colloidal state can be translated directly into the properties of protoplasm.

Properties of Colloids

adsorption

Finely divided substances provide a tremendous surface area. Consequently, surface phenomena become very apparent in colloidal particles. Colloids selectively *absorb ions from a solution,* and as a result in a given suspension they *become charged alike.* Since like charges repel, the particles are continually kept apart. In this way their coagulation is prevented—another reason for the permanency of a colloidal suspension. Placed in an electrolytic cell, the colloids of a suspension will migrate to one of the electrodes. In this way the polarity of the colloidal particle may be determined. It is now possible to deposit rubber, a colloid, electrolytically on a surface or on a form that is the anode of the cell. Activated charcoal is used in gas masks to remove poisonous gases from the inhaled air. The principle that the more easily a gas is liquefied, the more readily it is adsorbed was responsible for the fact that the chlorine gas of the First World War could be removed by the charcoal and yet the air could pass through the mask. Charcoal is also used to decolorize sugar solutions in the production of granulated sugar. Fuller's earth is used to decolorize petroleum preparations such as petrolatum and Nujol. Adsorption by fibers is responsible for the dyeing of fabrics. It must be evident by now that the property of adsorption is an important property of the colloids.

coagulation

If the charges on the colloids in a suspension could be destroyed, the particles would cease to repel each other. They could then come together to form aggregates that would settle out. The desired result can be easily accomplished by adding an electrolyte. The ions neutralize the charges on the colloids, and precipitation results. Since salts are usually used for this purpose, the process is called *salting out.* The formation of a delta at the mouth of a river, where the colloidal matter is precipitated upon coming in contact with the ions of sea water, is a natural illustration of this process.

dialysis

Substances that form true solutions and that crystallize upon the evaporation of water are called *crystalloids*. Conversely, substances that do not form crystals but form instead an amorphous or gluelike mass are the colloids.

In order to stabilize a colloidal suspension it is necessary that crystalloids be removed. This is done by dialysis. Pharmacologic laboratories employ dialysis extensively during the preparation of antitoxin. The process of digestion of food in the body is partly explained when we consider that only soluble substances may pass through the membrane of the intestinal wall; colloidal food cannot be absorbed.

osmotic pressure

The effect of colloids on the osmotic pressure of a solution is small in comparison with the effect of a comparable weight of crystalloid material. Osmotic pressure is due to the number of particles involved. Since colloidal particles are very large, their numbers are relatively small.

In the blood, however, the presence of colloids exerts a very important osmotic pressure. This pressure, which is due to the blood proteins, is called *oncotic pressure.* In comparison with globulin molecules, serum albumin molecules are small and simple and play a most important role in maintaining the osmotic pressure of the blood.

Protective Colloids

Such substances as gelatin and casein interfere with the precipitation of many insoluble compounds. When added to a colloidal suspension, they protect that colloid from the precipitating action of electrolytes. The mechanism of the protection is debated, but the action is probably due to the formation of a film around each colloidal particle. Many antiseptics contain very finely divided silver or silver oxide in suspension, protected by a protein. For instance, Argyrol contains silver oxide protected by vitellin of egg yolk; Collargol contains finely divided silver in albumin. In the manufacture of ice cream a small amount of gelatin is added as a protective colloid to give a smooth texture. It prevents the formation of gritty ice crystals and hard crystals of lactose.

Gels and Sols

The ameba that the biology student watches under the microscope is a blob of protoplasm. As it moves along, part of it appears to become more free flowing while other parts become more resistant to flow. Protoplasm is protein material, and proteins are extremely large molecules. They cannot form solutions but exist as colloids. Colloids that do not alter the viscosity of water are free-flowing colloids and are called *suspensoids* or sols; those that greatly increase the viscosity are called *emulsoids*. These can assume a more rigid appearance and form gels. Jellies are gels, as are custards and agar. Gels can adsorb large quantities of water. They lose this ability as they age. This process is called *syneresis.* One can observe the agar in a test tube shrink and pull away from the sides of the tube as the days pass. Could this process be partly responsible for some of the changes in bones and connective tissues that sometimes occur in persons of advanced age?

Even the clotting of blood is the formation of a gel. In a test tube, venous blood will become firm in a few minutes, so much so that it will not run out of the tube when it is inverted. Here again, syneresis occurs. The clot gradually shrinks, and a clear yellow fluid, serum, exudes from it. Calcium ions are necessary in the clotting of blood. Therefore, to prevent clotting, these calcium ions must be removed by adding oxalate ions, which remove the calcium through formation of calcium oxalate.

$$Ca^{++} + C_2O_4^{=} \rightarrow CaC_2O_4 \downarrow$$

Enzymes and some hormones are proteins and therefore colloidal. When giving an injection of insulin, one is working with a sol.

Colloidal gold is another example of a sol. It is used in the treatment of cancer and certain skin diseases.

SUSPENSIONS AND EMULSIONS

Remember that all the "shake-well" medications are suspensions of either solid in liquid (suspension) or tiny drops of liquid in liquid (emulsion). These substances are simply mixed together and can be easily separated, the suspension either by filtration or by gravity, the emulsion by gravity, or, in the case of a permanent emulsion, by "breaking" the emulsion.

Since the stability of the emulsion depends upon the presence of the emulsifying agent, it must follow that anything that will destroy the agent will break the emulsion. An emulsion may be broken by the conversion of the agent into another compound; e.g., an acid added to an emulsion stabilized by soap will form a fatty acid, which is not an emulsifying agent. It may also be broken by agitation; e.g., mayonnaise sometimes breaks in shipment. There are other methods too numerous to mention. The churning of butter and the whipping of cream depend on breaking an emulsion. Sugar added to cream before the whipping speeds up the process, for the crystals tend to aid in breaking the emulsion.

Many common examples of emulsions are worthy of mention. The proteins and lecithin of egg yolk keep the 30 per cent fat in the yolk in emulsified form. These agents can stabilize still more oil, as evidenced in the making of mayonnaise. The proteins in soup stock emulsify the fat present. Cooked starch from flour serves to emulsify the fat in gravies. Today butter fat obtained during the time that milk is plentiful is stored to be used later during a shortage. Then, by adding the proper amount of water and dried skim milk, either cream or milk can be produced through the use of the homogenizer. This machine forces the mixture through extremely fine holes under great pressure, reducing the fat globules to about one-tenth the normal size in milk. Because of this great dispersion, there is no *creaming* of the emulsion. During the digestion of fats, emulsification is an important step, for the great increase in the surface area aids in bringing the fat, water, and lipase in contact.

Study Exercises

1. Define solution, suspension, emulsion, colloid, permeable membrane, semipermeable membrane, osmosis, diffusion, dialysis, and oncotic pressure.
2. What is a 1M solution? a 2N solution? What weight of solute is dissolved in 100 Gm 10 per cent solution of sodium chloride? of glucose?
3. One liter of a calcium chloride solution contains 220 Gm solute. Describe its concentration as molar, normal, and per cent.
4. What weight of solute is dissolved in 500 ml 1N solution of sulfuric acid? What weight of hydrogen does this solution contain?
5. One liter of a solution contains 100 mEq sodium ion and 100 mEq chloride ion. How many grams of sodium ion and chloride ion are dissolved in 1 liter of this solution?
6. Correct each of the following statements: (*a*) a molar solution contains 1 Gm solute per liter of solution; (*b*) passage of dissolved material through a membrane is called osmosis; (*c*) blood cells may swell and rupture when suspended in a hypertonic solution; (*d*) colloidal particles can pass through a permeable membrane but not through a semipermeable one.

chapter eleven ELECTROLYTES

Ideas to preview or review

Pressure is defined as force per unit area. Vapor pressure is the force exerted in the space above the surface of a liquid by molecules that have escaped from the liquid.

Boiling point is the temperature at which the vapor pressure of a liquid equals the atmospheric pressure.

Freezing point is the temperature at which a liquid changes to a solid.

Osmotic pressure is the pressure that must be applied to a solution to prevent the flow of water into it through a semipermeable membrane.

A semipermeable membrane is one through which solvent (water) particles can pass but dissolved molecules cannot.

Current electricity is a flow of electrons.

Atomic number is the number of protons in the nucleus of an atom.

Electrodes are the terminals of a source of electricity. They are sometimes called poles, e.g., positive pole, or anode, and negative pole, or cathode.

Substances that can transmit an electric current are called *conductors*. The best conductors are metals such as copper, platinum, and silver; but some chemical compounds, either dissolved or melted, can also conduct an electric current. These are called *electrolytes*, and they are classified as acids, metal hydroxides (bases), and salts. Body fluids contain electrolytes and can therefore conduct electric currents. The transmission of nerve impulses and the tracings obtained with the electrocardiograph and the electroencephalograph depend, in part, upon the ability of body tissues to generate and conduct electrical energy.

PROPERTIES COMMON TO WATER SOLUTIONS OF ACIDS, HYDROXIDES, AND SALTS

Conductivity

If we examine an ordinary electric light bulb, we will find two contacts for the transmission of electric current. One is at the bottom of the base, and the other is the threaded section that is screwed into the socket. For the lamp to light, one contact must be attached to a wire coming from the source of electric current, and the other contact must be attached to a wire returning to it. This arrangement makes a closed circuit along which electrons can flow (current electricity). If either of the wires is broken, there will be no closed circuit, no flow of electrons, and the lamp will not light. When the separated wires are submerged below the surface of certain liquids or solutions, the lamp will light (Fig. 11–1). These liquids or solutions are electrolytes. When other liquids or solutions are used, the lamp does not light, indicating that these are nonelectrolytes or too weak to carry the current. The intensity of the light gives us an indication of the relative amounts of elec-

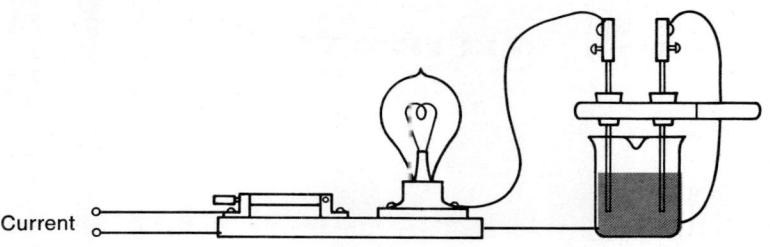

Fig. 11-1 Apparatus to determine the conductance of liquids.

tricity conducted by the various liquids. The ability of electrolytes to conduct an electric current is called conductivity.

Abnormal Properties of Solutions

When a nonelectrolyte such as table sugar (sucrose) is dissolved in water, the sucrose molecules replace some of the water molecules on the surface of the liquid, with the result that the vapor pressure of the solution is lower than that of pure water. This means also that the boiling point of the solution will be higher than that of water and the freezing point lower. Actually, when 1 gram-molecular weight of a nonelectrolyte is dissolved in 1,000 Gm water, the boiling point is elevated 0.52°C and the freezing point is depressed 1.86°C. The extent of this effect is proportional to the number of particles dissolved. Two gram-molecular weights have twice the effect, and so on. Gram-molecular weights of all nonvolatile nonelectrolytes dissolved in 1,000 Gm water have the same effect. The explanation is, of course, that gram-molecular weights represent the same number of molecules. Equal numbers of particles will cause equal vapor-pressure depressions, boiling-point elevations, and freezing-point depressions.

The osmotic pressure of a solution depends upon the number of solvent particles present in a given volume. In osmosis only solvent particles pass through the membrane. The presence of dissolved particles reduces the number of solvent molecules at the membrane, and this will increase the osmotic pressure of the solution.

We are stressing the fact that the vapor pressure, boiling point, freezing point, and osmotic pressure of a solution are proportional to its concentration.

A solution of an electrolyte has a greater effect on these properties than a solution of a nonelectrolyte of the same concentration. This is evident in the data shown in Table 11-1. The important point to note in these data is that the depression caused by potassium chloride is *nearly twice* that caused by nonelectrolytes and that the depression caused by potassium sulfate is nearly three times as great.

Summary of Properties

In addition to their ability to conduct an electric current, solutions of electrolytes have abnormally low vapor pressures and freezing points and abnormally high boiling points and osmotic pressures. It would appear that solutions of electrolytes contain greater numbers of dissolved particles than solutions of nonelectrolytes.

Table 11-1 FREEZING-POINT DEPRESSION OF MOLAL SOLUTIONS

Solute	Formula	Depression, °C
Sugar	$C_{12}H_{22}O_{11}$	−1.86
Glycerol	$C_3H_5(OH)_3$	−1.86
Potassium chloride	KCl	−3.61
Potassium sulfate	K_2SO_4	−5.51

ARRHENIUS' THEORY OF IONIZATION

In 1887 Arrhenius proposed a theory to account for these abnormal properties. He suggested that the molecules of an acid, hydroxide, or salt upon being dissolved in water broke up into smaller particles which he called *ions*. These ions could reunite, forming molecules again. For instance, sodium chloride could form sodium ion and chloride ion, which could then recombine to form molecules of sodium chloride in a reversible reaction. Arrhenius believed that ions could exist only in solution.

These ions bore electric charges, the number of charges being equal to the valence. Since metals, hydrogen, and ammonium groups were attracted to the negative pole, the cathode, they were called *cations*, and they were believed to carry a positive charge. Nonmetals, hydroxyl groups, and acid groups were attracted to the positive pole, the anode. They were called *anions*, and they were thought to carry a negative charge.

Different electrolytes varied in their abilities to conduct an electric current because they formed ions to a greater or lesser degree. An acid such as hydrochloric acid was a good conductor because it possessed a high concentration of ions and relatively few molecules. Such a solution was called a *strong electrolyte*. Conversely, a substance that produced few ions in solution was called a *weak electrolyte* and a poor conductor of electricity.

Arrhenius' Theory and the Properties of Electrolytes

If we think of the effect on the vapor pressure as being caused by individual particles, molecules, or ions, we can account for the abnormal increase in the boiling point and the extra-large depression in the freezing point of solutions of electrolytes. A solution of potassium chloride, KCl, forms 2 ions:

$$KCl \rightarrow K^+ + Cl^-$$

If it were 100 per cent dissociated, there would be twice as many particles and hence twice the depression in the freezing point.

Since potassium sulfate, K_2SO_4, forms 3 ions,

$$K_2SO_4 \rightarrow 2K^+ + SO_4^=$$

there would be three times the effect, if the dissociation were complete. It was shown in Table 11-1 that potassium sulfate has nearly three times the effect of a nonelectrolyte. According to Arrhenius' theory this is explained by incomplete dissociation, and the degree of dissociation can be calculated from the amount of the depression in the freezing point. The abnormally high osmotic pressures of electrolytes can be accounted for on the same grounds.

If we ask the questions: How do the electrolytes conduct an electric current? Do electrons flow from the conducting wires through the solution? Arrhenius' theory offers the answers. The negative pole, the cathode, is more richly supplied with electrons than the positive pole, the anode. The positively charged cations migrate to the cathode and pick off electrons, forming neutral molecules or atoms. The negatively charge anions are attracted to the anode and give up their electrons. In this way they too form neutral particles.

When the wires from a lamp are dipped into a solution of hydrochloric acid, positively charged hydrogen ions pick up electrons from the cathode to form hydrogen atoms (Fig. 11-1). Two of these combine to give a molecule of hydrogen gas. The negatively charged chloride ions give up electrons to the anode, forming chlorine atoms, and two of these combine to form a molecule of chlorine gas. Hydrogen gas bubbles off at the cathode, and chlorine gas bubbles off at the anode. Since electrons are removed from one end of the wire and supplied at the other, they are able to flow through the wire in the lamp, and the filament glows.

MODERN CONCEPTS OF IONIZATION

X-ray analysis, which was not available to Arrhenius, shows us that a sodium chloride crystal consists not of molecules but of a lattice arrangement in which sodium ions are surrounded on all sides by chloride ions and in which chloride ions are in turn surrounded by sodium ions. In other words, there are no molecules of the salt sodium chloride. Even in the solid state, the salt is 100 per cent ionized. This is in agreement with our concept of the valence bonds in sodium chloride.

Debye-Hückel Theory

If the salt potassium chloride is 100 per cent ionized, why is not its freezing-point depression double that of a nonelectrolyte of the same concentration? We find our answer in the Debye-Hückel theory. They suggested that in a concentrated solution positively charged ions attract negatively charged ions and repel those of a similar charge. The cation then finds itself in a cloud of negative ions which exerts a drag effect on it and reduces its mobility. In addition, the surrounding cloud of negative ions may mask the presence of the cation. Anions will also experience a drag effect caused by surrounding positive ions.

Because of these forces between ions, the number of ions on the surface of the solution is decreased somewhat, and the effect on the freezing point, vapor pressure, boiling point, and osmotic pressure is less than twice that of the nonelectrolyte. In addition, the dragging effect caused by the atmosphere of oppositely charged ions in their migration to the electrodes causes a corresponding reduction in their ability to conduct an electric current. The more dilute the solution, the less pronounced are these interionic forces and drag effects.

Brönsted and Lowry Theory

The discovery and use of new tools, such as the X ray, open to the scientist an ever-widening field of information. Brönsted and Lowry have proposed a theory which appears to be most acceptable in the light of today's knowledge. They classify electrolytes as acids and bases, and they define an acid as a *proton donor* and a base as a *proton acceptor*.

The proton in solution becomes attached to at least 1 molecule of water:

$$H^+ + H_2O \rightarrow H_3O^+$$

This ion, H_3O^+, is usually called the *hydronium* ion. It has also been named *oxonium* or *hydroxonium* ion. Since nursing and medical literature contains the old as well as the newer notations, students should be familiar with both. In this text, as the occasion arises, we use the different symbols: +, meaning proton; H^+, meaning hydrogen ion; and H_3O^+, meaning hydrated hydrogen ion, or hydronium ion.

DEGREE OF IONIZATION

Solutions of acids, bases, and salts showing a high degree of ionization are strong electrolytes. They exist completely or almost completely in the form of ions and very slightly, if at all, in the form of undissociated molecules. Hydrochloric acid appears to be 90 per cent ionized. Actually, its degree of ionization is higher than this, but interionic attractive forces mask the true ionization. Of any 100 molecules of hydrochloric acid, 90 appear to be ionized. It would seem that there are 90 hydrogen ions, 90 chloride ions, and 10 undissociated hydrogen chloride particles, giving a total effect of 190 particles. This figure is almost double the original 100 particles.

$$HCl \rightleftharpoons H^+ + Cl^-$$
$$\underset{10}{(100-90)} \quad 90 \quad 90$$

Comparable conditions prevail with the other strong electrolytes HNO_3, H_2SO_4, NaOH, KOH, and almost all the salts.

Weak electrolytes are poorly ionized; they exist in solution almost completely as undissociated molecules. Very few molecules dissociate into ions. Substances with a degree of ionization 1 per cent or less in a specified dilution (0.1N),[1] are classified as weak.

Acetic acid is 1.34 per cent ionized. Ignoring the decimal, we can say that for every 100 molecules of acetic acid, 1 dissociates, yielding 99 undissociated molecules, 1 hydrogen ion, and 1 acetate ion, giving a total of 101 particles.

$$CH_3COOH \rightleftharpoons H^+ + CH_3COO^-$$
$$\underset{99}{(100-1)} \quad\quad 1 \quad\quad 1$$

Carbonic acid is 0.17 per cent ionized. About 1 out of every 1,000 molecules of carbonic acid ionizes to yield 1 hydrogen ion and 1 bicarbonate ion, for a total of 1,001 particles.

$$H_2CO_3 \rightleftharpoons H^+ + HCO_3^-$$
$$\underset{999}{(1000-1)} \quad 1 \quad\quad 1$$

At any one time, very few ions are present in a solution of a weak electrolyte. If these ions are removed, more molecules will dissociate until eventually all the molecules are changed into ions.

In the laboratory, a student may spill sodium hydroxide (a strong electrolyte) on his hand. The skin should be washed with copious amounts of water, not to dilute the material, but to remove it by the mechanical effect of flowing water. To neutralize any remaining hydroxide, one should choose a weak acid such as acetic, citric, or lactic. As soon as the very few available hydrogen ions unite with the hydroxyl ions present to form molecules of water, more molecules of the weak acid will dissociate to furnish more hydrogen ions. In this way sufficient numbers of hydrogen ions are available, but never in a high concentration. There is a fairy tale of a magic purse that always contained three gold coins—no more, no less. No matter how much was spent, three gold coins remained. However, the comparison is imperfect because the undissociated molecules can change into ions and be used up.

ACIDS, PROTON DONORS

Acids may be either inorganic or organic compounds. Examples of inorganic acids are hydrochloric acid, HCl (strong); nitric acid, HNO_3 (strong); sulfuric acid, H_2SO_4 (strong); and phosphoric acid, H_3PO_4 (moderately strong). Examples of organic acids are acetic acid, CH_3COOH (weak); lactic acid, $C_3H_6O_3$ (weak); citric acid, $C_6H_8O_7$ (weak); and carbonic acid, H_2CO_3 (weak). Concentrated solutions of strong acids destroy tissues and fabrics.

Common Properties

The properties common to acids are the properties of the hydronium ion.

a. We have learned without studying chemistry to associate acids with a sour taste. We are familiar with the taste of such organic acids as citric acid in lemons, lactic acid in sour milk, and acetic acid in vinegar. Inorganic acids also taste sour.
b. Indicators, as their name implies, tell us something. They are usually organic dyes which change color in acidic, neutral, or alkaline solutions. Litmus, methyl orange, and phenolphthalein are commonly used indicators. In acid solutions litmus turns red, methyl orange turns red, and phenolphtalein (used medicinally as a laxative) remains colorless.
c. Solutions of acids react with active metals to form hydrogen gas and a salt:

---

[1] One-tenth gram-equivalent weight of solute per liter of solution.

$$2HCl + Zn \rightarrow ZnCl_2 + H_2 \uparrow$$
$$\text{Zinc chloride}$$

$$H_2SO_4 + Mg \rightarrow MgSO_4 + H_2 \uparrow$$
$$\text{Magnesium sulfate (epsom salt)}$$

This is one way of preparing these salts, especially if one can use or sell the by-product, hydrogen gas.

d. Acids react with metal hydroxides and metal oxides to form salts and water (neutralization):

$$2HCl + Mg(OH)_2 \rightleftharpoons MgCl_2 + 2H_2O$$
$$\text{Hydrochloric acid} \quad \text{Magnesium hydroxide} \quad \text{Magnesium chloride}$$

This reaction occurs in the stomach when milk of magnesia is taken as an antacid.

e. Acids react with carbonates and bicarbonates to form carbon dioxide, water, and a salt:

$$2HNO_3 + Na_2CO_3 \rightarrow$$
$$\text{Nitric acid} \quad \text{Sodium carbonate}$$

$$2NaNO_3 + H_2O + CO_2 \uparrow$$
$$\text{Sodium nitrate}$$

$$CH_3CHOHCOOH + NaHCO_3 \rightarrow$$
$$\text{Lactic acid} \quad \text{Baking soda}$$

$$CH_3CHOHCOONa + CO_2 \uparrow + H_2O$$
$$\text{Sodium lactate} \quad \text{Carbon dioxide} \quad \text{Water}$$

This second reaction occurs in making biscuits with sour milk. Comparable reactions occur with baking powders. Some effervescent medications such as Seidlitz powders react in the same way.

If a person swallows a corrosive acid, such as concentrated hydrochloric, sulfuric, or nitric acid, one should immediately administer as an antidote magnesium hydroxide, aluminum hydroxide, or calcium hydroxide rather than bicarbonate of soda. The above chemical reaction gives us the reason. The carbon dioxide gas generated in the reaction between the acid and sodium bicarbonate may exert pressure on the tissues of the stomach wall which, already damaged by the acid, may perforate. Some persons have developed the practice of administering bicarbonate of soda to themselves to correct gastric distress. In these and similar cases, the old adage applies: "He who diagnoses himself has a fool for a doctor."

Ionization of Acids

An acid has been defined as a proton donor. In solution the proton unites with a molecule of water to form hydronium ion. The reaction is reversible, and we can indicate the relative strengths of the acids in solution as follows:

Strong acid:

$$HCl + H_2O \rightleftharpoons H_3O^+ + Cl^-$$

Weak acid:

$$HC_2H_3O_2 + H_2O \rightleftharpoons H_3O^+ + C_2H_3O_2^-$$
$$\text{Acetic acid}$$

Some acids can donate 2 protons; these are called *diprotic* acids.

$$H_2SO_4 \rightleftharpoons H^+ + HSO_4^-$$
$$\text{Sulfuric acid} \quad \text{Bisulfate ion [acid or base (amphoteric)]}$$

$$HSO_4^- \rightleftharpoons H^+ + SO_4^=$$
$$\text{Sulfate ion (base)}$$

$$H_2CO_3 \rightleftharpoons H^+ + HCO_3^-$$
$$\text{Carbonic acid} \quad \text{Bicarbonate ion (acid or base)}$$

$$HCO_3^- \rightleftharpoons H^+ + CO_3^=$$
$$\text{Carbonate ion (base)}$$

Some can donate 3 protons; these are called *triprotic* acids.

$$H_3PO_4 \rightleftharpoons H^+ + H_2PO_4^-$$
$$\text{Phosphoric acid} \quad \text{Dihydrogen phosphate ion}$$

$$H_2PO_4^- \rightleftharpoons H^+ + HPO_4^=$$
$$\text{Hydrogen phosphate ion}$$

$$HPO_4^= \rightleftharpoons H^+ + PO_4^\equiv$$
$$\text{Phosphate ion}$$

When ionizations proceed in more than one step, the first or primary ionization predominates because neutral molecules will give up a proton more readily than negatively charged ions. If, however, the products of a reaction are removed, the reactions go all the way to completion. The removal of protons from a solution of phosphoric acid shifts the equilibrium toward the formation of phosphate ion, $PO_4^{\equiv}$.

BASES, PROTON ACCEPTORS

Bases may be neutral molecules or ions. They are able to combine with hydronium ions in solution. Strictly speaking, carbonate and sulfate ions are bases. However, the metal hydroxides and ammonia and its related compounds, the amines, comprise the substances most frequently considered to be bases. Hydroxyl ion, present in water solutions of these substances, imparts properties which we call *basic*. Strong hydroxides such as sodium hydroxide and potassium hydroxide are called *caustics*, while substances possessing basic or alkaline properties to a lesser degree are called *alkalies*.

Hydroxides

Some important metal hydroxides are sodium hydroxide, NaOH (strong); potassium hydroxide, KOH (strong); calcium hydroxide, $Ca(OH)_2$ (poorly soluble); and magnesium hydroxide, $Mg(OH)_2$ (poorly soluble). Concentrated solutions of strong soluble hydroxides destroy tissues and fabrics.

common properties

a. Solutions of hydroxides have a bitter taste. This taste is seldom encountered in foods but can be detected in soap and in some medicines, particularly in those containing alkaloids, such as codeine.
b. Solutions of hydroxides react with indicators, turning litmus to blue, methyl orange to yellow, and phenolphthalein to red.
c. Hydroxides produce a soapy, slippery feeling on the fingers.
d. Hydroxides react with acids to form salts and water, as in the neutralization reaction mentioned in the discussion of acids. Medications that can be used to neutralize acids are called *antacids*.

$$2HCl + Ca(OH)_2 \rightleftharpoons \underset{\text{Salt}}{2CaCl_2} + 2H_2O$$

e. Metal hydroxides can react with salts to form other metal hydroxides and other salts if one of the products of the reaction is insoluble or a weak electrolyte:

$$NaOH + \underset{\text{Ammonium chloride}}{NH_4Cl} \rightarrow NaCl + \underset{\text{Ammonia gas}}{NH_3 \uparrow} + H_2O$$

This reaction is used to test for the presence of ammonium ion, NH_4^+.

SALTS

Solutions of salts show no common properties, contrary to acids and metal hydroxides. They can conduct an electric current, and most soluble salts are strong electrolytes. An examination of the formulas of some salts tells us why this class of electrolytes has no common properties: sodium chloride, NaCl; potassium nitrate, KNO_3; copper sulfate, $CuSO_4$. There is no ion common to all salts. Of course, solutions of salts with a common anion or cation will have its properties in common. Sodium chloride, sodium carbonate, and sodium sulfate all exhibit the properties of their common cation, sodium. Calcium chloride, sodium chloride, and magnesium chloride have the properties of their common anion, chloride.

To some people salt means table salt, sodium chloride. As we have indicated, salts are a class of electrolytes in which compounds are composed of a positively charged ion other than hydrogen and a negatively charged ion other than hydroxyl.

Minerals

Inorganic salts are correctly described as *minerals,* chemical compounds found in nature resulting from inorganic processes. In works on nutrition and diet therapy, we read of the necessity of minerals for normal growth and development. These minerals are usually mentioned as elements: calcium, phosphorus, sodium, potassium, iodine, iron, magnesium, sulfur, and traces of cobalt, copper, manganese, fluorine, etc. Some of these minerals exist in the body and in foodstuffs as inorganic salts, some as organic salts, and some are part of protein material. Iron is bound up in the hemoglobin molecule, while sulfur and phosphorus are widely distributed in proteins from both plant and animal sources. In other words, sodium may be supplied to the body in the form of the inorganic salt sodium chloride or the organic salt sodium citrate. Sulfur may be found in the inorganic salt sodium sulfate or in the protein egg albumin.

The salts that are dissolved in the body are so dilute that the ions lead independent lives. We speak of the role of sodium ion or phosphate ion rather than of the function of sodium phosphate in the body. Some salts exist as insoluble crystals. The slender, interlocking, needle-shaped crystals of insoluble calcium phosphate give rigidity and strength to bones and teeth. Abnormal accretions composed principally of insoluble calcium salts form gallstones, kidney stones, and bladder stones.

The principal ions in intracellular and extracellular fluids are considered in Chapter 12. In addition to these, iron and iodine are essential dietary constituents. These may be in short supply in the diet, or they may not be properly utilized in the body. Some cases of iron deficiency can be alleviated by the administration of iron salts. Ferrous carbonate, ferrous sulfate, ferric chloride, and ferric ammonium citrate are used for this purpose.

Iodine is a constituent of the hormone thyroxin and must therefore be supplied in the diet for the synthesis of this substance. Iodide ion exists in sea water, and salt-water fish and plants grown in areas near the sea are suppliers of this mineral. In land-locked areas where the food is of local origin, iodide may be lacking in the diet. To correct this, sodium iodide or potassium iodide may be administered. Sodium iodide has been added to the table salt sodium chloride and marketed under the name of iodized salt. Where the supplies of iodide are ample, the indiscriminate use of iodized salt may be unwise for some people.

There appears to be a relationship between the presence of small amounts of *fluorides* in the diet or drinking water and subsequent resistance to tooth decay. This holds only if fluorides are supplied in early childhood: they seem to have little or no effect after the age of 10. Some communities have added small amounts of fluoride to the municipal water supply for the benefit of their youngsters. In other communities, violent controversies have developed over the advisability of fluoridation. It would be well for nurses to be informed of scientific evidence for or against such community projects and to lend their influence and support to the application of scientific advances for community betterment.

Some important electrolytes and their uses are given in Table 11–2.

Table 11–2 is by no means complete. It is interesting to compare the uses of some of these salts, particularly those with common ions, such as acetates, bromides, etc. In dilute solutions, ions live independent lives. Sodium ion has the same properties whether it is formed from sodium chloride, sodium citrate, sodium sulfate, or any other soluble salt. This is true of all other ions.

IONIZATION OF WATER

We found during our conductance experiment that water did not conduct current. This is not strictly true, in spite of our observations. It must be remembered that we used an electric light bulb to indicate the flow of current. Such a bulb requires a relatively large amount of electricity to make the

Table 11-2 SOME IMPORTANT ELECTROLYTES

Electrolyte	Formula	Where Found or Use
Acids:		
Hydrochloric acid	HCl (strong)	Normal constituent of gastric juice; administered in cases of hypoacidity; used to wipe metal.
Nitric acid	HNO_3 (strong)	A caustic and disinfectant; used to cauterize animal bites.
Sulfuric acid (oil of vitriol)	H_2SO_4 (strong)	Important in industrial processes; very dilute solutions sometimes used as astringents.
Phosphoric acid	H_3PO_4 (moderately strong)	Salts of this acid (phosphates) are important constituents of blood, bones, and coenzyme systems, and they are vital to normal metabolism.
Boric or boracic acid	H_3BO_3 (weak)	Mild antiseptic and eyewash; poisonous.
Carbonic acid	H_2CO_3 (weak)	In carbonated beverages; a normal constituent of blood along with bicarbonate ion.
Lactic acid	$C_3H_6O_3$ (weak)	In sour milk; formed in the muscles during exercise; formed by bacterial action on lactose (milk sugar).
Citric acid	$C_6H_8O_7$ (weak)	In citrus fruits—lemons, limes, oranges, etc.; intermediate product in carbohydrate and fat metabolism; a flavoring agent.
Acetic acid	CH_3COOH (weak)	In vinegar.
Hydroxides:		
Sodium hydroxide (lye and soda lye)	NaOH (strong)	A caustic; used to cut grease in drainpipes.
Potassium hydroxide (potash lye)	KOH (strong)	A caustic; used to cut grease in drainpipes.
Calcium hydroxide (slaked lime or milk of lime)	$Ca(OH)_2$ (poorly soluble, few ions)	Used in burn ointments, as a source of calcium, and as an antacid.
Magnesium hydroxide (milk of magnesia)	$Mg(OH)_2$ (poorly soluble, few ions)	An antacid and laxative.
Aluminum hydroxide (alumina gel)	$Al(OH)_3$ (weak)	An antidote for internal acid burns.

Table 11-2 SOME IMPORTANT ELECTROLYTES *(Continued)*

Electrolyte	Formula	Where Found or Use
Salts:*		
Sodium chloride	NaCl	Used as a gargle and mouth wash; a 0.9% solution has the same osmotic pressure as body fluid; used for irrigations, intravenously as a replacement fluid, and in more concentrated solutions as an emetic.
Aluminum acetate	$Al(CH_3COO)_3$	A local astringent and antiseptic.
Ammonium chloride	NH_4Cl	A diuretic and expectorant.
Barium sulfate†	$BaSO_4$	Insoluble and opaque to X rays; given as "barium meal" or "barium enema" to outline the gastrointestinal tract for diagnostic X rays.
Bismuth subnitrate	$BiONO_3$	An astringent and antiseptic.
Calcium carbonate (chalk)	$CaCO_3$	An antacid and an antidote for corrosive acids.
Calcium chloride	$CaCl_2$	Used to increase the calcium concentration of blood and to decrease the clotting time of blood.
Calcium lactate	$Ca(C_3H_5O_3)_2$	Used for the same purpose as calcium chloride and as an antacid.
Calcium oxalate	CaC_2O_4	A constituent of kidney stones.
Calcium phosphate	$Ca_3(PO_4)_2$	Forms part of bones and teeth; used as an antacid for hyperacidity.
Calcium sulfate (plaster of paris)	$(CaSO_4)_2 \cdot H_2O$	A partially hydrated powder.
Calcium sulfate (gypsum)	$CaSO_4 \cdot 2H_2O$	A fully hydrated crystal that forms the rigid or set plaster cast.
Copper sulfate	$CuSO_4$	An astringent and emetic.
Lead acetate (sugar of lead)	$Pb(CH_3COO)_2$	Used externally as an astringent and in the treatment of poison ivy.
Magnesium carbonate	$MgCO_3$	An antacid and laxative.
Magnesium citrate	$Mg_3(C_6H_5O_7)_2$	A purgative.
Magnesium sulfate (epsom salts)	$MgSO_4$	A cathartic.
Mercuric chloride (bichloride of mercury)	$HgCl_2$	A disinfectant; very poisonous.
Mercurous chloride (calomel)	$HgCl$	Used as a powder for skin rashes and ulcers and in dilute solutions as a mild laxative.
Potassium acetate	KCH_3COO	An alkaline diuretic.
Potassium bicarbonate	$KHCO_3$	Reduces the acidity of the stomach and of the urine.
Potassium bitartrate (cream of tartar)	$KHC_4H_4O_6$	A diuretic and cathartic.
Potassium bromide	KBr	A nerve sedative.
Potassium chloride	KCl	Used as a replacement fluid when potassium ion is lost from the body.
Potassium chromate	K_2CrO_4	A cauterizer.

* Soluble salts with rare exceptions are strong electrolytes.
† Barium sulfate is commonly but incorrectly called barium, which is a metal.

Table 11–2 SOME IMPORTANT ELECTROLYTES *(Continued)*

Electrolyte	Formula	Where Found or Use
Salts *(Continued)*:		
Potassium citrate	$K_3C_6H_5O_7$	An alkaline diuretic.
Potassium iodide	KI	Increases bronchial secretions; used in the treatment of certain types of metallic poisonings.
Potassium nitrate (saltpeter)	KNO_3	A diuretic.
Potassium permanganate	$KMnO_4$	An astringent, germicide, fungicide, and deodorant; used as an antidote for snakebite and phosphorous poisoning.
Potassium sulfite	K_2SO_3	An antiseptic used to check internal fermentation.
Radium sulfate	$RaSO_4$	Used in the treatment of cancer and some skin diseases.
Silver nitrate	$AgNO_3$	A germicide and local astringent.
Sodium acetate	$NaCH_3COO$	A diuretic and a laxative.
Sodium benzoate	NaC_6H_5COO	A food preservative; used in the treatment of rheumatism.
Sodium bicarbonate	$NaHCO_3$	Used as an antacid in acidosis and to clean teeth.
Sodium biphosphate	Na_2HPO_4	A cathartic.
Sodium bisulfite	$NaHSO_3$	Used in the treatment of gastric and intestinal fermentation.
Sodium borate (borax)	$Na_2B_4O_7$	A water softener, detergent, mild antiseptic, and astringent.
Sodium bromide	$NaBr$	A nerve sedative and cerebral depressant.
Sodium carbonate (washing soda)	Na_2CO_3	Used to soften water and for alkaline baths.
Sodium citrate	$Na_3C_6H_5O_7$	A diuretic; used to prevent the formation of stones in the urinary tract, to facilitate their removal, and as an intravenous fluid to overcome acidosis. It is also used to prevent drawn blood from clotting.
Sodium iodide	NaI	Increases bronchial secretions; used in the manufacture of iodized table salt.
Sodium nitrate (Chile saltpeter)	$NaNO_3$	Used in the treatment of diarrhea.
Sodium nitrite	$NaNO_2$	A vasodilator.
Sodium oleate	$NaC_{17}H_{33}COO$	Used to stimulate the flow of bile.
Sodium phosphate	Na_3PO_4	A cathartic.
Sodium salicylate	NaC_6H_4OHCOO	Used to reduce pain and fever.
Sodium sulfate (glauber's salt)	Na_2SO_4	A cathartic.
Sodium tartrate	$Na_2C_4H_4O_6$	A laxative and diuretic.

filament glow. Had we used a more delicate instrument, such as a milliameter or a galvanometer, we would have observed a flow of current. Water does ionize somewhat, and the reaction producing the ions is:

$$HOH + H_2O \rightleftharpoons H_3O^+ + OH^-$$

Since both ions are present in equal numbers, neither one can predominate, and an indicator shows no reaction. Water, then, is *neutral*.

According to Brönsted and Lowry, water is an acid since it gives up a proton, and hydroxyl ion is a base. Hydroxyl ion is such a strong base that we think of metal hydroxides as bases. If in a water solution hydronium ions are in excess, the solution manifests properties we recognize as acidic; if hydroxyl ions are in excess, the solution has alkaline properties. In pure water, however, the hydronium and hydroxyl ions are present in equal numbers. Their concentrations have been determined experimentally. At 25°C, the concentration of hydronium ion in pure water is 0.0000001 gram-ionic weight[1] per liter, and the concentration of hydroxyl ion is the same. These decimal numbers are cumbersome to use and can be expressed as negative powers of 10. We can change the expression of the concentration of hydronium ion from 0.0000001 to 1 in the seventh decimal place, or 1×10^{-7}. In a neutral solution, the concentration of hydroxyl ion is also 1×10^{-7} gram-ionic weight per liter.

If we multiply these two figures, we obtain the ion-product constant for water designated as K_w:

$$[H_3O^+] \times [OH^-] = K_w$$
$$1 \times 10^{-7} \times 1 \times 10^{-7} = 1 \times 10^{-14}\ [1]$$

This equation tells us that in a liter of pure water we have one ten-millionth gram-ionic weight of hydronium ion and the same amount of hydroxyl ion. It also indicates that in solutions of electrolytes the concentrations of these ions may vary, but the product of their concentrations cannot exceed the constant number 1×10^{-14}.

This means that if the concentration of one ion increases, the concentration of the other must decrease. It is impossible to have high concentrations of both hydronium and hydroxyl ions in the same solution. If the concentration of one is high, that of the other must be very low to satisfy the equation.

By knowing the ion-product constant for water and the concentration of one ion, we can compute the concentration of the other ion. In a given solution the concentration of hydronium ion is 0.001 gram-ionic weight per liter. What is the concentration of hydroxyl ion in this solution?

$$[H_3O^+] \times [OH^-] = K_w = 1 \times 10^{-14}$$

Substituting the concentration of H_3O^+,

$$1 \times 10^{-3} \times x = 1 \times 10^{-14}$$
$$x = 1 \times 10^{-11}$$

The concentration of hydroxyl ion is 1 in the eleventh decimal place, 0.00000000001 gram-ionic weight per liter. Since the concentration of hydronium ion is greater than that of hydroxyl ion, this solution is acidic.

In another solution the concentration of hydroxyl ion is 0.00001. The hydronium ion

---

[1] Gram-ionic weight is the atomic weight of the atom or atoms of which the ion is composed multiplied by 1 Gm. The atomic weight of hydrogen is 1. The gram-ionic weight is 1 Gm. Hydroxyl ion is composed of oxygen weight 16 plus hydrogen weight 1. Its gram-ionic weight is therefore 17 Gm. Seventeen grams of hydroxyl ion equals 1 gram ionic weight of hydroxyl ion.

[1] To multiply two terms containing exponents, the exponential numbers are added:

$$2 \times 10^{-2} \times 4 \times 10^{-3} = 8 \times 10^{-5}$$
$$0.02 \times 0.004 = 0.00008$$

To divide terms containing exponents, the exponential numbers are subtracted:

$$6 \times 10^{-4} \div 3 \times 10^{-3} = 2 \times 10^{-1}$$
$$0.003\overline{)0.000.6}\ \ 0.2$$

concentration is therefore 1×10^{-9} gram-ionic weigh per liter. Since hydroxyl ions are in excess, the solution is alkaline.

Since the concentration of one ion implies the concentration of the other, body fluids are usually described in terms of hydronium ions only, even if the concentration of hydroxyl ion exceeds that of hydronium ion.

pH

All cells and tissues depend upon the proper concentration of hydronium ion for normal function. This applies to bacteria as well as to red blood cells, to the function of respiration as well as to digestion. The behavior of colloidal solutions and enzymes alters when the hydronium ion concentration is changed. In a word, protoplasm needs a carefully controlled medium in which to live. Variations beyond certain very narrow limits cause death. In the light of these important facts, we needed a more convenient method of expressing the hydronium ion concentration. Before we knew about hydronium ions, Sørensen, as a result of his studies on the effect of acids on enzymes, suggested the pH system to describe hydrogen ion concentration. The symbol pH is defined as *the power (exponent) of the hydrogen ion concentration of the solution.*

We have indicated above that concentrations of 0.1, 0.01, 0.001, etc., can be expressed as 1×10^{-1}, 1×10^{-2}, 1×10^{-3}, etc. The exponents $-1, -2, -3$ are logarithms of these numbers. Sørensen proposed the use of the negative of logarithms to express concentration, so that $-(-1)$ would be 1, $-(-2)$ would be 2, etc.

$$pH = -\log \text{ concentration } H_3O^+$$

Mathematically, if we put the concentration of H_3O^+ in the denominator and use positive logs, we get the desired result:

$$pH = \log \frac{1}{[H_3O^+]}$$

We observe that the pH number gives hydronium ion concentration directly and hydroxyl ion concentration indirectly, since the product of these concentrations always equals 1×10^{-14}. Low pH numbers indicate solutions with a heavy concentration of hydronium ions and a low concentration of hydroxyl ions. Any pH number less than 7 denotes a solution in which there are more hydronium ions than hydroxyl ions. A solution with pH 6.99 is very slightly acidic. Solutions with pH numbers greater than 7 have more hydroxyl ions than hydronium ions, and they are alkaline. A solution with pH 7.01 is barely alkaline. Only at pH 7.00 is there an equal concentration of hydronium and hydroxyl ions. This is the neutral point. As the concentration of hydronium ions decreases, the pH number increases. We note further that a change of one whole pH number indicates a tenfold change in the concentration of the ions (Table 11–3). A solution with pH 2 has an hydronium ion concentration ten times greater than one with pH 3 and one hundred times greater than one with pH 4.

Table 11–3 gives pH values for concentrations expressed as one-tenth, one-hundredth, etc. A solution whose hydronium ion concentration is 0.0001 has a pH of 4, but a solution whose hydronium ion concentration is 0.0003 has a higher concentration. We would expect the pH number for this solution to be numerically smaller. Consulting log tables, we find that the pH of this solution is 3.52. Solutions whose pH numbers contain a decimal part are those in which the hydronium ion concentration is a number other than one-tenth, one-hundredth, etc.

In Table 11–4 vinegar has a pH number of 3, while that of grapefruit juice is 3.2. Which of these fluids contains the greater concentration of hydronium ion? The smaller number represents the greater concentration. Vinegar's pH 3 is smaller than 3.2 and therefore represents a higher degree of acidity.

Table 11-3 pH VALUES

Concentration H_3O^+, gram-ionic weight		Concentration OH^-, gram-ionic weight		pH	
1.0	1×10^{-0}	0.00000000000001	1×10^{-14}	0	Strongly acidic
0.1	1×10^{-1}	0.0000000000001	1×10^{-13}	1	
0.01	1×10^{-2}	0.000000000001	1×10^{-12}	2	
0.001	1×10^{-3}	0.00000000001	1×10^{-11}	3	Moderately acidic
0.0001	1×10^{-4}	0.0000000001	1×10^{-10}	4	
0.00001	1×10^{-5}	0.000000001	1×10^{-9}	5	Slightly acidic
0.000001	1×10^{-6}	0.00000001	1×10^{-8}	6	
0.0000001	1×10^{-7}	0.0000001	1×10^{-7}	7	Neutral
0.00000001	1×10^{-8}	0.000001	1×10^{-8}	8	Slightly alkaline
0.000000001	1×10^{-9}	0.00001	1×10^{-5}	9	
0.0000000001	1×10^{-10}	0.0001	1×10^{-4}	10	Moderately alkaline
0.00000000001	1×10^{-11}	0.001	1×10^{-3}	11	
0.000000000001	1×10^{-12}	0.01	1×10^{-2}	12	
0.0000000000001	1×10^{-13}	0.1	1×10^{-1}	13	Strongly alkaline
0.00000000000001	1×10^{-14}	1.0	1×10^{-0}	14	

The normal range of pH values for blood serum is 7.35 to 7.45. While both these figures represent solutions in which the hydroxyl ion concentration exceeds the hydronium ion concentration, pH 7.35 indicates a greater hydronium ion concentration than pH 7.45. At the lower pH value, there is about 25 per cent more hydronium ions than at the higher value.

Do not confuse the terms *neutral* and *normal*. The blood is normally slightly alkaline. If the hydronium ion concentration of the blood increases and the pH drops to 7 (neutral), which may happen, for example, in untreated diabetes and in certain kidney diseases, the patient will suffer very severe acidosis and will be in danger of death. (Acidosis is a pathologic condition in which the blood pH drops below 7.35.)

Table 11-4 pH VALUES OF SOME AQUEOUS SOLUTIONS

	pH
Pancreatic juice	7.5–8.0
Intestinal juice	7.0–8.0
Blood serum	7.35–7.45
Cerebrospinal fluid	7.35–7.45
Saliva	6.8–7.2
Urine	4.8–7.5
Pure water	7.0
Grapefruit juice	3.2
Vinegar	3
Gastric juice	1

INDICATORS

The exact determination of pH involves electrical measurements on instruments called pH meters, but sufficiently accurate determinations can be made with indicators. We have mentioned litmus, methyl orange, and phenolphthalein. Indicators, whether natural or synthetic dyes, are generally weak acids or weak bases which change from un-ionized to ionized forms (or vice versa) with an accompanying color change. There is no single indicator that covers the complete pH range from 0 to 14, but by using several indicators, the complete scale can be covered (Table 11–5).

By properly comparing solutions, one can determine the pH within an accuracy of 0.2 of a unit. Test papers have been prepared that change to different colors, depending upon the pH of the solution. The pH can be determined by comparing the color of the paper with a standard chart. Nitrazine paper has a range from 4.5 to 7.5, and it is a useful indicator for determining the pH of urine specimens.

When sulfadiazine is used in treatment, it is important to be certain that the urine is alkaline. Sulfadiazine and some of the products produced by its action are soluble in alkaline urine but precipitate out and cause blockage in acidic urine. Some compounds, such as the urinary antiseptic methenamine, require an acidic urine. When frequent pH measurements are necessary, the nurse may be required to determine them, and a test paper such as nitrazine paper is usually used.

NEUTRALIZATION

Neutralization is a reaction between an acid and an alkali. The hydroxyl ion is one of the strongest bases, and when it accepts a proton, it becomes a molecule of water which ionizes to a very slight degree. The reaction between hydrochloric acid and sodium hydroxide is

$$H^+ + Cl^- + Na^+ + OH^- \rightleftharpoons H_2O + Na^+ + Cl^-$$

HCl is a strong electrolyte and exists almost completely in the form of ions, as does NaOH. NaCl is completely ionized. Therefore the only particles reacting are

$$H^+ + OH^- \rightleftharpoons H_2O$$

The reverse reaction here is the very slight ionization of water. The sodium and chloride ions take no part in the reaction; they are merely *spectator ions*, though upon evaporation of the water, crystals of sodium chloride are formed.

A diprotic acid such as carbonic acid can be neutralized in stages, depending upon the amount of hydroxide used:

$$Na^+ + OH^- + H_2CO_3 \rightleftharpoons Na^+ + HCO_3^- + H_2O$$
(Weak)   Sodium bicarbonate

$$Na^+ + HCO_3^- + Na^+ + OH^- \rightleftharpoons 2Na^+ + CO_3^= + H_2O$$
Sodium carbonate

Essentially neutralization is

$$H^+ + OH^- \rightarrow H_2O$$
Proton   Ion   Molecule

Table 11–5 COMMON INDICATORS

Indicator	pH Range	Color Change (Acid-Alkaline)
Congo red	3.0–5.2	Blue to violet red
Methyl orange	3.1–4.4	Red to orange yellow
Alizarin red	3.7–4.2	Yellow to pink
Bromocresol purple	5.2–6.8	Yellow to purple
Litmus	5.0–8.0	Red to blue
Bromothymol blue	6.0–7.6	Yellow to blue
Neutral red	6.8–8.0	Yellow to orange
Cresol red	7.2–8.0	Yellow to red
Phenolphthalein	8.3–10.0	Colorless to red
Thymolphthalein	9.3–10.5	Yellow to blue

HYDROLYSIS OF SALTS

Hydrolysis is the chemical reaction of a substance with water that forms new substances. Since body fluids are water solutions, the nurse must have an understanding of the reactions of salts with water. When we test water solutions of sodium chloride, sodium bicarbonate, and copper sulfate with litmus, we find that the sodium chloride solution has no effect on litmus; the sodium bicarbonate solution turns litmus blue, indicating excess hydroxyl ions; and the copper sulfate solution turns litmus red, indicating excess hydronium ions. Sodium bicarbonate contains no hydroxyl, and copper sulfate contains no hydronium. Therefore these ions must have been formed from water. Each water molecule in dissociating forms 1 hydroxyl and 1 hydronium ion. For a basic reaction, hydronium ions must be removed from the solution in some fashion; for an acidic reaction, hydroxyl ions must be removed.

Sodium bicarbonate supplies a high concentration of sodium and bicarbonate ions. The bicarbonate ions unite with hydronium ions to form molecules of the weak electrolyte carbonic acid. As hydronium ions are removed from the solution as molecules of carbonic acid, more water molecules dissociate, forming more hydronium and hydroxyl ions. Since molecules of sodium hydroxide do not form, hydroxyl ions build up in the solution and impart to it alkaline properties:

$$\underset{\text{Ion}}{Na^+} + \underset{\text{Ion}}{HCO_3^-} + \underset{\text{Molecule}}{H_2O} \rightleftharpoons \underset{\text{Ion}}{Na^+} + \underset{\text{Ion}}{OH^-} - \underset{\text{Molecule}}{H_2CO_3}$$

When copper sulfate dissolves in solution, molecules of the weak electrolyte copper hydroxide are formed. These molecules remove hydroxyl ions from the solution. Molecules of the highly ionized sulfuric acid are not formed. As hydroxyl ions are removed from the solution, more water molecules ionize. Hydronium ions accumulate in the solution and give it acidic properties:

$$\underset{\text{Ion}}{Cu^{++}} + \underset{\text{Ion}}{SO_4^=} + 4H_2O \rightleftharpoons \underset{\text{Ion}}{2H_3O^+} + \underset{\text{Ion}}{SO_4^=} + \underset{\text{Molecule}}{Cu(OH)_2}$$

Sodium chloride in solution forms sodium and chloride ions. The only substances that could be formed in a reaction with water would be sodium hydroxide and/or hydrochloric acid. Both of these are strong electrolytes which exist as ions rather than molecules. Hydronium ions are not removed to form molecules of hydrochloric acid, and hydroxyl ions are not removed to form molecules of sodium hydroxide. The ionization of water is not disturbed. A solution of sodium chloride contains sodium ions, chloride ions, molecules of water, and the same concentration of hydronium and hydroxyl ions as in water alone. In other words, sodium chloride plus water forms salt water. There is no chemical reaction. This is a salt that does not hydrolyze:

$$\underset{\text{Ion}}{Na^+} + \underset{\text{Ion}}{Cl^-} + \underset{\text{Molecule}}{H_2O} \rightleftharpoons \underset{\text{Ion}}{Na^+} + \underset{\text{Ion}}{Cl^-} + \underset{\text{Molecule}}{H_2O}$$

One way in which salts are formed is the reaction between an acid and an hydroxide. If the acid is strong and the hydroxide weak, the resulting salt will react with water to give acidic properties. If the hydroxide is strong and the acid weak, the salt will hydrolyze to impart alkaline properties. If both are strong, the resulting salt will not hydrolyze.

Let us use this knowledge to predict the reaction of potassium citrate with water. This is a salt of a strong hydroxide, potassium hydroxide, and a weak acid, citric acid. It will hydrolyze to give alkaline properties. In medicine it is used as an alkaline diuretic.

$$\underset{\text{Ion}}{3K^+} + \underset{\text{Ion}}{C_6H_5O_7^=} + \underset{\text{Molecule}}{3H_2O} \rightleftharpoons \underset{\text{Ion}}{3K^+} + \underset{\text{Ion}}{3OH^-} + \underset{\text{Molecule}}{C_6H_8O_7}$$

Ammonium chloride is also used as a diuretic. A water solution of this salt has acidic properties:

$$\underset{\text{Ion}}{NH_4^+} + \underset{\text{Ion}}{Cl^-} + \underset{\text{Molecule}}{H_2O} \rightleftharpoons \underset{\text{Molecule}}{NH_3} + \underset{\text{Ion}}{H_3O^+} + \underset{\text{Ion}}{Cl^-}$$

If a salt is derived from a weak acid and a weak hydroxide, we would need to know the ion-product constants of the two electrolytes to predict whether the reaction with water would be acidic, alkaline, or neutral. Sugar of lead, lead acetate, is a soluble salt

Table 11-6 HYDROLYSIS REACTIONS OF SOME SALTS

Salt	Acid Component	Hydroxide Component	pH of Solution		
			Less than 7 (Acidic)	7 (Neutral)	More than 7 (Alkaline)
Sodium chloride............	HCl (strong)	NaOH (strong)		X	
Sodium bicarbonate.........	H_2CO_3 (weak)	NaOH (strong)			X
Sodium citrate..............	$C_6H_8O_7$ (weak)	NaOH (strong)			X
Ammonium chloride.........	HCl (strong)	NH_3 (weak)	X		
Potassium sulfate...........	H_2SO_4 (strong)	KOH (strong)		X	
Aluminum chloride..........	HCl (strong)	$Al(OH)_3$ (weak)	X		
Cupric sulfate...............	H_2SO_4 (strong)	$Cu(OH)_2$ (weak)	X		
Sodium carbonate..........	H_2CO_3 (weak)	NaOH (strong)			X
Potassium nitrate...........	HNO_3 (strong)	KOH (strong)		X	

used in the laboratory to test for the presence of sulfur. Lead acetate undergoes hydrolysis, but one has to know the ion-product constants for lead hydroxide and acetic acid to predict the degree of hydrolysis.

We can generalize from Table 11-6 that:

a. Salts of strong acids and strong hydroxides do not hydrolyze. The solutions are approximately neutral.
b. Salts of strong acids and weak hydroxides hydrolyze to form solutions with acidic properties.
c. Salts of weak acids and strong hydroxides hydrolyze to form solutions with alkaline properties.

BUFFERS

In nontechnical language, a buffer is something that deadens a blow, softens a shock, or bears the brunt of a collision. The normal chemical reactions that occur in the body can take place only within very narrow limits of pH variation. The pH range of the blood compatible with health lies between 7.35 and 7.45. When we consider the variety of foods in the human diet (some of which form acid residues and others alkaline residues), the differences in rates of metabolism (varying from the body at rest to the body under stress), and the amount of metabolic wastes transported by the blood, we recognize that the very small variations in blood pH that occur in health indicate that the body is a master chemist. The blood pH is controlled by the action of *blood buffers* which resist changes in the hydrogen ion concentration. Recall that the concentrations of hydrogen and hydroxyl ions are related and that a change in one causes a change in the other. If the concentration of one is controlled, the other is controlled also.

A buffering system to keep the pH of a solution above 7, that is, on the alkaline side, would consist of a combination of the salt of a weak acid and a strong hydroxide plus the weak acid. The bicarbonate ⇌ carbonic acid system is the most important buffering system in the blood. Here we have a mixture of carbonic acid, H_2CO_3, a weak acid, and sodium bicarbonate, $NaHCO_3$, a salt of a strong base. This mixture of sodium bicarbonate and carbonic acid in water has an alkaline reaction. The solution contains a high concentration of sodium ions, a high concentration of bicarbonate ions, molecules of undissociated carbonic acid, hydrogen ions, and hydroxyl ions (the concentration of hydroxyl ions being greater than that of hydrogen ions).

As hydrogen ions are formed in metabolism and are added to the blood, bicarbonate ions unite with them to form molecules of carbonic acid:

$$H^+ + HCO_3^- \rightarrow H_2CO_3$$

In this way hydrogen ions are removed from the blood, and the pH does not drop. Excess carbonic acid breaks up into carbon dioxide and water. Carbon dioxide is excreted through the lungs, and water through the lungs, the skin, or the kidneys. By this arrangement, hydrogen ions formed during metabolism are excreted as water molecules, and an increase in hydrogen ion concentration is prevented.

Bicarbonate ions can also react with water (hydrolyze) to form molecules of undissociated carbonic acid and hydroxyl ions:

$$HCO_3^- + H_2O \rightarrow H_2CO_3 + OH^-$$

The hydroxyl ions so formed can either react with hydrogen ions to form water or accumulate to establish the proper alkaline blood pH.

Conversely, if the concentration of hydrogen ions tends to decrease and the concentration of hydroxyl ions to increase (a threatened alkalosis, a pathologic condition in which the blood pH is higher than 7.45), molecules of undissociated carbonic acid can ionize, thus increasing the concentration of hydrogen ions to bring the blood pH back to normal:

$$H_2CO_3 \rightarrow H^+ + HCO_3^-$$

In addition to the bicarbonate ⇌ carbonic acid system, which is the most important, there are several other buffering systems in the blood which act on the same principle (Table 11-7). These are the dihydrogen phosphate ⇌ monohydrogen phosphate system, the hemoglobin ⇌ reduced hemoglobin system, the oxyhemoglobin ⇌ reduced oxyhemoglobin system, and the serum proteins. At normal blood pH, the serum proteins—albumin, globulin, and fibrinogen —exist as negatively charged particles, or anions. In cases of threatened acidosis they

Table 11-7 SUMMARY OF BUFFERS

Reactions to Remove H^+ or Increase OH^-	Reactions to Increase H^+ or Reduce OH^-
Bicarbonate: $\quad HCO_3^- + H^+ \longrightarrow H_2CO_3$ $\quad HCO_3^- + H_2O \longrightarrow H_2CO_3 + OH^-$ $\quad H^+ + OH^- \longrightarrow H_2O$	$H_2CO_3 \longrightarrow H^+ + HCO_3^-$ $HCO_3^- \longrightarrow H^+ + CO_3^=$
Phosphate: $\quad HPO_4^= + H^+ \longrightarrow H_2PO_4^-$	$H_2PO_4^- \longrightarrow H^+ + HPO_4^=$
Hemoglobin*: $\quad Hb^- + H^+ \longrightarrow HHb$	$HHb \longrightarrow H^+ + Hb^-$
Oxyhemoglobin: $\quad HbO_2^- + H^+ \longrightarrow HHbO_2$	$HHbO_2 \longrightarrow H^+ + HbO_2^-$
Serum proteins: $\quad Albuminate^- + H^+ \longrightarrow albumin$	$Albumin \longrightarrow H^+ + albuminate^-$

*The symbol Hb stands for hemoglobin.

are considered the body's last line of defense. They can react with hydrogen ions to form undissociated protein molecules, or they can form positively charged particles, or cations:

$$\text{Albuminate}^- + \text{H}^+ \rightleftharpoons \text{albumin}$$
Anion

$$\text{Albumin} + \text{H}^+ \rightleftharpoons \text{H albumin}^+$$
Cation

It is interesting to note that while the blood is alkaline, the urine is usually acidic. This is part of the buffering system in which the kidneys control the type of cation excreted. Sulfates and phosphates formed during the oxidation of protein can be excreted as sodium sulfate and sodium phosphate. When this occurs, the urine tends to have a higher pH. On the other hand, the kidneys can return sodium ions to the blood to help maintain the alkalinity, and they can excrete sulfates and phosphates as ammonium sulfate and ammonium phosphate. These salts hydrolyze to make the urine more acidic.

From a chemical viewpoint, these buffering systems are marvelous to contemplate. Only after the buffers have been exhausted does acidosis or alkalosis develop. The body is truly the master chemist.

Study Exercises

1. List five properties of hydrogen ion, and give four properties of hydroxyl ion.
2. What is an indicator? Give the reactions of three indicators in acid solutions.
3. What is meant by conductance?
4. Define acid and base according to the Brönsted and Lowry theory.
5. Name two acids formed in the body.
6. What are the formulas for soda lye, potash, and lime water?
7. What is a strong electrolyte? What is a weak electrolyte?
8. Name three strong acids and three weak acids.
9. Name two strong hydroxides and two weak hydroxides.
10. Name 10 strong salts.
11. Account for the fact that a 60 per cent solution of acetic acid is described as concentrated and weak, while a 1 per cent solution of hydrochloric acid is dilute and strong.
12. What antidotes that might be available in the average household would you choose if (a) a child swallowed lye; (b) a child swallowed hydrochloric acid; (c) a child spilled sulfuric acid on his hand; (d) a child spilled potash on his hand. Explain your choice in each case.
13. Why do acid fruits such as cranberries and rhubarb affect aluminum pots in which they are cooked?
14. Why is it inadvisable to choose a strong electrolyte as an antidote?
15. Citrate of magnesia (magnesium citrate) in a 10 per cent solution is given by mouth as a saline cathartic. This solution has a higher osmotic pressure than body fluids. Account for the resulting watery fecal evacuation.
16. Why is a paste of sodium chloride rubbed on the skin area where the terminals of the electrocardiograph are to be attached?
17. Define ion-product constant, pH, neutral point, hydrolysis, indicator, buffer salt, anion, weak acid, and neutralization.
18. List five salts that do not hydrolyze, five that form acid solutions in water, and five that form alkaline solutions in water.
19. What is the normal pH range of the blood compatible with health? What is acidosis? What is alkalosis?
20. What is the principal buffering system in the blood? How does it decrease the hydrogen ion concentration? How does it increase the hydrogen ion concentration?

21. What is the reaction when NaCl is added to water? Predict the pH of each of the following solutions as acidic, neutral, or alkaline: cupric nitrate, potassium citrate, ammonium sulfate, sodium sulfate, and ammonium phosphate.

22. The pH readings for urine are given as 4.8 to 7.5. What do these figures mean? How do you account for such a wide variation, the widest of all body fluids?

chapter twelve ELECTROLYTES IN THE BODY

Ideas for preview or review

Equivalent weight is the atomic weight divided by the valence. A milli-equivalent weight (mEq) is one-thousandth of an equivalent weight.

Proteins, being colloids, tend to be retained within intact membranes.

The osmotic pressure of a solution increases as the solution becomes more concentrated and decreases as the solution becomes more dilute.

Dialysis is the passage of dissolved substances through membranes.

Living membranes are selective. Of two very similar particles, e.g., sodium ion and potassium ion, a cell membrane may permit one to pass into the cell but not the other.

Passage of dissolved particles through living membranes appears to be a cooperative venture between the particle and the membrane, whereby the particle is transported into the cell.

Has it ever occurred to you that all the visible parts of the body are dead? The skin, the nails, the hair, the corneas in the eyes are all lifeless. The living body exists inside, enclosed in an air-tight casing. All the living cells are surrounded by and submerged in a solution of salts more dilute than ocean water. This solution makes up 70 per cent of the adult body weight. Part of it, 50 per cent, is enclosed in the cells of the body inside the cell membranes. This portion is called *intracellular fluid*. The rest, 20 per cent, is outside the cells either in blood or lymph vessels or flowing over the cells. This solution is called *extracellular fluid*. Of the extracellular fluid, an amount equal to 5 per cent of the body weight is in the closed circulatory system. We refer to this portion of extracellular fluid as *plasma*.[1] The rest, 15 per cent of body weight, has been called by various names: body fluid, cell fluid, tissue fluid, and lymph. We shall refer to this portion of extra cellular fluid as *interstitial fluid,* which means fluid in the tiny spaces between the cells (Fig. 12–1).

Let us consider a muscle cell surrounded by its cell membrane. The membrane separates the cell contents from the environment. The fluid content of the cell is intracellular fluid. The fluid immediately outside the membrane is interstitial fluid, part of the extracellular fluid. All the substances reaching the cell reach it through the medium of interstitial fluid. All the substances that leave the cell do so by way of interstitial fluid. The interstitial fluid is in contact with the outside of the capillary walls. Substances can pass from the interstitial fluid through the walls of the capillaries into the plasma.

Glucose, carried by the general circulation dissolved in the plasma, can dialyze through the capillary walls into the interstitial fluid. From the interstitial fluid, it can dialyze through the cell membrane into the cell. The glucose is then part of the intracellular fluid.

If a man weighed 200 lb, 140 lb would be fluid. Of this, 100 lb would be in the cells (intracellular fluid), 30 lb would be inter-

---

[1] Fluid enclosed in the circulatory system is sometimes called intravascular fluid.

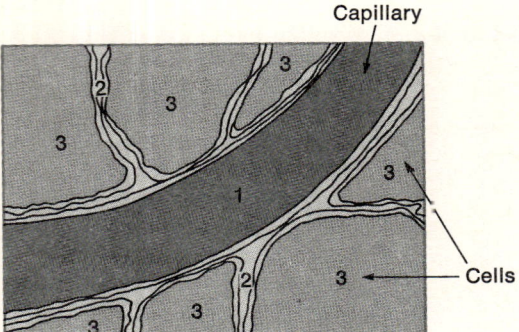

Fig. 12-1 Body fluids: (1) plasma; (2) interstitial fluid; (3) intracellular fluid.

stitial fluid, and 10 lb would be plasma. We can make a rough translation from weight to volume. Since we have mentioned that the fluid is very dilute, we can figure what the volume would be if it were water. A liter of water weighs 1 kg, or 2.2 lb. This would mean 13.6 liters interstitial fluid and 4.5 liters plasma. Recall that a liter is slightly larger than a quart (1.06 qt). Plasma constitutes 55 per cent of the total blood volume and the cells, 45 per cent. A 200-lb man would have about 8 liters of blood.

FLUID EXCHANGES

Fluids are contantly lost from the body and are replaced by ingested liquids, foods, and the water formed in metabolism. The oxidation of 1 molecule of glucose, for example, yields 6 molecules of water. Even foods that appear to be solid contain bound water. Looking at a raw egg, it is obvious that it is fluid and contains considerable water. What happens when it is cooked into a hard-cooked egg? Does it lose its water? In the cooking process the water becomes bound to the protein in the egg, and the resultant hard-cooked egg appears to be a solid. It would be safe to say that all foods contain fluid and that oxidation of all foods produces water in the cells.

Under normal conditions of body temperature and environmental temperature, about 1 liter fluid is lost through the lungs and skin in a 24-hr period. Generally, between 1 and 2 liters urine are eliminated in the same period of time. The amount varies with fluid intake and temperature. When the environmental temperature is high, the amount of fluid lost by the skin increases and the amount lost in urine decreases. In addition, about 150 ml fluid escapes in fecal material.

The body gains fluids through (a) ingestion of liquids, (b) ingestion of what appears to be solid food, and (c) water formed through chemical changes in the body.

The body loses fluids through (a) urination, (b) exhalation, (c) perspiration, and (d) defecation.

The amount of fluid gained must equal the amount of fluid lost. If more fluid is gained than lost, the excess fluid will remain in the tissues, giving rise to *edema*. If more fluid is lost than gained, the eventual result will be *dehydration*.

COMPOSITION OF FLUIDS

Body fluids are principally water. They have an osmotic pressure equal to that of 0.9 per cent sodium chloride solution. They contain glucose, amino acids, proteins, urea and other nitrogen compounds, fats and other lipids, dissolved gases, antibodies, etc. In addition, they contain small amounts of vitally important electrolytes. The concentration of electrolytes is expressed as milliequivalents per liter.

At the dilutions at which these electrolytes exist in body fluids, the ions lead independent existences. We speak of sodium ion and chloride ion, not of sodium chloride. Observe from Table 12-1 that the cation in largest supply in extracellular fluids, both plasma and interstitial fluid, is sodium. Within the cell, the most abundant cation is potassium. Observe also that the principal anion in extracellular fluids is chloride and that in intracellular fluid is biphosphate. Within the normal pH range of blood, 7.35 to 7.45, proteins, which are amphoteric, act as anions.

In the living cells, the electrolytes are compartmentalized as indicated. The question arises: Why do not the sodium and potassium ions migrate across the membranes and equalize the concentrations? When cells die, that is exactly what happens, but living membranes are able to discriminate and, to a certain extent, to control the kinds and amounts of substances entering and leaving the cells. In the case of sodium and potassium ions, the cell membrane acts as a pump, pumping out extra sodium ions and allowing potassium ions to diffuse back into the cell. This action requires work, and only living membranes can avail themselves of the energy generated by the cell to accomplish this. When a cell is injured, potassium ions leak out into extracellular fluid. Some authorities think that this action is involved in the occurrence of shock in the body.

The proper concentrations of electrolytes in the proper compartments of the body are essential for health and for life itself.

In muscle and nerve action, sodium, potassium, and hydroxyl ions are stimulating ions and calcium, magnesium, and hydrogen ions are relaxing ions. A decrease in the concentration of calcium ions causes a decrease in its relaxing effects and muscles may go into spasm. Alterations in the concentration of potassium have very dramatic effects; this is particularly true in heart muscle.

Since protoplasm is protein in nature, we would expect to find proteins in intracellular fluid. The proteins in the circulating blood, i.e., serum albumin, serum globulin, and fibrinogen, account for the proteins in plasma. Note that interstitial fluid, the intermediary between plasma and intracellular fluid, is particularly poor in proteins. Proteins, being colloids, tend to remain within intact membranes. To become part of the interstitial fluid, proteins would need to migrate through either the membranes of the cells or the membranes of the capillary walls. The presence of proteins within the capillaries is partly responsible for the differences in osmotic pressure between the arterial and venous portions of the capillaries. One of their functions is to facilitate the return flow

Table 12-1 ELECTROLYTE COMPOSITION OF BODY FLUIDS*

Electrolytes	Extracellular Fluid		Intracellular Fluid
	Plasma	Interstitial Fluid	
Cations:			
Sodium, Na^+	142	145	15
Potassium, K^+	5	4	157
Calcium, Ca^{++}	5	3	5
Magnesium, Mg^{++}	3	2	27
Total cations	155	154	204
Anions:			
Chloride, Cl^-	103	116	4
Bicarbonate$^-$, HCO_3^-	27	27	10
Sulfate$^=$, $SO_4^=$	1	2	18
Biphosphate$^=$, $HPO_4^=$	2	3	100
Proteins$^-$	16	1	72
Organic acid$^-$	6	5	
Total anions	155	154	204

*All values are expressed as milliequivalents per liter.

of interstitial fluid into the venous portion of the capillary. Osmotic pressure due to plasma proteins is called *oncotic* pressure.

The concentrations of body fluids described in Table 12-1 are typical for an adult. Although different investigators report slightly dissimilar figures, the general picture is about the same. Body fluids in newborn infants and children are more dilute than in the adult. A higher proportion of their body weight is water. Infants have much less intracellular fluid than do adults: 20 to 25 per cent as compared with 50 per cent for adults. But they have a much greater volume of fluid in the interstitial space: 45 per cent as compared with 15 per cent for adults. In comparison with adults, infants have a higher fluid intake and a higher urinary output. One might conclude that they are more fluid.

ACID-BASE BALANCE

In medical language acid means anion and base means cation. Chlorides, bicarbonates, sulfates, and phosphates are referred to as acid; potassium, sodium, calcium, and magnesium are called base. Recalling the hydrolysis of salts, we realize that any sodium, potassium, or calcium salt dissolved in water will give either a neutral or alkaline reaction. Therefore, from these cations we get either no reaction or basic reactions. Naturally, hydrogen ion is not called base. It and hydroxyl ions are always present in body fluids in varying amounts. When we discuss electrolytes in the body, we are speaking of these cations and anions. In discussing acid-base balance, we refer primarily to the ratio of hydrogen to hydroxyl ions.

We have established that body fluids shift from compartment to compartment and that body fluid can be lost or be in oversupply. Fluid loss can mean loss of only water or it can mean loss of water and electrolytes. Electrolytes can leak from a compartment in which they are normally in rich supply to another in which they are usually dilute. In addition to enzymes, water, partially digested food, and undigestible substances, fluid lost in diarrhea contains intestinal and pancreatic juice and bile. These juices contain sodium, potassium, and calcium ions as well as chloride, phosphate, and bicarbonate ions. Ordinarily, though, most of these ions would not be excreted but would be reabsorbed by the body and redistributed in the fluid compartments. Calcium is normally excreted in the feces, but although some potassium is lost by this route, usually the kidneys regulate the excretion and retention of ions in the body. Recall that the ions of the strong electrolytes do not react with hydrogen or hydroxyl ions to form acids or hydroxides but that ions involved in the formation of weak electrolytes react with available hydrogen ions to form molecules of poorly dissociated acids and with hydroxyl ions to form molecules of hydroxides. Equivalent weights of sodium ion and chloride ion, in water, give a neutral reaction, pH 7. Equivalent amounts of sodium ion and bicarbonate ion form an alkaline solution with a pH greater than 7. A solution composed of sodium as the positive ion and a mixture of chloride and bicarbonate as the negative ions is alkaline. As the bicarbonate ions decrease and the chloride ions increase, the solution becomes less alkaline. It approaches pH 7. In other words, as the number of hydroxyl ions decreases, the number of hydrogen ions increases. The solution becomes more acid.

Respiratory Acidosis and Alkolosis

Acidosis is a medical term, not a chemical term. It refers to the acid-base balance in the blood. Normally, the blood pH is between 7.35 and 7.45. As we know, this indicates a condition in which there are more hydroxyl than hydrogen ions. The solution is slightly alkaline. If the hydrogen ions increase and the hydroxyl ions decrease, even slightly, the solution, although it may still be alkaline, is more acidic than originally. Acidosis is a condition in which there are more than the normal number of hydrogen ions in the blood. Any pH number less than 7.35 indicates acidosis. A pH reading of 7, which is the neutral point (equal concentrations of hydrogen and hydroxyl ions), would mean that the patient was in coma and, unless the condition could be corrected quickly, not long for this world. Do not think that the word acidosis means that the blood is acidic. No one has been known to survive even a very slightly acid condition (pH 6.8) for any length of time. Let us say then, if we are speaking of living persons, that the blood is never acidic.

A very important concept to grasp, in understanding acidosis, is that the pH depends on the *relative amounts,* or the ratio of concentration, of certain ions. For all practical purposes, we can explain the principle involved, as far as extracellular fluids are concerned, by referring to the relative quantities of only sodium, chloride, and bicarbonate ions and undissociated molecules of carbonic acid. The kidneys retain or excrete ions depending upon the available supply. If sodium is in good supply, waste products of oxidation are excreted in the form of the sodium salt; e.g., sulfur in protein is oxidized to sulfate and excreted in the urine as sodium sulfate. If sodium is needed in the body, the amino group in the amino acids is split off and changed to ammonia. The wastes are excreted as ammonium salts. This increase of salts of a weak base and a strong acid makes the urine more acid (lowers the pH). Patients who have been on a sodium-free diet for some time may excrete no sodium in the urine. Thus, we see that the body tends to *compensate* for changes in ionic concentration.

Carbonic acid decomposes to form carbon dioxide and water. Carbon dioxide is excreted by the lungs. An increase in the respiratory rate or breathing much deeper than normally tends to remove carbon dioxide from the blood. As the carbon dioxide concentration in the blood decreases, the amount of undissociated carbonic acid decreases also.

$$H_2CO_3 \rightarrow H_2O + CO_2$$

If carbon dioxide escapes from the blood via the lungs, this reaction is not reversible. Unless there is a proportional drop in the concentration of bicarbonate ions, the ratio of bicarbonate ions to undissociated carbonic acid will rise and the pH will rise. If the bicarbonate ions adjust to the changed concentration of carbonic acid, the pH will remain normal.

Conversely, when the patient has difficulty breathing, as in some types of pneumonia, or where an obstruction occurs in the respiratory passages, the concentration of undissociated carbonic acid increases. Here again, unless the concentration of bicarbonate increases also, the ratio of bicarbonate to carbonic acid will drop and the pH will drop. If it does adjust, the pH will stay in the normal range. Conditions in which the adjustment is successfully achieved are *compensated.*

At pH 7.4, which is the middle of the normal range, the ratio of bicarbonate to carbonic acid is 20.[1]

$$\frac{HCO_3^-}{H_2CO_3} = \frac{28\,mEq/liter}{1.4\,mEq/liter} = 20$$

Let us say that in the condition of hyperpnea[2] the carbonic acid concentration drops to 1 mEq per liter.

$$\frac{HCO_3^-}{H_2CO_3} = \frac{28}{1} = 28$$

With a decrease in carbonic acid, we have what amounts to an increase in bicarbonate to make the solution more alkaline. To compensate, the bicarbonate must be decreased to 20 mEq per liter.

$$\frac{HCO_3^-}{H_2CO_3} = \frac{20}{1} = 20$$

The kidneys help to make this adjustment by

---

[1] The figures used in the following illustrations are not important. They have been selected to illustrate the discussion.
[2] Hyperpnea is a condition characterized by a very rapid respiratory rate.

excreting a highly alkaline urine and retaining in the body substances that will help neutralize the alkalinity of the blood.

In the patient with labored breathing, the carbonic acid concentration may rise to, let us say, 2 mEq per liter.

$$\frac{HCO_3^-}{H_2CO_3} = \frac{28}{2} = 14$$

A drop in this ratio means a drop in pH; in other words, acidosis. To compensate, the bicarbonate concentration must rise to 40 mEq per liter.

$$\frac{HCO_3^-}{H_2CO_3} = \frac{40}{2} = 20$$

In this situation, the kidneys excrete a urine with a high concentration of acids and ammonia. If the respiratory impairment was due to an obstruction, breathing will return to its normal rate once the obstruction is removed. In the case of the patient with very rapid and deep breathing (hyperpnea), we have indicated that, unless the body can compensate for the decrease in carbonic acid, the blood pH will rise. This condition is called *respiratory alkalosis*. Alkalosis may arise from other causes than respiratory difficulties, but in any case the blood pH rises. Alkalosis is a condition in which the concentration of hydrogen ions is less than normal. A blood pH higher than 7.45 is indicative of alkalosis. As with acidosis, the pH range beyond normal that is compatible with life is very narrow. The upper limit is pH 7.8. At pH 7.0, which indicates a very severe acidosis, the blood would contain one-ten-millionth of a gram of hydrogen ions; at pH 8.0, the blood would contain 100-millionth of a gram of hydrogen ion. There are only ten times more hydrogen ions at pH 7.0 than at pH 8.0; yet this relatively small range is beyond the limits compatible with life, i.e., pH 7.0 to 7.8. We begin to appreciate the marvelous job done by the blood buffers and the kidneys.

metabolic acidosis and alkalosis

Marked changes in the amount of carbon dioxide in the blood, as we have seen, with the resulting disturbance of the bicarbonate-carbonic acid ratio, give rise to *respiratory acidosis* (pH decrease, increase in H^+) and *respiratory alkalosis* (pH increase, decrease in H^+). The accumulation in the blood of nonvolatile acids, which exhaust the blood buffers, is called *metabolic acidosis*. In diabetes mellitus, if uncontrolled, β-hydroxybutyric acid and β-ketobutyric acid (also called acetoacetic and diacetic acid), formed from abnormal oxidation of fat, require neutralization by the buffer salts in the blood. When the amount of these acids is high, a large quantity of buffer is used up in the neutralization process. The net effect of this reaction is to decrease the concentration of bicarbonate in the blood. (The sodium bicarbonate buffer system is the major buffer in the blood.)

Since the blood pH is regulated by the ratio of bicarbonate ion to carbonic acid, the decrease in bicarbonate amounts to an effect similar to an increase in carbonic acid. This action tends to make the fluids more acidic. Even though the kidneys then excrete a more acidic urine by eliminating ammonium salts in place of sodium salts, as the buffers are used up, the blood pH falls.

This condition triggers another set of reactions. As the pH falls, respiration is stimulated to rid the body of carbon dioxide. This is a compensatory reaction in an attempt to raise the pH to normal. In metabolic acidosis, such as exists in some cases of diabetes, we observe an increased respiratory rate. This increased respiration would, by the more rapid removal of carbon dioxide, make body fluids more alkaline. However, this does not solve the original problem of neutralizing the ketone acids, and the effect is transient.

In general, the same types of reactions are observed in other conditions in which abnormally large amounts of nonvolatile acids must be neutralized. Such conditions include (a) failure of the kidneys to excrete the acids normally formed in metabolism,

(b) overtreatment with ammonium chloride or hydrochloric acid, and (c) intended or accidental ingestion of nonvolatile acids or substances that form nonvolatile acids in the body.

In addition to the accumulation of acid substances in the body, metabolic acidosis may be caused by the loss of alkaline body fluids. By losing alkaline fluids, the body is, in effect, more acid. The intestinal fluids are alkaline and, in normal digestion, much of this fluid is resorbed by the body before the fecal material is excreted, thus maintaining normal body pH. In diarrhea, these intestinal juices are lost. When we realize that these fluids contain sodium ions, potassium ions, chloride ions, and bicarbonate ions, in addition to enzymes, hormones, bile salts, and other substances, and that the pH of intestinal juice is about 8.3 and the pH of bile is between 7.8 and 8.6, we can predict that continued loss of these fluids will cause the blood pH to fall.

Until recent years, diarrhea in infants was a very serious, often fatal disease. Infants have a more rapid exchange between the fluid compartments and also do not have the margin of safety that adults do. Diarrhea used to lead to dehydration and metabolic acidosis which was the cause of death. Using sodium chloride as a replacement fluid could, for a very short time, delay the dehydration effects, but it would not alter the acidosis because sodium chloride is neutral in reaction. With increasing knowledge of the electrolyte composition of the body, it was realized that a salt with an alkaline reaction was needed. Such salts as sodium lactate, sodium citrate, and sodium bicarbonate all hydrolyze to form alkaline solutions. The use of these electrolytes in this condition has truly been a lifesaver. Potassium would also be included, if needed, to prevent disturbances in heart rhythm and muscular weakness.

In metabolic alkalosis, the blood pH rises. This can be brought about by the addition of strong bases or bicarbonate ions to the blood or by the loss of acid. Examples of addition of alkaline substances would be overdosages of sodium bicarbonate or of acids that yield bicarbonate on oxidation, thus increasing the bicarbonate concentration in the body. Certain medications can bring about the same effect by causing a loss of hydrogen ions. Vomiting and prolonged gastric suctioning can cause a loss of hydrochloric acid thus leaving the body more alkaline. This is the reverse of the condition arising in diarrhea. The kidneys attempt to compensate by excreting a more alkaline urine. Metabolic wastes are excreted as sodium salts rather than as ammonium salts, as happens when acidosis threatens. The respiratory rate is depressed in an attempt to build up the concentration of carbonic acid in the blood to counteract the alkaline condition.

Carbon Dioxide Combining Power

We have been emphasizing primarily changes in pH in our discussion of acid-base balance. However, a knowledge of both pH and the amount of bicarbonate, carbonic acid, and carbon dioxide in the plasma is important in understanding the patient's status in metabolic acidosis and alkalosis. These concentrations can be equated to the partial pressure of carbon dioxide. In other words, these substances can be changed to carbon dioxide and their concentrations reported as such.

A brief description of the technique will help in understanding the results. A sample of the patient's blood is drawn and centrifuged, separating the plasma. The plasma is treated with carbon dioxide gas at the same pressure as carbon dioxide in the lungs. To this treated plasma is added an acid which liberates all the carbon dioxide that can be obtained from the original sample plus whatever carbon dioxide was absorbed by the sample. This technique is, in a way, a measure of the alkaline reserve or the amount of available base in the plasma.

In metabolic acidosis, the sodium ions

are tied up, neutralizing the nonvolatile acids in the blood. They cannot react with the carbon dioxide introduced into the plasma, nor have they been combined with bicarbonate in the plasma. When the added acid liberates carbon dioxide in this condition, the amount will be low.

In metabolic alkalosis, there is a high concentration of bicarbonate or of base which can react with the carbon dioxide. In this case the combining power will be high. Although this technique is not as reliable in the cases of respiratory acidosis and alkalosis, let us predict what would be indicative. In respiratory acidosis, the concentration of carbonic acid is high. The amount of carbon dioxide obtained in this test will be high. In respiratory alkalosis, when the amount of carbonic acid is low, the combining power is low.

The carbon dioxide combining power is but an indication of the abnormal acid-base balance. The total picture is obtained when, in addition to a knowledge of the partial pressure of carbon dioxide, one measures also the pH and the concentration of either bicarbonate ion or excess base. A test that is replacing the carbon dioxide combining power determination, because of its greater accuracy, is the plasma total carbon dioxide content. In this determination, a blood sample is collected in the absence of air, the plasma separated, strong acid added, and the amount of carbon dioxide that bubbles off is collected, measured, and the volume corrected for pressure differences. This test is a measure of the amount of dissolved carbon dioxide plus the amount of bicarbonate, carbonic acid, carbonates, and carbamino compounds in the blood.

FUNCTIONS OF BODY ELECTROLYTES

The electrolytes make up the buffer systems in the body and, under normal conditions, maintain the pH between 7.35 and 7.45. They supply the ionic conditions necessary for nerve and muscle action and enable tissues to function normally. The proper concentration of electrolytes is necessary to maintain the osmotic pressure of body fluids. Electrolytes are involved in the control of the total volume of body fluid as well as of the volume of fluid in the various body compartments.

The proper concentrations of sodium, potassium, calcium, and chloride ions are essential for the regulation of normal heart action. Potassium relaxes heart muscle and calcium stimulates it. Potassium and calcium are involved in nerve conductivity. In addition, calcium is necessary for coagulation of blood and the formation of bones and teeth. Abnormal concentrations of calcium can cause tetany, a condition characterized by muscle spasms.

Phosphate is necessary for normal muscle contraction and is also a constituent of bones and teeth. Phosphate is part of enzyme systems involved in oxidation. Two very important compounds are adenosine diphosphate (ADP) and adenosine triphosphate (ATP). ATP is involved in the synthesis of proteins in the body, in the manufacture of urea and ammonia in the liver and the kidneys, respectively, in the absorption of glucose from the intestines, and in muscle contraction and nerve function.

Choride and bicarbonate ions cooperate in the transport of carbon dioxide and oxygen in the blood. The *chloride-bicarbonate shift,* in which chloride and bicarbonate ions exchange places from the plasma into the red cells and vice versa, enables the blood to carry carbon dioxide or oxygen with minimum changes in pH.

Principles of Electrolyte Balance

a. Water intake must equal water output.
b. Edema results if intake exceeds output.
c. Dehydration results if output exceeds intake.
d. Dilution of fluids results when water is retained without concomitant electrolyte retention.

e. Dilution of fluids results when electrolytes are lost without concomitant water loss.
f. Concentration of fluids results when water is lost without loss of electrolytes.
g. Concentration of fluids results when electrolytes are retained without comparable retention of water.
h. Fluids become more acidic through loss of cations, particularly sodium and potassium.
i. Fluids become more acidic through addition of anions, particularly chloride. (Bicarbonate and biphosphate ions are exceptions.[1])
j. Fluids become more alkaline through addition of cations.
k. Fluids become more alkaline through loss of anions.
l. Replacement fluids are selected to compensate for the electrolyte imbalance.
m. Fluids can pass from one body compartment to another.
n. Retention of carbon dioxide tends to make fluids more acidic.
o. Loss of carbon dioxide tends to make fluids more alkaline.

ELECTROLYTE AND FLUID DISTURBANCES

The discussion that follows relates situations that the nurse will meet. A knowledge of the principles of electrolyte balance will lead to an understanding of the pathology and treatment of these conditions.

Electrolyte Losses

Electrolytes may be lost along with proportional amounts of fluid. Fluid may be retained while electrolytes are lost, in which case the remaining body fluid may be more dilute.

[1] Salts of strong hydroxides and weak acids give alkaline reactions with water.

Electrolytes may be lost along with larger amounts of fluid. Here the remaining body fluids would become more concentrated.

losses from the digestive tract

Loss of gastric juice, as in persistent vomiting, causes the body to lose fluid as well as electrolytes. The fluid loss may lead to dehydration. Gastric juice contains hydrochloric acid and sodium ions along with traces of potassium, calcium, and magnesium. It is thought that gastric juice is manufactured in the body from sodium chloride. Loss of gastric juice removes from the body hydrogen and chloride ions and leaves in the body sodium and bicarbonate ions. The result, as we expect, is alkalosis. The same condition may arise as a result of prolonged gastric suction. Both fluid and electrolytes are lost. The appropriate replacement fluids will compensate for these losses.

Loss of fluid from lower down in the digestive tract involves the loss of pancreatic and intestinal juices. Intestinal fluid contains more bicarbonate than chloride. Since chloride is associated with the strong hydrochloric acid and bicarbonate with the weak carbonic, loss of bicarbonate without sufficient compensation leads to acidosis. In this condition, there is also the possibility of dehydration. Recall the description of infant diarrhea. When this disorder is unchecked, the child succumbs to dehydration and acidosis.

fistular drainage

Sometimes a considerable amount of fluid is lost through an opening to a body cavity or through drainage from an abscess. The loss usually involves both fluid and electrolytes. Depending upon the ions lost, either acidosis or alkalosis may develop. If there is much loss of fluid, there is always the chance that the patient will become dehydrated. In situations like this, the proper amounts and kinds of replacement fluids work wonders.

burns

Burns cause loss of fluid and electrolytes, with a proportionately larger loss of electrolytes than fluid. However, when the burned area is extensive, the fluid loss becomes critical. Loss of interstitial fluid causes plasma to leak from the capillaries and intracellular fluid to move out of the cells. The loss of plasma means a reduction in plasma volume and a closer packing together of blood cells (hemoconcentration). The reduced plasma volume interferes with the normal function of the kidneys. It may also cause a drop in blood pressure and the danger of shock. In these cases, the plasma volume must be restored to normal by the administration of plasma and/or plasma expanders. Physiologic salt solution, given by itself, may leak right out of the blood vessels. Expanders are colloids, and because they do not dialyze easily through the capillary walls, they tend to hold the fluid inside the blood vessels. Sometimes, in severe burns, the patient may have attacks of vomiting. This involves a further loss of electrolytes. In addition to saline solution and glucose, which are commonly administered in these conditions, electrolytically balanced fluids are used to restore the disturbed electrolyte balance.

salt[1] depletion

In heat exhaustion, sodium chloride from extracellular fluid is lost in profuse perspiration. As the concentration of sodium and chloride decreases in the plasma, the kidneys try to compensate by retaining these ions. There are two ways to increase the concentration of a solution. One is to increase the amount of solute. This the kidneys do by retaining sodium and chloride.

[1] Salt here means sodium chloride.

The other way is to reduce the amount of solvent. The kidneys do this by excreting water. Thus the kidneys work both ways to restore the concentration of sodium and chloride to normal. We observe that although the appropriate concentrations are achieved, it is at the expense of a reduction of the *volume* of plasma. As the volume of plasma decreases, the kidneys begin to conserve water. The decreased concentration of sodium in the interstitial fluid causes potassium ions to shift out of the cells to replace the sodium normally present in extracellular fluid. To restore the electrolytes to normal, sodium chloride solution is given, orally if possible or by the intravenous route. Frequently potassium is administered along with the saline solution.

The situation described above may also occur in other situations. A very low sodium chloride intake, as in a "low-salt" or "salt-free" diet, might amount to the same thing: salt depletion. Kidney dysfunction, in which excretion of sodium and chloride ions is high, or conditions in which salt is lost, as in Addison's disease, can establish the same electrolyte and fluid derangement.

hemorrhage

In hemorrhage, whole blood is lost. This means not only reduced blood volume and reduced pressure but also reduced oxygen-carrying capacity due to loss of cells. Cells fail to receive sufficient oxygen and nutrients, producing a stress situation in the cells. Injured cells or cells under stress lose potassium from their intracellular fluid. Sodium from interstitial fluid moves into the cells to replace the potassium, and chloride ions move along with the sodium. The patient may go into shock. Whole blood is given to replace the blood loss; it is given rapidly. To accomplish the rapid replacement of blood volume big needles are used, or the blood container is raised during the flow (raising the container increases the pressure), or the blood may be pumped into the blood vessel. Oxygen is sometimes given to increase the oxygen-carrying capacity of the blood.

Electrolyte Retention

Retention of electrolytes in the body is usually accompanied by edema. The kidneys, by conserving water, attempt to dilute the excess ions to normal concentration.

sodium retention

In congestive heart failure and kidney disorders such as nephritis and nephrosis as well as cirrhosis of the liver, sodium is retained along with water. Therefore, in these conditions we find edema. Although the diseases are different, the mechanisms for sodium and water retention seem to be somewhat similar. In these conditions the volume of blood plasma is smaller than normal, and the kidneys have less plasma flowing through them. The rate at which the kidney glomeruli filter plasma is reduced. This slower rate allows the kidneys to retain in the body proportionately more sodium and water. Hormones may aggravate the condition. Aldosterone, from the adrenal glands, stimulates the absorption of sodium from the plasma in the kidneys to the blood. Patients with these problems are put on "low-salt" or "salt-free" diets. Usually diuretics are administered to induce the excretion of sodium and water (Table 12-2).

potassium retention

Potassium is retained in the body in kidney failure, in Addison's disease, in conditions in which the blood is hemolyzed (red blood cells rupture, liberating hemoglobin), in severe burns, and as a result of dehydration or decreased filtration by the glomeruli in the kidneys. In severe cases, the excess potassium can cause paralysis of the muscles and even disturbances of the heartbeat. If there are no lesions in the kidneys, large infusions of sodium chloride are given. Since potassium is usually excreted by the kidneys in lieu of sodium, this treatment tends to reduce the amount of potassium retained. The artificial kidney is sometimes used to remove the excess potassium.

chloride retention

In kidney defects in which the amount of urine is reduced, chloride may be retained in the body. We have mentioned that sodium excess is accompanied by edema. In sodium depletion, as may occur with a low salt intake, the body may not be able to retain sufficient water, and it becomes dehydrated. Dehydration causes reduced urine output and chloride retention. Ammonium chloride is used as a diuretic. If the kidneys do not excrete sufficient fluid, chloride may be retained. Can you anticipate the condition that arose as a result of excess chloride ions? It is acidosis. The excess chloride can be balanced by sodium ions. This is accom-

Table 12-2 SUGGESTED COMPENSATION FOR ELECTROLYTE DISTURBANCES

Substances in Imbalance	Compensation for Loss	Compensation for Retention
Water	Water or isotonic sodium chloride solution	Diuretics
Sodium	Sodium lactate	Ammonium chloride
	Sodium bicarbonate	Decrease salt intake
		Diuretics
Chloride	Ammonium chloride	Sodium lactate
		Sodium citrate
		Sodium bicarbonate
Potassium	Potassium chloride	Sodium chloride

plished by the administration of sodium lactate or sodium citrate. Recall that both lactate and citrate ions are oxidized to carbon dioxide and water, yielding energy. The sodium ions are left to counterbalance the chloride ions and correct the acidosis.

FLUID INTAKE AND OUTPUT

Normally, in the adult, the volume of fluid ingested is balanced by the volume of fluid lost from the body. Fluid is lost in the urine, in the feces, in expired air, and from the skin. The amounts of fluid lost from the skin and the bladder are related. If much fluid is lost through perspiration the volume of urine decreases. On the other hand, when the environmental temperature is low, smaller amounts are lost through the skin and larger amounts through the urine. At low temperatures only water is lost through the skin. This is referred to as *insensible* water loss. As we have noted, in heat exhaustion perspiration contains sodium and chloride.

It is important, for the proper treatment of the patient, to know fluid intake and output and the volumes of fluid lost from the different body areas and to recognize the physiologic effects of fluid and electrolyte imbalance (Table 12-3). The patient may lose abnormally large amounts in diarrhea, in hemorrhage, in vomitus, through suction drainage, through profuse sweating, through salivation, through bronchial secretions, etc. The nurse who realizes the effects on the patient of these losses will consider the intake-output record for the patient as important as it truly is. The nurse will recognize that it is possible for fluids to shift from one compartment in the body to another, sometimes very rapidly, and this may require frequent measurements of fluid volumes. If a patient is receiving fluids intravenously and the urinary output increases noticeably, the nurse may consider the advisability of slowing down the intravenous administration. The informed nurse can be an invaluable asset to both the physician and the patient by intelligent observation and faithful, reliable, dependable performance.

Table 12-3 SOME PHYSIOLOGIC EFFECTS OF FLUID AND ELECTROLYTE IMBALANCE*

Substance in Imbalance	Lack	Excess
Water....................	Dehydration	Edema (accumulation of fluid in tissues)
	Thirst	
	Wrinkling of skin and tongue	Increased production of fluids, e.g., tears, saliva
	Dry mucous membranes	
	Reduced urine output	Increased blood pressure and spinal pressure
	High specific gravity of urine	
Sodium, Na⁺	Weakness	Edema
	Dry tongue	Pulmonary edema
	Low blood pressure	
	Low temperature	
Potassium, K⁺	Muscular weakness	Muscular paralysis
	Decreased diastolic blood pressure (heart relaxation)	Slow pulse
		Heart block (loss of every second or third beat)
	Rapid heart rhythm	
Chloride, Cl⁻	Alkalosis	Acidosis
	Decreased respirations	Increased respirations
	Muscle spasms	Decreased responsiveness
	Coma	Coma

*An imbalance in one electrolyte causes disturbances in others and in the total amount of any body fluid.

Electrolytes in the Body

Study Exercises

1. What is meant by extracellular fluid, intracellular fluid, interstitial fluid, and plasma?
2. What are three ways that the body obtains water?
3. List five avenues through which the body loses water.
4. What is the principal cation in extracellular fluid?
5. What is the principal cation in intracellular fluid?
6. What is the principal anion in intracellular fluid and in extracellular fluid?
7. How can we account for the very minute amount of protein in interstitial fluid?
8. Discuss the relationship between plasma proteins and the osmotic pressure in the capillaries.
9. Compare the distribution of body fluids in infants and adults.
10. How does the body compensate for a threatened shortage of sodium ions?
11. What is alkalosis? What is respiratory alkalosis? What is acidosis? What is metabolic acidosis?
12. List five functions of body electrolytes.
13. What are two conditions in which gastric fluids are lost? What might be the primary effect of this fluid loss on the pH of the body?
14. What would be the effect of persistent diarrhea on the pH of the body?
15. What kinds of substances are blood expanders?
16. Give three results of fluid loss in extensive burns.
17. What are three effects on the body of salt depletion?
18. Explain why retention of electrolytes is usually accompanied by edema.
19. Why would chloride retention with sodium depletion lead to acidosis?
20. Why are accurate fluid intake-output measurements essential in the care of patients with fluid and electrolyte disturbances?

chapter thirteen ORGANIC CHEMISTRY

Ideas for preview or review

We have included in this section several rather involved formulas. The student is expected to compare formulas for similarities and dissimilarities and to relate these differences to differences in properties. The student is not expected to memorize involved formulas.

Oxidation can mean the loss of electrons, the addition of oxygen, the loss of hydrogen, or an increase in valence. Reduction can mean the gain of electrons, the loss of oxygen, the addition of hydrogen, or a decrease in valence.

Graphic or structural formulas indicate the arrangement of atoms in the molecule.

Nonpolar valence involves a sharing of electrons rather than a loss or gain, as in electrovalent compounds.

Unsaturated compounds contain double or triple bonds.

Catalysts alter the rate of chemical reactions. Catalysts cannot cause a reaction to occur that would not take place, however slowly, in the absence of the catalyst.

A substance is considered to be soluble if it can mix with the solvent in the form of average-sized molecules. Liquids that are not mutually soluble are said to be immiscible.

For many years organic chemistry was the chemistry of compounds formed in the organs and in organisms; i.e., the only source of these compounds was considered to be the animal and vegetable kingdoms. It was believed that they were formed in contact with the vital force, in vivo, and could not be formed without this force, in vitro. It was not until 1828 that the significance of organic chemistry changed. In that year—quite by accident, to be sure—*Wöhler* made urea in the laboratory in vitro. The laboratory production of urea was of no importance in itself, for there was no particular demand for it. However, the experiment opened up the whole field of synthetic organic chemistry.

To appreciate the significance of this discovery, we merely have to look around us. Synthetic dyes of every hue, new fabrics, medicinals for almost every use, perfumes, plastics, insecticides, and hosts of other compounds have all resulted from that change in viewpoint. Not only has the chemist been able to reproduce natural substances, but he has taken natural substances, determined their composition, and then altered them to better suit his purposes. It would be useless to try to state the actual number of organic compounds that have been prepared, for new ones are being made each day. More than 300,000 have been described in scientific journals.

Organic chemistry is best described by saying that it is the *study of carbon and its compounds*. However, carbon compounds belonging to the mineral kingdom, oxides, carbonates, etc., are still considered inorganic compounds, even though many of them are formed in living things: carbon dioxide, produced in respiration, bicarbonate ion in the plasma and other body tissues.

COMPARISON OF ORGANIC AND INORGANIC CHEMISTRY

organic reactions

The bulk of inorganic compounds are electrolytes, existing in solution in the form of ions. With these substances, reactions are practically instantaneous. Most organic compounds are nonelectrolytes, existing in the form of molecules. Here the molecules must come into contact and the atoms must be rearranged or redistributed to form new compounds. As a result, most organic reactions are slow. Some of them may take hours or even days to come to equilibrium. We shall find that many organic reactions require the use of catalysts to speed up the reactions.

isomeric compounds

The organic and inorganic compounds differ in other respects. In organic chemistry, we do not find more than one compound with the same percentage composition and the same molecular weight; in other words, with the same molecular formula. If the formula is H_2SO_4, you know that it is sulfuric acid. Not so with organic compounds. The formula C_2H_6O may represent ethanol (ethyl alcohol), or it may represent dimethyl ether, CH_3OCH_3. This variation is possible because different arrangements of atoms in the molecule form substances with different properties, therefore different compounds. Compounds with the same kinds and numbers of atoms, but with different arrangements of atoms in the molecule, are isomers. One can readily appreciate that as the number of atoms in a molecule increases, there are greater and greater numbers of possible arrangements and increasing numbers of isomers. Understanding this, we can realize that molecular formulas are inadequate. The formula of an organic compounds must show not only the elements and the number of each present but also the arrangement of the atoms within the molecule. These formulas, structural or semistructural, make it possible to show the different arrangements existing between isomers.

polymers

We encounter few inorganic compounds that have the same percentage composition but different molecular weights (compounds made up of several duplicating units), but this condition is common in organic chemistry. These compounds can, in many cases, have very high molecular weights. Examples of polymers are glycogen, dextrin, and starch. All are composed of the fundamental building unit, $C_6H_{10}O_5$, but there are different numbers of each of these units in each of the three different compounds. We identify these differences in the formula by indicating in a subscript the number of units in the molecule, if known, e.g. $(C_6H_{10}O_5)_x$. We can define polymers as compounds made up of repeating structural units.

solubility

Most inorganic compounds are either soluble or insoluble in water; very few are soluble in other solvents. While some organic compounds are water soluble, a great proportion are soluble in so-called organic solvents such as acetone, ether, alcohol, and the hydrocarbon derivatives, benzine, kerosene, naphthalene, etc.

Characteristic Groups

During the discussion of organic chemistry, we shall continually emphasize the grouping of atoms characteristic of a particular class of compounds. These groups are important in organic chemistry, for each grouping contributes its properties to those of the whole compound and the properties of the compound are usually the sum of the properties of the groups contained in it. As we study complex compounds, we shall point out these constituent groups. If we stop to think

Organic Chemistry

about it, we remember that we encountered a similar phenomenon in inorganic chemistry. Ammonium sulfate is used to test for an ammonium salt and also for a sulfate, and the properties of the salt are the sum of those of the constituent parts. We did not emphasize this fact because there is no molecule of ammonium sulfate as there is of a non-electrolyte. Instead, we discussed the properties of the constituent ions. In organic chemistry these groups are part of a molecule.

GENERAL CLASSIFICATIONS

Organic molecules are of two types. There are molecules in which the atoms are arranged as the links of a chain. To be sure, there may be other links or even chains attached to the main chain, but fundamentally there are two ends to the molecule. There are other molecules in which the atoms are arranged to form a ring. These molecules are called *cyclic* and must, of course, have at least 3 atoms so joined.

Based on these structural differences, organic compounds are divided into two large classifications. The open-chain compounds are called *aliphatic* compounds, a name derived from the Greek word for "fat" or "oil," because representative compounds are the fatty acids derived from fats. The other large group is composed of *cyclic* compounds. Aromatic compounds, so named because many of those first investigated had pleasant odors, are ring compounds derived from coal tar. There are aliphatic compounds with pleasant odors and aromatic compounds that have decidedly unpleasant odors or are odorless.

Hydrocarbons

It is convenient to group organic compounds according to their structures, and hence according to the similarity of their properties.

methane series

This series is named for the first member, methane. Other names will be given it as we proceed.

Methane. The simplest organic compound is methane, which has the formula CH_4. This makes a good compound to use in developing the structural formula. Such formulas are based on the fact that an atom of carbon has a valence number of 4 and is written

Each of the valence bonds represents a pair of shared electrons. On this page, a flat surface, all 4 bonds must be shown in the same plane. It must be remembered, however, that this is merely a plan like the floor plan of a house and serves much the same purpose. Actually, the carbon atom is better represented by the three-dimensional figure, the tetrahendron, a solid having four sides and four apexes. Each valence bond should be visualized as protruding from an apex of the tetrahendron. This concept can be demonstrated only with models. If we distribute the 4 atoms of monovalent hydrogen around the carbon atom, we get

Properties of Methane. Methane is a colorless gas that is formed during the decay of carbohydrate material in the absence of air. The bubbles of gas that rise when the ooze at the bottom of a pool of stagnant water is stirred are bubbles of methane. Accordingly, this gas is sometimes called *marsh gas;* this name is also applied to the series. It is a constituent of coal gas and of natural gas. It is often found in coal mines and is the *firedamp* of the miners, which often is responsible for explosions in the mines. It burns with a slightly yellow flame, the equation for the complete combustion being

$$CH_4 + 2O_2 \rightarrow CO_2 + 2H_2O$$

This is typical of the combustion of all hydrocarbons; they all burn to form carbon dioxide and water.

Ethane. The next member of this series is ethane, C_2H_6. We may develop the structural formula of this compound from that of methane by substituting a CH_3 group for 1 of the hydrogen atoms of methane.

```
    H   H
    |   |
H — C — C — H
    |   |
    H   H
```

The substitution might have been made for any one of the other hydrogen atoms, since they are all alike. Note that the actual difference between the formulas of methane and ethane is a CH_2 group. This will be the difference between successive members of the series, which, because of this constant difference, is called a *homologous series*.

Propane. The next member of the series is propane, C_3H_8, and the structural formula can be developed from that of ethane by substituting CH_3 for any 1 of the hydrogen atoms.

```
    H   H   H
    |   |   |
H — C — C — C — H
    |   |   |
    H   H   H
```

Butane. The fourth member of the series is butane, C_4H_{10}. Following the same procedure as before, we get the structural formula

```
    H   H   H   H
    |   |   |   |
H — C — C — C — C — H
    |   |   |   |
    H   H   H   H
```

Isobutane. Observe that in the formula for propane 6 of the hydrogen atoms are attached to end carbon atoms, and in producing the formula for butane the CH_3 group might have been substituted for any one of these 6 hydrogen atoms. Observe further that the other 2 hydrogen atoms are attached to a carbon atom that is not the end carbon. Substitution of the CH_3 group for one of these hydrogen atoms will produce the structural formula of an entirely different compound.

```
        H   H   H
        |   |   |
    H — C — C — C — H
        |   |   |
        H   |   H
            |
        H — C — H
            |
            H
```

This compound has the same empirical or molecular formula, C_4H_{10}, as butane. It, therefore, has the same percentage composition and the same molecular weight, but it has a different arrangement of the atoms in the molecule. This compound is an isomer of butane, or, better, it is one of a pair of isomers. It is called isobutane, and its properties are different from those of butane.

The General Formula. This series goes on, successive members differing by CH_2, until we reach $C_{60}H_{122}$. In order to write these molecular formulas easily we merely remember the general formula for this series, C_nH_{2n+2}, where n represents the number of carbon atoms in the hydrocarbon. For instance, the formula of the hydrocarbon in this series containing 32 carbon atoms would be $C_{32}H_{66}$.

It should be obvious that as the number of carbon atoms increases, the number of possible isomers also increases, for if there are many carbon atoms in the molecule, there should be many different arrangements of the atoms. The straight-chain compounds are called normal compounds; i.e., they have a carbon-to-carbon-to-carbon linkage. The branch-chain compounds are described as isocompounds. The aliphatic hydrocarbons with 5 carbon atoms are called pentanes; those with 6, hexanes; 7, heptanes; 8, octanes; 9, nonanes; 10, decanes; etc.

Saturation. An examination of the structural formulas of the compounds in this series shows that there are no unsatisfied valence bonds. Nothing can be added to the formula, for there is no place for an atom to be attached to a carbon atom while the carbon valence number of 4 is maintained. For this reason, such compounds are called *saturated compounds.* This series is also called the *saturated hydrocarbon series.* Their names end in *-ane.*

Nomenclature. The naming of complex substances presents many difficulties. In 1892, an international congress of chemists adopted the Geneva system of nomenclature, which is now universally used. By this system it is possible to name any organic compound regardless of complexity, and anyone familiar with the rules can translate the names into structural formulas. However, the compounds that we shall study are not sufficiently complex to warrant learning the rules of the system. From time to time basic names used will be given in case they are encountered in collateral reading.

A common method of naming compounds considers parts of compounds as named after the hydrocarbons. If we remove 1 hydrogen atom from methane, CH_4, we have a group of atoms with 1 valence, CH_3-. This group is called *methyl.* In the compounds methyl alcohol, methyl chloride, etc., we will find this CH_3-group. From ethane, C_2H_6, we get ethyl group, C_2H_5-, and, as we might expect, the names ethyl alcohol, ethyl ether, ethyl iodide, etc. As the length of the carbon chain increases, we find more than one possible arrangement of atoms and, therefore, isomeric groups. From propane, C_3H_8, we have a normal propyl group and an isopropyl group, both C_3H_7-, but *n*-propyl is arranged $CH_3CH_2CH_2-$ and isopropyl, CH_3CHCH_3. These open-chain or aliphatic groups are called *alkyl groups* or *alkyl radicals* and are sometimes symbolized by the capital letter R. Aromatic groups may be formed in like manner. The group

$$\begin{array}{c} H \\ C \\ HC \diagup \diagdown C- \\ \| \quad \| \\ HC \diagdown \diagup CH \\ C \\ H \end{array}$$

C_6H_5-, is called *phenyl,* named for the compound phenol, which is C_6H_5OH. The aromatic groups are called *aryl groups* or *aryl radicals* (Table 13–1).

ethylene series

The second hydrocarbon series to be discussed is the ethylene series. It also gets its name from the first member of the series. The names of all the compounds in this series end in *-ene.* According to the Geneva system, ethylene is called ethene. It will be evident as we proceed that there can be no methylene or methene.

Structural Formula. The molecular formula of ethylene is C_2H_4. If we write the 2 carbon atoms and then distribute the 4 hydrogen atoms symmetrically, we get

$$\begin{array}{c} H \quad H \\ | \quad | \\ H-C-C-H \\ | \quad | \end{array}$$

We now discover 2 bonds that are unsatisfied; yet we have used all the hydrogen atoms. This problem is solved by allowing these valence bonds to satisfy each other, whereupon we arrive at the correct formula for ethylene.

$$\begin{array}{c} H \quad H \\ | \quad | \\ H-C=C-H \end{array}$$

The formula was introduced this way to emphasize the fact that all the valence bonds are not satisfied in the same manner as in ethane. This compound is described as un-

Table 13-1 HYDROCARBONS

Number of Carbon Atoms	Hydrocarbon	Structural Formula	Semistructural Formula	Group	Structural Formula	Semistructural Formula
1	Methane CH_4	H—C(H)(H)—H	CH_4	Methyl	H—C(H)(H)—	Alkyl Groups CH_3—
2	Ethane C_2H_6	H—C(H)(H)—C(H)(H)—H	CH_3CH_3	Ethyl	H—C(H)(H)—C(H)(H)—	CH_3CH_2—
3	Propane C_3H_8	H—C(H)(H)—C(H)(H)—C(H)(H)—H	$CH_3CH_2CH_3$	n-propyl	H—C(H)(H)—C(H)(H)—C(H)(H)—	$CH_3CH_2CH_2$—
				Isopropyl	H—C(H)(H)—C(H)—C(H)(H)—H	$CH_3CH\ CH_3$
4	Butane C_4H_{10}	H—C(H)(H)—C(H)(H)—C(H)(H)—C(H)(H)—H	$CH_3CH_2CH_2\text{-}CH_3$	Butyl	4 isomers	$CH_3CH_2CH_2CH_2$— $CH_3CHCH_2CH_3$ $\|$ H CH_3—C(H)—CH_2— $\|$ CH_3 CH_3 $\|$ CH_3—C— $\|$ CH_3
		CH_3—C(H)(CH_3)—CH_3	$CH_3CH_3CH_3\text{-}CH$	Isobutyl		
6	Benzene C_6H_6	(benzene ring with H,C,CH,CH,CH,CH,C,H)	C_6H_6	Phenyl	(benzene ring with one substitution point)	Aryl Group C_6H_5—

saturated. The general formula for the series is C_nH_{2n}.

Uses. Ethylene has been used as a general anesthetic since 1923. It has many points in its favor. However, the explosive limits of ethylene-oxygen mixtures are broad, the most dangerous mixture containing 25 per cent ethylene. Ethylene has another very interesting use: it is employed to color green oranges and lemons. Such fruits can be ripe even though the skin may still be green. They

Organic Chemistry

ship well in this condition. In a warehouse a remarkably low concentration of ethylene in the air will rapidly give the fruit its characteristic color. Ethylene glycol, the popular antifreeze sold under the name Prestone, is made from ethylene.

acetylene series

This series is named for its first member, acetylene. According to the Geneva system the compounds have the suffix *-yne;* thus, acetylene becomes ethyne. It should be evident by now that, in the naming of the various hydrocarbon series, the same prefixes indicating the number of carbon atoms are used throughout. We may, for example, have hexane, hexene, and hexyne, each meaning a hydrocarbon with 6 carbon atoms and the suffix indicating the series. The names of the first few members of the series do not indicate the number of carbon atoms, but little difficulty is encountered with them.

Structural Formula. The molecular formula of acetylene is C_2H_2. If we try to develop the structural formula as we did with ethylene, we may write the 2 carbon atoms with their bonds and then distribute the hydrogen atoms symmetrically.

$$H-\overset{|}{\underset{|}{C}}-\overset{|}{\underset{|}{C}}-H$$

We now find 4 bonds with nothing to satisfy them. This formula is completed by joining the bonds between the carbon atoms to produce a triple bond.

$$H-C\equiv C-H$$

Once more we have an unsaturated compound, indicated by this triple bond. The general formula for this series is C_nH_{2n-2}.

cyclic hydrocarbons

Cyclopropane. Cyclopropane has the molecular formula C_3H_6 and its structural formula is as follows:

$$\begin{array}{c} H \quad\quad H \\ H-C---C-H \\ \diagdown \;\; \diagup \\ C \\ \diagup \;\; \diagdown \\ H \quad\quad H \end{array}$$

Cyclopropane is of interest because of its use as a general anesthetic. It is a colorless gas with a mildly pungent but not unpleasant odor. In the concentrations used in anesthesia, it is odorless. It is heavier than air and very inflammable and explosive. The explosive limits when mixed with air are 3.0 to 8.5 per cent; with oxygen, 2.5 to 50.0 per cent. The gas has many advantages to recommend its use, but explosiveness is a great disadvantage. Every precaution must be taken to avoid an explosion. In some cases nurses have not been allowed to wear sharkskin or nylon uniforms because of the great tendency of these fabrics to develop static electricity. For 6 to 8 min after administration of the gas, the patient exhales an inflammable mixture.

Benzene. Aromatic chemistry is based on benzene,[1] which has the molecular formula C_6H_6. Many attempts have been made to devise a formula for benzene. Kekulé suggested that it is composed of 6 CH groups joined in a ring:

$$\begin{array}{c} H \\ C \\ HC \quad\quad CH \\ HC \quad\quad CH \\ C \\ H \end{array}$$

---

[1] Benzene should not be confused with benzine. The later compound is one of the lower-boiling fractions in the fractional distillation of petroleum. Benzene, on the other hand, is obtained from coal tar, which results from the destructive distillation of soft coal in the making of illuminating gas, coke, and ammonia. Benzene may be prepared in the laboratory by the polymerization of acetylene.

$$3C_2H_2 \xrightarrow{500°} C_6H_6$$

An examination of this formula shows that the carbon atoms have a valence of 3, which we have learned is not true. Kekulé took care of this by inserting alternating double linkages, as follows:

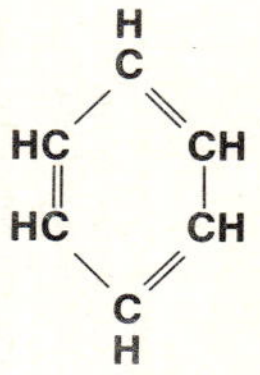

$$\begin{array}{c} H \\ C \\ HC \diagup \diagdown CH \\ HC \diagdown \diagup CH \\ C \\ H \end{array}$$

but benzene seems to be a type of compound in some sense intermediate between a saturated and an unsaturated compound. To account for this behavior, Kekulé theorized that the double linkages were continually oscillating back and forth between two positions as indicated by the two formulas

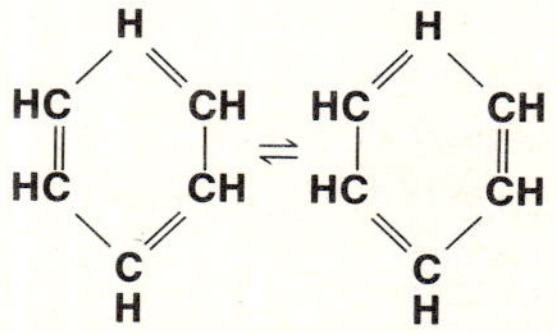

$$\begin{array}{cc} H & H \\ HC = CH & HC - CH \\ HC - CH \rightleftharpoons HC = CH \\ C & C \\ H & H \end{array}$$

For convenience in writing structural formulas in aromatic chemistry, a conventional outline formula has been adopted for the benzene ring.

It must be remembered in using this symbol for C_6H_6 that there is a CH at each corner of the hexagon.

Toluene. Toluene has the formula $C_6H_5CH_3$, or

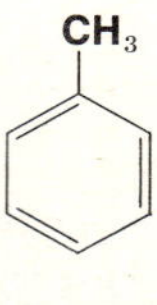

Toluene is obtained from coal tar, boils at 111°C, and is lighter than water. These properties are given to explain a use of toluene encountered by nurses. A urine sample can be kept for some time if it is covered with a layer of toluene. It seals the urine from the air, and—more important—it prevents the evaporation of water from the urine, which would change its density. Samples of the urine may still be withdrawn by means of a pipet.

polynuclear aromatic hydrocarbons

This impressive title refers to those substances that result from the condensation of two or more rings. Of these we shall only discuss *naphthalene*. This compound has the formula

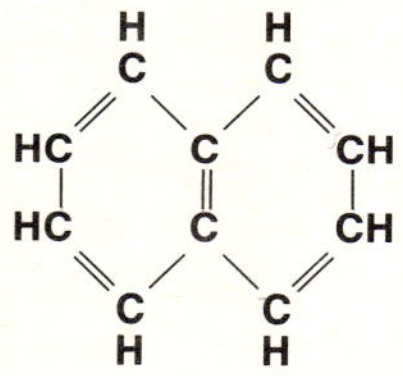

or

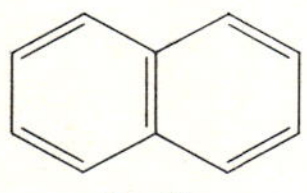

$C_{10}H_8$

It is a white crystalline solid that readily sublimes. Naphthalene has a pungent odor and is sold in flakes or balls for use as protection against moths (moth balls). It is obtained from coal tar. About one hundred thousand pounds are used annually, for it is a constituent of a great number of important dyes. The carbon atoms in this hydrocarbon are numbered

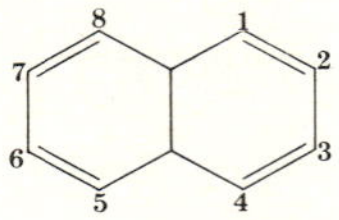

The numbers are used to indicate the positions of groups when substituted in the ring.

There are two other condensed carbocyclic rings widely distributed in nature.

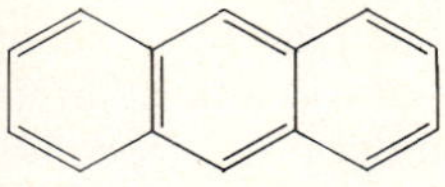

Organic Chemistry

Anthracene $C_{14}H_{10}$

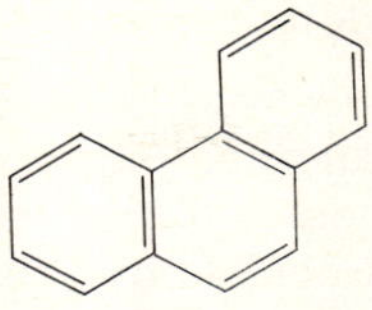

Phenanthrene $C_{14}H_{10}$

Anthracene is similar to naphthalene in its reactions. The red dye alizarin is one of its important derivatives and is the parent of many other dyes. Phenanthrene is an isomer of anthracene. This nucleus is common to one family of narcotics, the morphine family, and is part of the structure of cholesterol and a class of hormones called steroids.

Properties of Hydrocarbons

substitution

There are two important types of chemical reaction in organic chemistry: substitution and addition. Substitution takes place with saturated compounds, and addition is a function of double and triple bonds.

The halogens (fluorine, chlorine, bromine, and iodine) are particularly reactive, and these nonmetals are capable of reacting with the saturated hydrocarbons.

If liquid bromine is added to a saturated hydrocarbon, an atom of bromine will very slowly substitute for a hydrogen atom and this atom will unite with the other bromine atom of the molecule to form hydrobromic acid, according to the following equation:

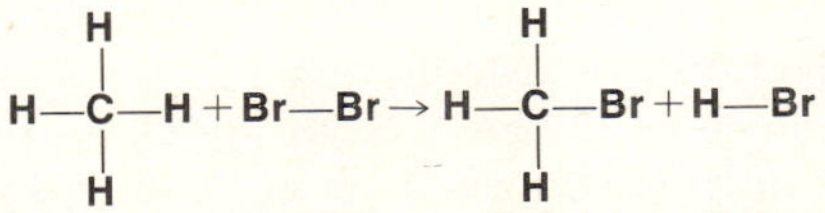

$$H-CH_3 + Br-Br \rightarrow H-CH_2-Br + H-Br$$

Methane + Bromine → Methyl bromide + Hydrobromic acid

Monosubstitution. Benzene can produce only one monosubstitution compound. For instance, if bromine is added to benzene, the following reaction takes place:

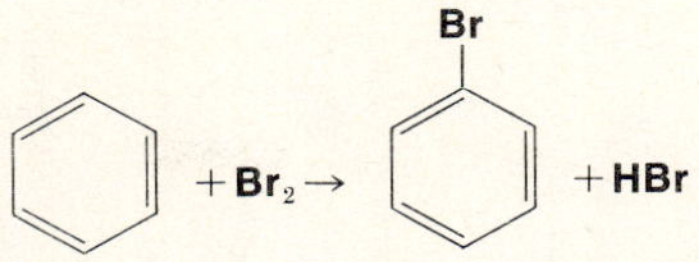

Benzene + Br_2 → Monobromobenzene, or bromobenzene + HBr

The reaction is exactly like substitution in aliphatic compounds. The bromine might have been substituted for any other hydrogen atom, for they are all alike.

Concentrated nitric acid in the presence of concentrated sulfuric acid will substitute the nitro group for one of the hydrogen atoms of the benzene ring.

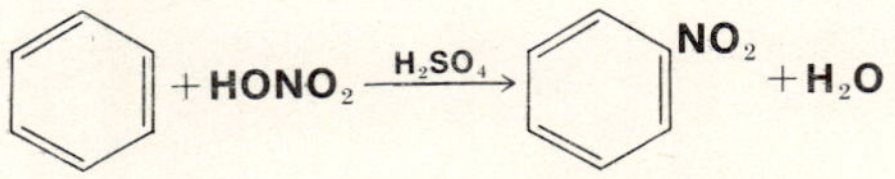

$$\text{Benzene} + HONO_2 \xrightarrow{H_2SO_4} \text{Nitrobenzene} + H_2O$$

Attention is called to this reaction, for it is the hydroxyl group of the nitric acid that is split out in forming the water, leaving the nitro group, $-NO_2$. In other words, this is not the nitrate of benzene. Further, this group should not be confused with the nitrite group, (NO_2^-). It will be seen that the valence of nitrogen in the nitrite group is 3. Nitroglycerin, $C_3H_5(NO_3)_3$, is really misnamed and should be called glyceryl nitrate, but the common name, though incorrect, is still used. Glyceryl nitrate is used in medicine as a vasodilator.

Disubstitution. The substitution of two atoms of bromine in the molecule for two atoms of hydrogen permits three possible arrangements. If the substitutions occur on adjacent carbons, carbons 1 and 2, the compound is called orthodibromobenzene. If the substituted bromine atoms are on carbons 1 and 3, the compound is metadibromobenzene, and if the substitutions occur on carbons 1 and 4, we have paradibromobenzene. The prefixes ortho, meta, and para are designated by the lowercase letters *o, m,* and *p,* respectively.

$C_6H_4Br_2$

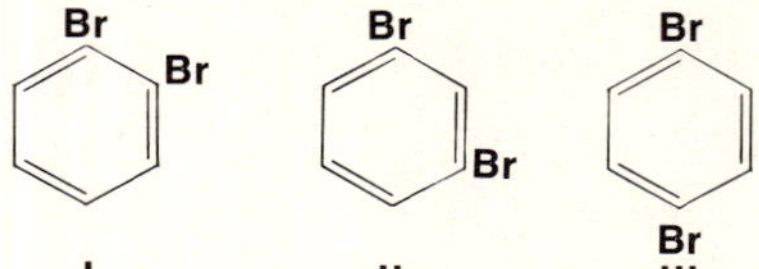

I o

II m

III p

It should be remembered that in naming compounds in this manner it is the relative positions of the groups that determine the name, not the actual positions. For instance, o-dibromobenzene (o for *ortho*) could be written

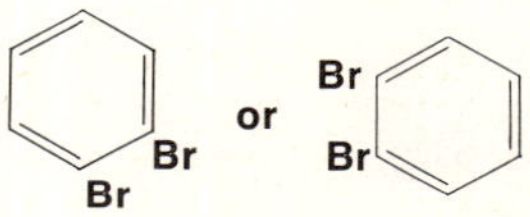

Trisubstitution. In the naming of trisubstitution compounds of benzene, we must resort to a third method. In this case the carbon atoms are numbered as follows:

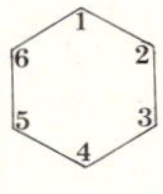

There are three isomers of a trisubstituted benzene of the formula $C_6H_3X_3$. For instance, the three tribromobenzenes are

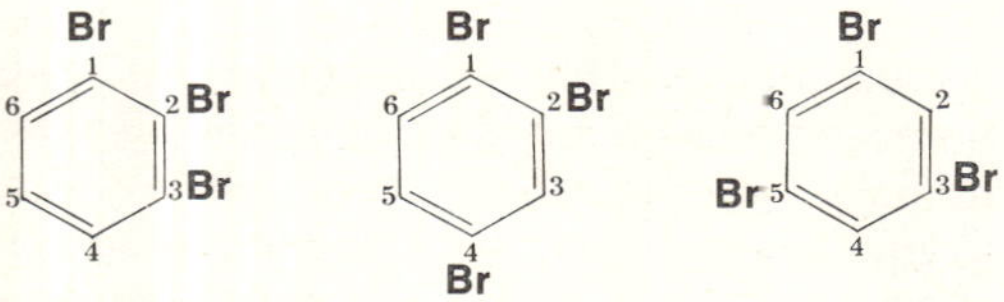

Isomer I is called 1,2,3-tribromobenzene; II is 1,2,4-tribromobenzene; and III is 1,3,5-tribromobenzene. We must use this method of naming when there are more than two substituents in the ring. Note that 1,2,5-tribromobenzene would be identical with II and that the 1,2,6-compound would be identical with I, etc.

addition

This type of reaction takes place with unsaturated compounds. As we may recall, there is always a double or triple bond in the structure of these compounds. Recalling further how we developed the structural formulas of ethylene and acetylene, we must be conscious of the vulnerable spot in these molecules. If ethylene is treated with bromine, the halogen breaks the double bond, leaving a single bond, and both atoms of the halogen add on directly to the carbon atoms, as shown in the equation

$$H_2C=CH_2 + Br-Br \rightarrow H_2C(Br)-C(Br)H_2$$

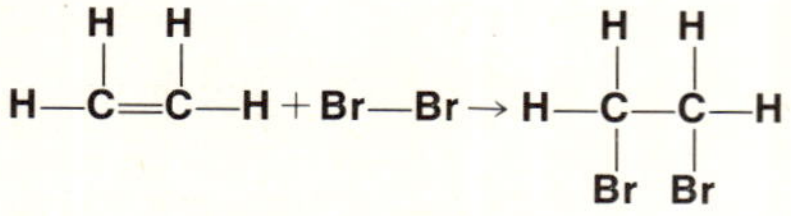

Let us pause now to examine this equation. Note that the compound formed, ethylene dibromide, is saturated; i.e., the double bond has been destroyed. Note also that no hydrobromic acid was formed.

This action of bromine may be used as a test for unsaturation. We add bromine to the compound; if the color disappears and no hydrobromic acid is formed, the compound must be unsaturated.

If bromine is added to a compound having a triple bond in the molecule, such as acetylene, 2 molecules of bromine will be required, as shown in the equation

$$H-C\equiv C-H + 2(Br-Br) \rightarrow HBr_2C-CBr_2H$$

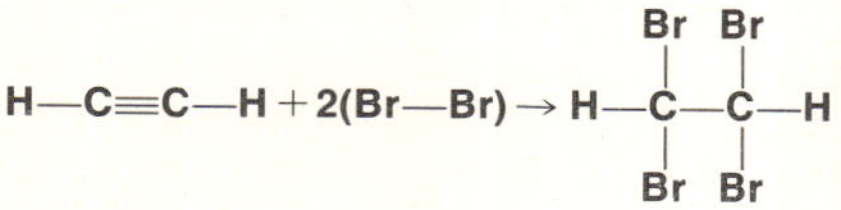

It should be evident now that the number of molecules of the halogen required in this addition reaction gives us an idea as to the degree of unsaturation of a compound, i.e., 1 molecule per double bond, 2 molecules per triple bond; in general, 1 molecule for each valence bond broken. This is the basis of the so-called iodine number of a fat or oil. Iodine is used because it can be more easily handled on a quantitative basis.

hydrogenation

In the presence of finely divided nickel as a catalyst, hydrogen may be added to an unsaturated compound. The hydrogen acts as the halogen did; for instance.

$$H-\underset{}{C}=\underset{}{C}-H + H-H \rightarrow H-\underset{H}{\overset{H}{C}}-\underset{H}{\overset{H}{C}}-H$$

In the case of a triple bond, 2 molecules of hydrogen are required as with bromine.

$$H-C\equiv C-H + 2(H-H) \rightarrow H-\underset{H}{\overset{H}{C}}-\underset{H}{\overset{H}{C}}-H$$

This process is called hydrogenation, the adding of hydrogen, and is a very important process.

Halogen Compounds

monochlor derivatives

Methyl chloride, or monochloromethane, and ethyl chloride, or monochloroethane, are important compounds. They are used as refrigerants because they have low boiling points. Ethyl chloride is also used as a local anesthetic. The liquid is sprayed on the area and as it evaporates so much heat is extracted from the tissues that the area becomes insensitive. Obviously, it can only be used for minor operations such as incising a boil or removing a splinter. The ethyl chloride is contained in a small glass or metal tube that fits the hand and has a pinhole orifice covered by a spring cap. In use, the cap is easily lifted with one finger. The heat of the hand vaporizes enough of the liquid to develop quite a pressure, and the liquid sprays out of the orifice with some force, to be directed onto the spot to be chilled. Such tubes should be stored at a low temperature, of course. Ethyl chloride is used also as a general anesthetic, but induction and recovery are so rapid that it is difficult to maintain an even anesthesia.

chloroform

Chloroform has long been used as a general anesthetic. It has the advantage of being noninflammable, but its use is accompanied by such danger to the patient that it has been passed over in favor of safer anesthetics. Chloroform is often used as an analgesic by the country doctor who must perform a delivery without skilled assistance. A small amount of it on cloth in the bottom of a glass held by the patient allows the inhalation of enough chloroform to relieve pain. Too much cannot be inhaled, for soon muscular coordination ceases and the hand drops. The process can be repeated as consciousness and pain return. It should not be used near a naked flame because it produces noxious gases. Chloroform is oxidized slowly by the air to form phosgene, a poisonous war gas, and hydrochloric acid.

$$2CHCL_3 + O_2 \rightarrow 2COCl_2 + 2HCl$$

Before chloroform is used in anesthesia it should be tested for the presence of chloride ions with silver nitrate solution. A turbidity of silver chloride indicates that oxidation has taken place, and the presence of phosgene may be inferred. This oxidation dictates the manner of storing chloroform; the container must be full of liquid to exclude the air. It is better to fill a small brown bottle with chloroform and discard a small amount than to use a larger bottle and not fill it completely. Chloroform is used somewhat as a counterirritant in liniments. Its chief use in the laboratory is as a solvent.

iodoform

Iodoform, a pale-yellow solid, has been used as an antiseptic in ointments and as a powder, which may be dusted on a wound. Its mildly antiseptic action is probably due to the iodine that is set free slowly. Iodoform is also used to test for ethyl alcohol.

carbon tetrachloride

Carbon tetrachloride is a heavy, volatile liquid. Its vapor is very heavy, as may be seen by calculating its molecular weight. This property, together with the noncombustibility of the liquid and its vapor, makes it a very efficient fire extinguisher. This type of extinguisher is especially efficient on chemical fires. It is not recommended for extinguishing burning alcohol because noxious fumes are formed. In no case should the carbon tetrachloride be vaporized in a small room, as its vapor is poisonous.

Carbon tetrachloride is an excellent solvent for gums, resins, and fats. That fact makes it useful in dry cleaning, and it is much safer than the inflammable hydrocarbons. A mixture of carbon tetrachloride and hydrocarbons is sold under the name Carbona. In spite of the presence of the hydrocarbons, the vapor of the mixture does not burn. Carbon tetrachloride is recommended for dissolving grease and oil in electric equipment, such as motors and generators. Since it is a nonelectrolyte, there is no danger of a short circuit. Administered in capsules, carbon tetrachloride has been remarkably successful in the treatment of hookworm. Its low cost—less than 1 cent per dose—is an important factor in its large-scale use.

Alcohols

The characteristic arrangement of atoms in an alcohol is —C—O—H.

The —O—H group is called *hydroxyl*. Alcohols with 1 hydroxyl group are called *monohydric* alcohols; those with 2, *dihydric* alcohols (glycol); those with 3, *trihydric* alcohols (glycerol). The suffix *-ol* in a name indicates that the compound is an alcohol.

monohydric alcohols

methyl alcohol   H—C(H)(H)—O—H

Methyl alcohol, CH_3OH, is also called methanol. This name is obtained by dropping the final *e* of the name of the 1-carbon hydrocarbon methane and adding the last two letters of "alcohol." For a great many years this alcohol has been prepared by the destructive distillation of wood; hence its other name; wood alcohol.

Properties. Methyl alcohol burns with a blue flame to form carbon dioxide and water.

$$2CH_3OH + 3O_2 \rightarrow 2CO_2 + 4H_2O$$

This reaction is typical of the combustion of all alcohols. Methyl alcohol is an excellent solvent for gums, resins, and shellac and has been used extensively in the paint industry. It is poisonous, attacks the optic nerve, and leads to blindness and death. Even its vapor will produce blindness in time. In the old days, painters could look forward to ending their days in blindness. Denatured alcohol has removed this industrial hazard.

ethyl alcohol   H—C(H)(H)—C(H)(H)—O—H

Ethyl alcohol, CH_3CH_2OH or C_2H_5OH, is also called ethanol. It is a colorless, volatile liquid, soluble in water in all proportions. This solubility results from the fact that both compounds are polar.

Preparation. Ethyl alcohol is prepared chiefly by the process of fermentation. If a solution of glucose is allowed to stand in contact with the enzyme zymase, which is produced by yeast, ethyl alcohol and carbon dioxide are formed by the reaction

$$C_6H_{12}O_6 \xrightarrow{\text{Zymase}} 2C_2H_5OH + 2CO_2$$
Glucose

Starch from grain or potatoes may be used if another enzyme, diastase, is used first. This enzyme is formed in the sprouting of barley, which is called malt after it has sprouted. Diastase promotes the hydrolysis of the starch to maltose.

$$2(C_6H_{10}O_5)_n + nH_2O \xrightarrow{\text{Diastase}} nC_{12}H_{22}O_{11}*$$
Starch — Maltose

Maltase, also produced by yeast, converts maltose into glucose.

$$C_{12}H_{22}O_{11} + H_2O \xrightarrow{\text{Maltose}} 2C_6H_{12}O_6$$
Maltose — Glucose

Then, of course, the glucose produces the alcohol, as shown in the equation above. Because much of our alcohol is made from grain, ethyl alcohol is also known as grain alcohol.

Properties and Uses. Alcohol burns with a slightly luminous flame according to the reaction

$$C_2H_5OH + 3O_2 \rightarrow 2CO_2 + 3H_2O$$

It is an excellent solvent and finds a wide use in many industries. Ethyl alcohol is used as an antiseptic and as a preservative of biologic specimens. It is also used as the solvent in such medicinal preparations as tinctures, spirits, and elixirs.

Physiologic Effect. Ethanol evaporates quickly, extracting heat rapidly from the environment. Alcohol baths and sponges are used to cool a patient with an elevated temperature. The alcohol, as it evaporates, removes heat from the patient's skin, thus lowering the body temperature. Alcohols, in general, are hygroscopic (absorb water), and for this reason ethanol is used as an astringent on the skin, preventing sweating. Because ethanol coagulates protein, it can be used to harden the skin and help prevent the development of decubitus ulcers. For the same reason, it is also used as an antiseptic and disinfectant. When the protoplasm of a microorganism is coagulated, the organism can no longer function. Ethanol is a vasodilator and is sometimes used in peripheral vascular disease. On oxidation, ethanol forms carbon dioxide, water, and energy. One gram ethanol produces 7 large calories of heat (7 Cal). Ethanol is a source of quick energy, but it is actually a depressant of the nervous system. What seems like stimulation is really unrestrained activity due to the removal of inhibitions. Many people believe that alcohol taken internally increases the body temperature. Actually, there is a temporary feeling of warmth due to the increased flow of blood to the skin, which also produces the flushed appearance, but by this very process the heat loss is greater and the body temperature falls. Large quantities of alcohol are irritating to the stomach, and continued use may result in gastritis.

Denatured Alcohol. Ethyl alcohol to which has been added some substance that cannot be easily removed and that renders the alcohol unfit for drinking purposes is known as denatured alcohol. The use of denatured alcohol permits manufacturers who use alcohol in their industrial processes to avoid the high excise tax on beverage alcohol.

Tribromoethanol. Tribromoethanol, CBr_3-CH_2OH (Avertin), is a basal anesthetic,

$$\begin{array}{c} H \quad H \\ | \quad | \\ H-C-C-O-H \\ | \quad | \\ H \quad H \end{array} \qquad \begin{array}{c} Br \quad H \\ | \quad | \\ Br-C-C-O-H \\ | \quad | \\ Br \quad H \end{array}$$
Ethanol — Tribromoethanol

instilled into the rectum. It produces unconsciousness, but other drugs must be used to bring about deep anesthesia.

---

*The subscript n appears in the formula for starch because its exact molecular weight is not known, though we do know that it is a multiple of that of the unit $C_6H_{10}O_5$. Other letters are often used for other polysaccharides to indicate that their molecular weights are different multiples of the unit weight.

isopropyl alcohol

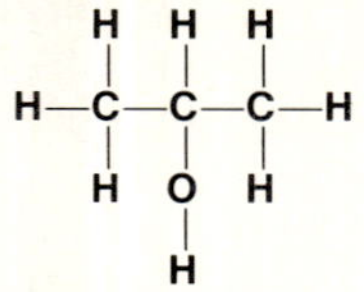

H H H
H—C—C—C—H $CH_3CHOHCH_3$
H O H   1   2   3
   H

Isopropyl alcohol is also called isopropanol or propanol-2, to indicate that the hydroxyl group is on the second carbon. Isopropanol has astringent and coagulating actions similar to those of ethanol and, because of its lower cost, is frequently used for these purposes in place of ethanol.

phenol

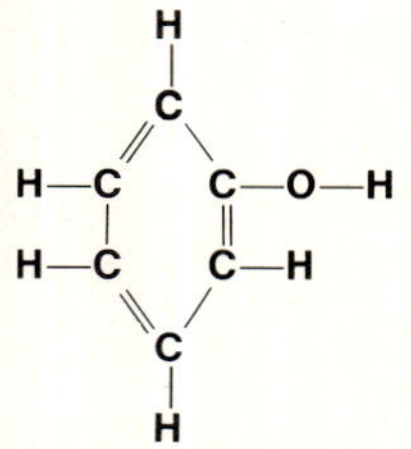

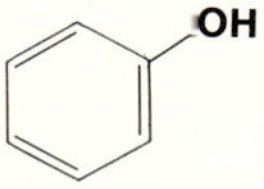

Phenol, C_6H_5OH, is the original disinfectant used by Lister. Other disinfectants and antiseptics are standardized by using phenol as a reference. They are then rated in terms of phenol coefficients. Although the name and formula tell us that phenol is an alcohol, it can react with metal hydroxides such as sodium hydroxide to form phenolates.

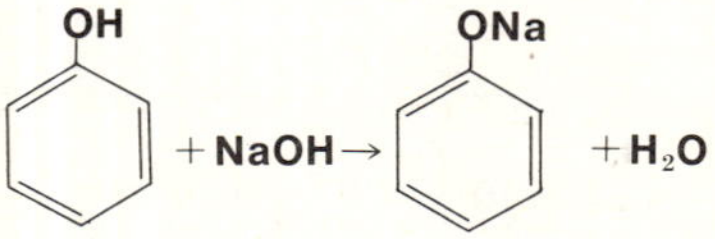

OH   ONa
 | |
(ring) + NaOH → (ring) + H_2O
Phenol   Sodium
                    phenolate

Its common name, carbolic acid, reminds us that phenol has weak acid properties.

Phenol is a solid in the form of shiny white needlelike crystals. It causes reddening and blistering of the skin on contact and is extremely toxic if taken internally. The antidote is ethyl alcohol, but it must be administered very quickly before too much destruction of tissue has taken place and then rapidly removed by tube.

Some of the popular disinfectants are derivatives of phenol. Lysol, for example, is a mixture of soap and the three isomeric cresols. The cresols are formed when an —OH group is substituted in toluene.

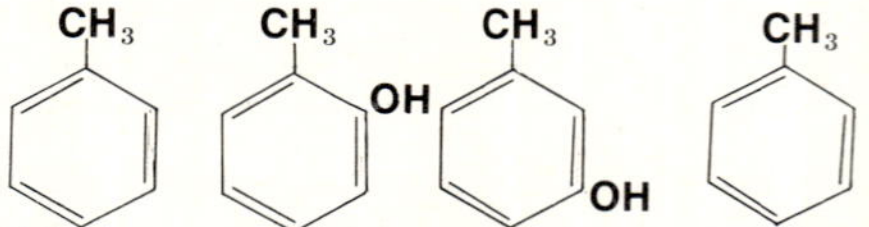

Toluene   o-cresol   m-cresol   p-cresol

Benzyl Alcohol. A more typical aromatic alcohol has the —C—O—H group outside the benzene ring. The simplest of these is benzyl alcohol.

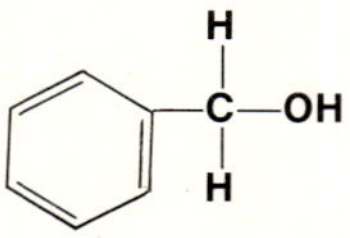

The benzyl group, $C_6H_5CH_2$—, is used in the name of related compounds in the same fashion as the phenyl group, C_6H_5—.

Cholesterol. As its name implies, cholesterol contains the alcohol arrangement, —C—OH, and belongs to a class of compounds called *higher alcohols,* or *sterols.* These substances have properties similar to the fats. Cholesterol is made of three 6-membered rings like phenanthrene

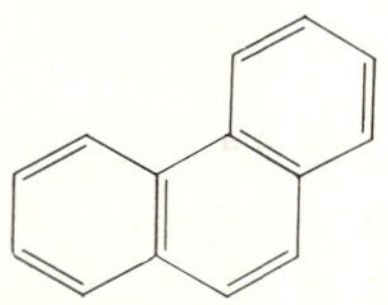

Phenanthrene

plus a 5-membered ring, cyclopentane,

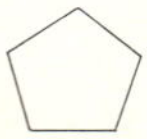

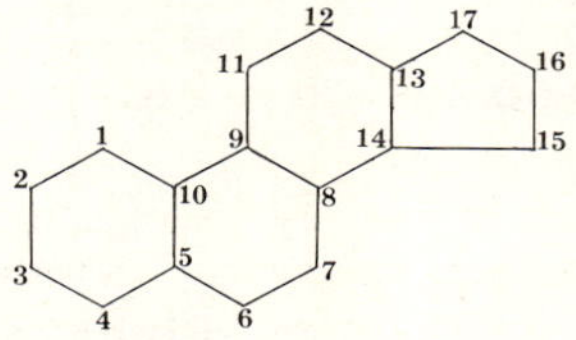

Organic Chemistry

giving what is called the *steroid nucleus*:

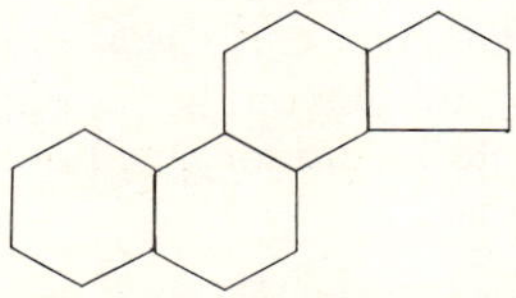

This skeleton is found in some hormones, e.g., estradiol and cortisone, and in some vitamins, e.g., ergosterol and calciferol.

The positions of the atoms in the rings are designated by number.

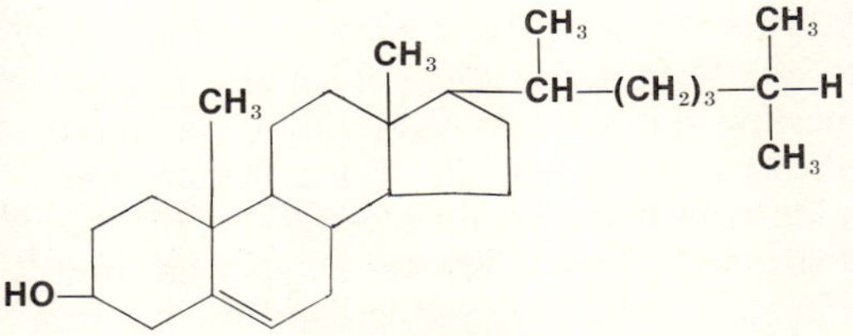

Cholesterol

polyhydric alcohols

Glycols. Glycols are dihydroxy alcohols, and the only one of interest is ethylene glycol,

$$\begin{array}{c} CH_2OH \\ | \\ CH_2OH \end{array}$$

There is little reason for discussing this compound except to bridge the gap between monohydric alcohols and glycerol. Incidentally, this compound is the base of the popular antifreeze known as Prestone.

Glycerol. Glycerol is a trihydroxy alcohol, which might be considered to result from the substitution of a hydroxyl group for a hydrogen atom on each of the carbon atoms in propane.

$$\begin{array}{c} CH_2OH \\ | \\ CHOH \\ | \\ CH_2OH \end{array}$$

It is a product of the digestion of fats and oils.

Glycerol is a viscous colorless liquid that boils at 290°C and has a sweet taste. It is an interesting fact that as the number of hydroxyl groups increases in compounds, the sweet taste increases. Ethylene glycol is slightly sweet and glycerol is sweeter.

Glycerol contains 3 alcoholic hydroxyl radicals, each of which should react as if it were alone. It should react with 3 molecules of an acid in the same manner as 3 molecules of a monohydric alcohol, e.g., ethyl alcohol. It reacts with concentrated nitric acid to form glyceryl nitrate, or nitroglycerin.

$$\begin{array}{c} CH_2O\;\boxed{H\quad HO}\;NO_2 \\ | \\ CHO\;\boxed{H\;+\;HO}\;NO_2 \\ | \\ CH_2O\;\boxed{H\quad HO}\;NO_2 \end{array} \xrightarrow[H_2SO_4]{conc.} \begin{array}{c} CH_2ONO_2 \\ | \\ CHONO_2 + 3H_2O \\ | \\ CH_2ONO_2 \end{array}$$

or

$$C_3H_5(OH)_3 + 3HNO_3 \xrightarrow[H_2SO_4]{conc.} C_3H_5(NO_3)_3 + 3H_2O$$

Nitroglycerin is an oily liquid that is a powerful explosive. It is sensitive to shock and dangerous to transport. Being a liquid, it finds explosive uses differing from those of a solid. Nobel found in 1867 that nitroglycerin absorbed in some absorbent material was much safer to handle and yet lost none of its explosive power. The resulting product was dynamite.

Nitroglycerin is also used medicinally to reduce the pain characteristic of angina pectoris because it relaxes arterial walls. It is usually dispensed in tablets containing minute amounts of the drug. These tablets become inert on standing or if exposed to the air. This fact, coupled with the development of a tolerance for such compounds, accounts for the disappointment experienced by many persons using the drug.

Aldehydes

The aldehydes and ketones are two important classes of compounds. Their characteristic groups are found in substances such as the sugars, whose properties are due, in part at least, to their being aldehydes or ketones. Both aldehydes and ketones contain the carbonyl group.

$$-\underset{\|}{\overset{O}{C}}-$$

The characteristic group of aldehyde is

$$-\underset{}{\overset{H}{C}}=O$$

or —CHO. It is important that this group be written —CHO and *not* —COH. (This latter symbol is used to designate alcohols.) Note that, since the carbon in the aldehyde group has only 1 valence free, this group must appear at the end of a chain and can never be in the middle.

nomenclature

A common method of naming aldehydes is to derive their names from the acids they form on oxidation; i.e., formaldehyde forms formic acid; benzaldehyde forms benzoic acid. A more logical method consists in substituting the letters *al*, the first letters of aldehyde, for the final *e* of the hydrocarbon. Thus formaldehyde with 1 carbon atom is methanal, and acetaldehyde with 2 carbon atoms becomes ethanal, etc.

formation

The aldehydes are formed by the slow oxidation of alcohols like methanol and ethanol.

$$H-\underset{\underset{H}{|}}{\overset{H}{|}}{C}-O-H + [O] \longrightarrow H-\underset{}{\overset{H}{C}}=C + H_2O$$

Formaldehyde or methanal

$$H-\underset{\underset{H}{|}}{\overset{H}{|}}{C}-\underset{\underset{H}{|}}{\overset{H}{|}}{C}-O-H + [O] \longrightarrow H-\underset{\underset{H}{|}}{\overset{H}{|}}{C}-\underset{}{\overset{H}{C}}=O + H_2O$$

Acetaldehyde or ethanal

This oxidation may also be called a dehydrogenation, or removal of hydrogen. The name *aldehyde* is derived from the words *al*cohol *dehyd*rogenation.

Benzaldehyde, C_6H_5CHO, an aromatic aldehyde, is employed as a perfume and flavoring material under the name of oil of bitter almonds.

$$C_6H_5-\overset{H}{\underset{}{C}}=O$$

properties

Polymerization. Aldehydes have the ability to link together to form polymers. The polymer has a molecular weight that is a multiple of the simple aldehyde. The polymer of acetaldehyde is called *paraldehyde* $(CH_3CHO)_3$. It is used medicinally as a hypnotic, is administered rectally, and is used for its sedative effect for patients with delirium tremens.

Oxidation. If an aldehyde is treated with an oxidizing agent, an acid is formed.

$$H-\overset{H}{\underset{}{C}}=O + [O] \longrightarrow H-\overset{O}{\underset{}{C}}=O$$
Formaldehyde Formic acid

It has been mentioned that the aldehydes are named for the acid that they form on oxidation. Acetaldehyde forms acetic acid.

$$H-\underset{\underset{H}{|}}{\overset{H}{|}}{C}-\overset{H}{\underset{}{C}}=O + [O] \longrightarrow H-\underset{\underset{H}{|}}{\overset{H}{|}}{C}-\overset{O}{\underset{}{C}}=O$$
Acetaldehyde Acetic acid

Benzaldehyde forms benzoic acid.

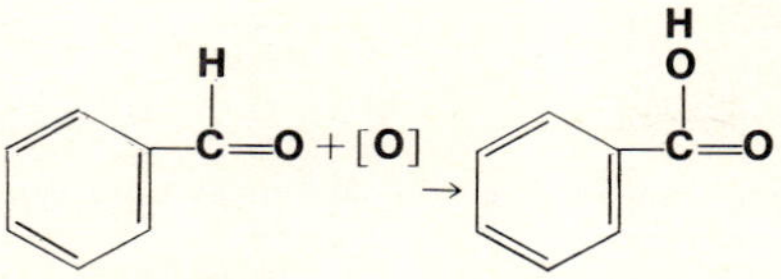

Benzaldehyde + [O] → Benzoic acid

Glucose forms gluconic acid.

Reduction. If an aldehyde is capable of being oxidized, it must follow that it can act as a reducing agent. Several simple tests have been devised to measure the reducing ability of aldehydes. Three of these, Fehling's test, Benedict's test, and Clinitest, use blue cupric ions (Cu^{++}) as oxidizing agent. In an alkaline medium, cupric ions exist mainly as cupric hydroxide.

If Fehling's solution, to which a few drops of an aldehyde solution has been added, is heated to boiling, the following reactions take place: reduction,

$$2Cu(OH)_2 \rightarrow 2CuOH + H_2O + [O]$$
Blue cupric hydroxide — Yellow cuprous hydroxide

oxidation,

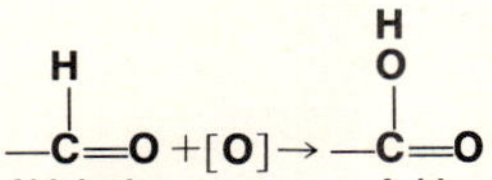

Aldehyde + [O] → Acid

followed by

$$2CuOH \rightarrow Cu_2O \downarrow + H_2O$$
Red cuprous oxide

The progress of the reaction can be followed by the formation and color of the cuprous oxide formed. Various colors from yellow-green through orange and brown to brick-red are produced. These colors are much more pronounced if a substance such as glucose is the source of the aldehyde group. The cause of these color changes is debatable. Cuprous hydroxide is yellow, and cuprous oxide is red. These colors, together with the blue of the solution in early stages of the reaction, could account for the transition from one color to the next. The amount of aldehyde present can be estimated from the color formed.

formaldehyde

Formaldehyde, HCHO, is a gas with a pungent odor. It is soluble in water; a 40 per cent solution is sold under the name Formalin. Formaldehyde hardens proteins and is used to preserve biologic specimens. It is used in embalming fluid and as part of a mixture to sterilize sharp instruments used in surgery. The disinfecting of sharp instruments poses a special problem since most of the effective disinfectants ruin the cutting edges of the instruments. Formaldehyde is used as a gas for fumigation.

acetaldehyde

Acetaldehyde, CH_3CHO, is an intermediary product of metabolism in the body. In industry it is used in the manufacture of paraldehyde and of plastics and resins.

glucose

Glucose, $C_6H_{12}O_6$, also called dextrose or grape sugar, is the principal sugar in the blood and contains an aldehyde group. The properties of the aldehyde group permit a simple method of determination of glucose in the blood and urine.

Other sugars—galactose, $C_6H_{12}O_6$, maltose, $C_{12}H_{22}O_{11}$, and lactose, $C_{12}H_{22}O_{11}$—are also aldehydes.

trichloroacetaldehyde

Trichloroacetaldehyde, CCl_3CHO, has been given the name *chloral*. It is an oily liquid, which adds a molecule of water to form chloral hydrate, $CCl_3CH(OH)_2$. Chloral is used as a hypnotic, its use being similar to that of paraldehyde. Both are irritating to the stomach and therefore are often given mixed with olive oil as a retention enema.

Ketones

Ketones have the characteristic formula

$$-\underset{|}{\overset{|}{C}}-\overset{O}{\overset{\|}{C}}-\underset{|}{\overset{|}{C}}-.$$ The carbonyl group, $-\overset{O}{\overset{\|}{C}}-$, obviously, cannot occur at the end of a carbon chain in ketones. *Acetone* is the most important ketone. It is formed through the oxidation of isopropyl alcohol.

$$\begin{array}{c} H_3C \\ \diagdown \\ C-O-H \; + \; [O] \\ \diagup \\ H_3C \end{array} \rightarrow \begin{array}{c} H_3C \\ \diagdown \\ C=O + H_2O \\ \diagup \\ H_3C \end{array}$$

It is a volatile liquid with a sweetish odor. Small quantities are found in the blood, breath, and urine of an uncontrolled diabetic patient and account for the expression "acetone breath." Acetone is an excellent solvent and finds many applications because of this property. It is included in nail polish remover, is the chief ingredient in paint remover, and is used in one of the processes for making rayon.

The sugar *fructose*, $C_6H_{12}O_6$, also called levulose, obtained from honey, is a ketone.

oxidation

Ketones may be oxidized by strong oxidizing agents. In this case, however, the action is very different from that with aldehydes. Here, decomposition of the ketone takes place; if an acid is formed, it will always contain less carbon than the ketone. The oxidation of acetone illustrates the point.

$$\begin{array}{c} CH_3 \\ \diagdown \\ C=O + 4[O] \rightarrow CH_3COOH + CO_2 + H_2O \\ \diagup \\ CH_3 \end{array}$$
Acetic acid

Fructose on oxidation also breaks at the ketone group to give us two acids.

Acids

The carbonyl group $-\overset{O}{\overset{\|}{C}}-$ plus the hydroxyl group $-O-H$ combined makes $-\overset{O}{\overset{\|}{C}}-OH$, which is called the *carboxyl group*. This group is characteristic of organic acids. Organic acids are moderately weak.

Those acids in which the carboxyl group is attached to an alkyl radical are called *fatty acids,* since some of them can be obtained from natural fats. There has been a tendency lately to include all acids that may be obtained from fats in this category.

It has been shown that these acids may be formed by the oxidation of an aldehyde or an alcohol. They may also be obtained through the general process of adding sulfuric acid to a salt of the acid desired.

nomenclature

Many of the organic acids have been known so long that common names have been assigned to them. According to the Geneva system, *-oic* replaces the final *-e* of the name of the parent hydrocarbon; for example, acetic acid, CH_3COOH, is called ethanoic acid. Note here that the carbon of the carboxyl group must be counted in deciding the parent hydrocarbon. This fact becomes evident if we consider the series

$$\underset{\text{Ethane}}{CH_3CH_3} \rightarrow \underset{\text{Ethanol}}{CH_3CH_2OH} \rightarrow \underset{\text{Ethanal}}{CH_3CHO} \rightarrow \underset{\text{Ethanoic acid}}{CH_3COOH}$$

Table 13–2 shows the acids whose formulas must be remembered for future use. A comparison of oleic and stearic acids should show that oleic acid contains 1 double bond in its structure; i.e., the difference between the two acids is 2 hydrogen atoms, the same difference that we found between C_2H_6 and C_2H_4. In the same manner we could decide that there were 2 double bonds in linoleic acid and 3 double bonds in linolenic acid. Of course, the shortage of 4 hydrogen atoms in linoleic acid could indicate a triple bond,

Organic Chemistry

Table 13-2 FATTY ACIDS AND UNSATURATED ACIDS

Acid	Formula
Formic	HCOOH
Acetic	CH_3COOH
Propionic	C_2H_5COOH
Butyric	C_3H_7COOH
Palmitic	$C_{15}H_{31}COOH$
Stearic	$C_{17}H_{35}COOH$
Oleic	$C_{17}H_{33}COOH$
Linoleic	$C_{17}H_{31}COOH$
Linolenic	$C_{17}H_{29}COOH$

but actually it has 2 double bonds. Also, in linolenic acid the shortage of 6 hydrogen atoms could indicate a triple and a double bond, but it actually has 3 double bonds.

formic acid

The first member of the fatty acid series was named from the Latin word *formica,* which means "ant," because the acid was originally prepared by the distillation of red ants. Formic acid is the irritating substance in the sting of many insects. For instance, the sting of a bee consists of the insertion of a "hypodermic needle" and the injection of a small amount of formic acid. The inflammation is caused by this acid in the tissues.

acetic acid

The acid of vinegar is acetic acid. If apple cider is allowed to stand in contact with the air, the sugar first ferments to form alcohol, which becomes oxidized to acetaldehyde and then to acetic acid. Pure acetic acid freezes to an icelike mass at 16.6°C, which is why it is called *glacial* acetic acid. It is a good solvent for many compounds, and its esters are used in large quantities as solvents.

butyric acid

The acids above propionic acid in the fatty-acid series have disagreeable odors unless they are so nonvolatile as to be practically odorless. Butyric acid, for instance, gives the odor to rancid butter. It is also one of the constituents of perspiration and is partially responsible for its odor.

other organic acids

Benzoic acid, C_6H_5COOH, occurs in gum benzoin and also to a small extent in cranberries.

Benzoin can be used as an inhalant because it is volatile with steam, and the vapors soothe the mucous membranes of the throat. Benzoic acid itself is used as a stimulant, an antiseptic, and a diuretic.

Some organic acids are di- or triprotic; i.e., they have 2 or more replaceable hydrogen atoms. *Oxalic acid,* found in small amounts in rhubarb and cranberries and used as a bleach and stain remover, is a diprotic acid.

Malonic acid, used in the manufacture of barbituric acid and the barbiturates, is another example.

Citric acid, from lemons, limes, etc., is a triprotic acid; i.e. it has 3 replaceable hydrogen atoms.

properties

The hydrogen atom of the carboxyl group can form the hydronium ion. This is the replaceable hydrogen of the acid. The acids can react with active metals or with metal hydroxides to form *salts;* acetic acid for example, reacts slowly with zinc to form hydrogen and zinc acetate.

$$Zn + 2CH_3COOH \rightarrow H_2 \uparrow + (CH_3COO)_2Zn$$

Zinc acetate, a salt, is used as an antiseptic and astringent.

$$NaOH + CH_3COOH \rightarrow \underset{\text{Sodium acetate}}{CH_3COONa} + H_2O$$

Solutions of sodium acetate are used as laxatives and diuretics. Benzoic acid reacts with bases to form salts.

$$C_6H_5COOH + NaOH \rightarrow \underset{\text{Sodium benzoate}}{C_6H_5COONa} + H_2O$$

Sodium benzoate is the common preservative of such foods as sweet cider and catchup. Labels stating a content of 0.1 per cent sodium benzoate are commonly found on containers of such foodstuffs.

The diprotic and triprotic acids can form more than one salt. From oxalic acid we can have

$$\begin{array}{c} COONa \\ | \\ COONa \end{array} \quad \text{and} \quad \begin{array}{c} COONa \\ | \\ COOH \end{array}$$

Oxalic acid forms an insoluble calcium salt.

$$\begin{array}{c} COO \\ | \\ COO \end{array}\!\!\!>\!\!Ca$$

Some oxalate may be formed in the body and is excreted by the kidneys. In the presence of calcium ion (Ca^{++}), insoluble calcium oxalate crystals may form. One kind of kidney stones is formed of this material. Calcium ions are a normal constituent of the blood and are essential to the blood-clotting process. By adding oxalic acid to drawn blood, the calcium ions in the blood are removed as calcium oxalate and the blood-clotting process is prevented. Blood treated in this way is called *oxalated* blood. The magnesium salt of citric acid is the well-known laxative marketed under the name of citrate of magnesia.

Sodium benzoate is used internally in the treatment of rheumatism.

Esters

The characteristic group in esters is

$$-\underset{|}{C}(=O)-O-\underset{|}{C}-$$

This arrangement of atoms is similar to the carboxyl group, with a carbon atom in place of the hydrogen atom in —COOH. Esters are formed from the reaction of acids and alcohols, with the elimination of a molecule of water.

$$\underset{\text{Methyl alcohol}}{H_3C-O-H} + \underset{\text{Acetic acid}}{H_3C-C(=O)-O-H}$$

$$\Downarrow$$

$$\underset{\text{Methyl acetate}}{H_3C-C(=O)-O-CH_3} + H_2O$$

This is a reversible reaction. The forward reaction is *esterification*. The reverse, the reaction of an ester with water, is *hydrolysis*. The reaction slowly reaches a state of equilibrium, but by use of a catalyst the equilibrium point is reached more rapidly.

The fats and oils are esters of the long-chain fatty acids and the trihydroxy alcohol, glycerol. Oleic acid and stearic acid reacting with glycerol yield, respectively, glyceryl oleate (in olive oil) and glyceryl stearate (in beef suet).

Cholesterol, being an alcohol, reacts with fatty acids to form cholesterol esters.

physical properties

The esters of the lower members of the fatty-acid series are volatile liquids that are insoluble in water. Many of them have pleasant odors; e.g., amyl acetate has the odor of pears; isoamyl acetate, that of bananas; methyl butyrate, that of pineapples; octyl acetate, that of oranges; and amyl butyrate, that of apricots. It is rather interesting that an acid such as butyric acid, which has an objectionable odor, can form esters with the pleasant odors described above. In general, the esters are excellent solvents, and enormous quantities are used in the lacquer industry. While modern nail polish and the finish on an automobile are far apart in application, there is little fundamental difference between them; both are lacquers. The fruity odor detected in the use of each indicates a similarity in the solvent used.

chemical properties

Hydrolysis. Esters react with water to form acid and alcohol. This reaction, being the reverse of esterification, does not go to completion but reaches an equilibrium point. In the stomach, in the presence of hydrochloric acid and the enzyme lipase, emulsified fats undergo hydrolysis.

$$C_3H_5(C_3H_7COO)_3 + 3H_2O \xrightarrow{HCl}$$
Glyceryl butyrate
(from cream)

$$C_3H_5(OH)_3 + 3C_3H_7COOH$$
Glycerol Butyric acid

Saponification. The metallic salts of the higher fatty acids are soaps, and the process of making soap is called *saponification.*

Glyceryl oleate + sodium hydroxide →

sodium oleate + Glycerol
Castile soap

medicinal uses

Amyl nitrite, $C_5H_{11}ONO$, is used to relieve the pain of angina pectoris. Its action is similar to that of nitroglycerin, but there is a great difference in the method of its use as well as in the speed with which it brings relief. Amyl nitrite is a volatile liquid dispensed in fragile glass ampules, or "pearls," that may be crushed in a handkerchief to allow the inhalation of the vapor. Relief is obtained in a dramatically short time. Nitroglycerin tablets, on the other hand, are dissolved under the tongue to be absorbed sublingually. The action of this drug requires 2 or 3 min. The similarity in the effect of these two drugs is believed to be due to the fact that they both produce the nitrite ion in the body.

Ethyl nitrite, C_2H_5ONO, is used as a vasodilator to increase sweating and so reduce fever. The preparation used contains about 4 per cent of the drug in 90 per cent alcohol and is called sweet spirit of niter. It is especially convenient to administer to children because the pleasant odor makes it attractive.

Benzyl benzoate, the ester of benzyl alcohol and benzoic acid, is used to kill head lice (pediculi).

Ethers

The characteristic group of the class of compounds called ethers is —C—O—C—. Ethers can be prepared by removing 1 molecule of water from 2 molecules of alcohol. Using 2 molecules of ethyl alcohol and concentrated sulfuric acid as a dehydrating agent,

Diethyl ether

When both carbon-hydrogen groups are the same, the ether is called a *simple* ether. If the groups are different, we have a *mixed* ether.

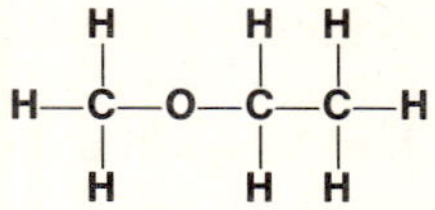

```
    H       H   H
    |       |   |
H — C — O — C — C — H
    |       |   |
    H       H   H
```

This is methylethyl ether.

diethyl ether

Diethyl ether, $(C_2H_5)_2O$, is commonly known in hospitals and in everyday life simply as "ether." It is also called sulfuric ether because of its method of preparation shown above. It is a colorless liquid that boils at 35°C. Both the liquid and its vapor are very inflammable. The danger of fire during anesthesia is very great if an open flame or cautery must be used. The danger is lessened if we remember that ether vapor is more than twice as heavy as air.

Diethyl ether is one of the most satisfactory of the volatile liquid anesthetics, as indicated by its general use over the last hundred years. The chief disadvantages of ether anesthesia are irritation of the respiratory passages, with the accompanying increase in bronchial secretions, and postoperative vomiting.

Ether is an excellent solvent for fats, oils, resins, and gums. For instance, ether is used for the extraction of fats in food analysis. Adhesive tape is easily removed after ether is applied to it, for the ether easily dissolves the adhesive. Ether is not satisfactory for spot cleaning because it evaporates so rapidly that a ring is invariably left.

divinyl ether

Divinyl ether, $(CH_2=CH)_2O$, has been successfully used as an anesthetic since 1930. It presents the same fire hazard as diethyl ether, but the other disadvantages are greatly reduced. It should not be used for operations lasting over an hour, and only a trained anesthetist can safely administer it even for such a period. However, for short operations it is most convenient, for it may be administered by the open-drop method. Divinyl ether is a colorless liquid with a boiling point of 39°C, which gives it about the same volatility as diethyl ether.

COMPOUNDS WITH MORE THAN ONE FUNCTIONAL GROUP

Hydroxy Acids

These compounds are at the same time acids and alcohols.

lactic acid

Lactic acid, $C_3H_6O_3$, α-hydroxypropionic acid, is formed when milk sours. It is found in muscles and in the brain and is an intermediate product in the oxidation of glucose in the muscles. Part of the lactic acid formed is subsequently converted to glycogen. One theory of muscle fatigue is that lactic acid accumulates in the muscle.

The word *alpha* in the name of the acid designates the position of the hydroxy group. In naming the substituted acids, the first carbon atom adjacent to the carboxyl group is alpha. The second carbon atom is beta, the third carbon atom is gamma, and so on through the Greek alphabet. The sodium salt of lactic acid is used as a replacement fluid in intravenous therapy, because it gives an alkaline reaction, which is sometimes desirable, and because the carbon-oxygen-hydrogen part of the molecule can be oxidized in the body to yield energy.

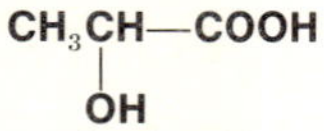

```
CH₃CH — COOH
    |
    OH
   Lactic acid
```

citric acid

Citric acid, $C_6H_8O_7$, contains three carboxyl groups and an alcohol or hydroxy group. It is found in all citrus fruits. Oranges, lemons, limes, and grapefruits are high in citric acid. Why, then, are they listed in nutrition texts as foods that yield an alkaline residue? The answer is, as we might guess, that they form the sodium salt in the body. This salt has an alkaline reaction.

β-hydroxybutyric acid

The aliphatic acid with 4 carbon atoms is butyric acid. An —OH group on the second carbon atom from the carboxyl group makes *β-hydroxybutyric acid.* This acid accumulates in the blood and urine of persons with uncontrolled diabetes. It is one of the substances referred to as ketone bodies (although it is not a ketone), the accumulation of which in the blood causes ketosis.

$$CH_3-CH-CH_2-COOH$$
$$|$$
$$OH$$

β-Hydroxybutyric acid has no trivial name; we must use its chemical name.

salicylic acid

This acid may be considered as benzoic acid with an alcohol group or as phenol with a carboxyl group.

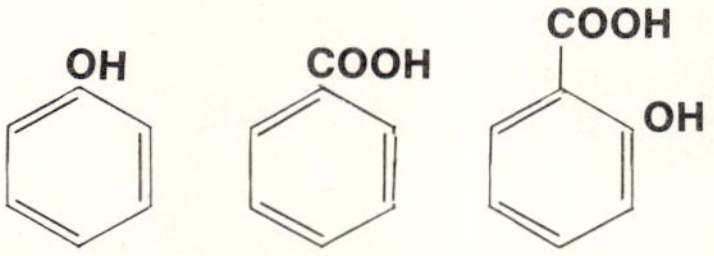

Phenol   Benzoic acid   Salicylic acid

Salicylic acid reacts with sodium hydroxide to form sodium salicylate, a salt that has long been used in the treatment of rheumatism. However, it is absorbed largely in the stomach, sometimes causing nausea and gastric irritation. The action of sodium salicylate seems to be due to the formation by hydrolysis of salicylic acid, which is the actual analgesic. The difficulties encountered with sodium salicylate prompted the search for a less irritating salicylate that could be taken orally. The result is acetylsalicylic acid, or aspirin. Since salicylic acid is an alcohol as well, it can react with an acid to form an ester.

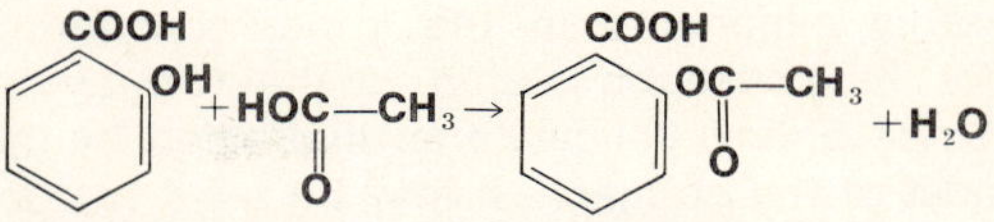

Salicylic acid + Acetic acid → Acetylsalicylic acid + H_2O

Salicylic acid, being an acid, can react with an alcohol to form an ester.

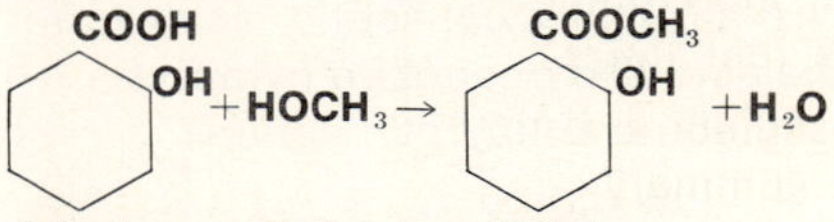

Salicylic acid + Methyl alcohol → Methyl salicylate + H_2O

Methyl salicylate is the artificial *oil of wintergreen.* The natural oil is gaultheria oil. This drug is commonly included in liniments and ointments to be rubbed on painful joints and muscles. It is readily absorbed through the skin and hydrolyzes to form salicylic acid. In this connection, there seems to be a difference in the rate of absorption, and therefore in the efficacy, of the natural and the artificial oils of wintergreen. Since the natural oil is absorbed faster, it is to be preferred over the artificial one. This ester is used for flavoring, to provide wintergreen or checkerberry flavor.

Another important medicinal substance formed from salicylic acid and phenol is phenyl salicylate, also called salol, which has the formula

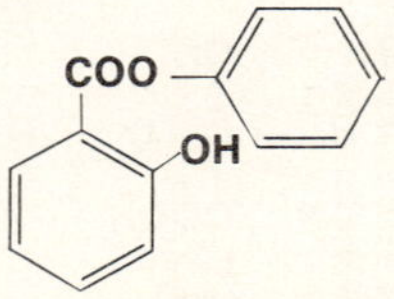

This drug is used as an intestinal and urinary antiseptic. In an alkaline medium it is hydrolyzed to form phenol and salicylic acid.

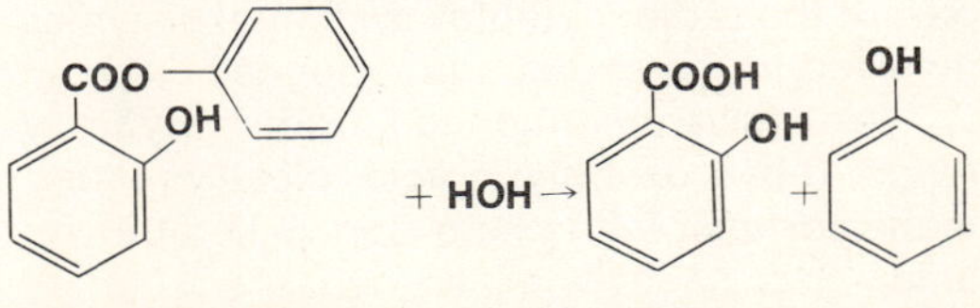

+ HOH →

Phenol is released in the intestines, and is

partly eliminated in the urine, accounting for its antiseptic action in the two areas. Since phenyl salicylate is unaffected by the acid of the stomach, it may be used to coat tablets that must act only in the intestines. Such tablets are said to be *enteric coated.*

In summary, all the salicylates hydrolyze to form salicylic acid, which is analgesic. They are all mildly antiseptic, salol forming both the acid and phenol on hydrolysis. They also have some antipyretic action.

Keto Acids

The typical arrangement of atoms in a ketone is

$$-\overset{|}{\underset{|}{C}}-\overset{O}{\underset{}{\overset{\|}{C}}}-\overset{|}{\underset{|}{C}}-.$$

The simplest acid that is also a ketone is *pyruvic* acid. It is α-ketopropionic acid. Its chemical name tells us that there are three carbon atoms in the acid, with a ketone group on the first carbon atom after the carboxyl group.

$$CH_3-\overset{O}{\underset{}{\overset{\|}{C}}}-COOH$$

Pyruvic acid is formed from lactic acid in the oxidation of glucose. Recall that oxidation can mean the addition of oxygen, the loss of hydrogen, or a numerical increase in the valence.

$$\begin{array}{c} CH_3 \\ | \\ HCOH \\ | \\ COOH \\ \text{Lactic acid} \end{array} \rightarrow \begin{array}{c} CH_3 \\ | \\ C=O \\ | \\ COOH \\ \text{Pyruvic acid} \end{array} + 2H$$

The hydrogen atoms lost by lactic acid in the formation of pyruvic acid are picked up by hydrogen acceptors. These acceptors are part of the rather complex system of enzymes involved in the oxidation of glucose.

We have mentioned, under Hydroxy Acids, β-hydroxybutyric acid; closely related is the corresponding keto acid, β-ketobutyric acid.

$$\begin{array}{c} H \\ | \\ O \\ | \\ CH_3CCH_2COOH \\ | \\ H \end{array} \qquad \begin{array}{c} O \\ \| \\ CH_3CCH_2COOH \end{array}$$

β-Hydroxybutyric acid   β-Ketobutyric acid

A comparison of the formulas shows that the difference between these two compounds is the same as the difference between lactic and pyruvic acid: the loss of 2 hydrogen atoms. The keto acid can be formed from the hydroxy acid, and it also accumulates in the blood and urine of diabetic persons. The keto acid sometimes loses a molecule of carbon dioxide from its carboxyl group and forms acetone. Now we have the three substances that go under the name of the ketone bodies: β-hydroxybutyric acid, β-ketobutyric acid, and acetone. These three substances can cause *ketosis.*

β-Ketobutyric acid has two other names, which are frequently used in the literature. It is called acetoacetic acid and diacetic acid. Both names reflect the relationship in structure between β-ketobutyric acid and 2 molecules of acetic acid.

$$CH_3-\overset{O}{\overset{\|}{C}}-\boxed{OH\ H}\ CH_2-\overset{O}{\overset{\|}{C}}-OH$$
Acetic acid   Acetic acid

$$CH_3-\overset{O}{\overset{\|}{C}}-CH_2-\overset{O}{\overset{\|}{C}}-OH$$
Diacetic acid
Acetoacetic acid
β-Ketobutyric acid

These acids, β-ketobutyric acid and β-hydroxybutyric acid, are formed in the body during the oxidation of fatty acids. A diet high in fats is called a *ketogenic* diet. In the past, a ketogenic diet was sometimes used in an attempt to control epileptic seizures. Most patients found a high-fat diet particularly unpalatable. Fortunately, today most seizures can be controlled with drugs, and the ketogenic diet is used but rarely. In starvation, the body oxidizes reserve fat supplies in the absence of carbohydrate. This is a highly ketogenic diet; one might expect ketosis to develop, and it does. Since these substances are acids, it is easy to see why uncontrolled ketosis may develop into acidosis.

Study Exercises

1. How does organic chemistry differ from inorganic chemistry?
2. What are isomers? Illustrate with examples.
3. How can you account for the difference in the speed of organic reactions compared with that of inorganic reactions?
4. What is meant by an unsaturated hydrocarbon?
5. How may an unsaturated compound be recognized by its structural formula?
6. How does bromine act with a saturated hydrocarbon? Illustrate by an equation using the general formula for the hydrocarbon.
7. What name is given to the type of reaction described in question 6?
8. How does bromine act with an unsaturated hydrocarbon? Illustrate by an equation using ethylene as the hydrocarbon.
9. What name describes the type of action illustrated in question 8?
10. What are aromatic compounds?
11. Write the formula for benzene structurally and symbolically.
12. How many monosubstitution products of benzene are there? How many disubstitution products are there? Give two ways of naming the isomers of dichlorobenzene.
13. What are polynuclear aromatic hydrocarbons? Name three of these and draw their structural formulas.
14. Why may ethyl chloride be used as a local anesthetic?
15. What are two uses of carbon tetrachloride? What properties dictate each use?
16. What is the test for unsaturation?
17. What is meant by hydrogenation? What practical use is made of this process?
18. What is the characteristic group of an alcohol? What is a monohydric alcohol? What is a dihydric alcohol? What is a trihydric alcohol? Give an example of each.
19. What is the physiologic effect of methyl alcohol?
20. What is fermentation? Write the equation for the fermentation of glucose.
21. What is denatured alcohol? Why is it denatured?
22. For what purposes is ethyl alcohol used in the hospital?
23. For what purposes is isopropyl alcohol used in the hospital?
24. What is the antidote for phenol poisoning?
25. What are sterols? Name three.
26. What are three uses of glycerol? What properties of glycerol are responsible for these uses?
27. What is the characteristic group of an aldehyde?
28. What is meant by polymerization?
29. What use is made of paraldehyde?
30. What is the action of an aldehyde on Fehling's or Benedict's solution?
31. Write the equation for the slow oxidation of an aldehyde.
32. What is the characteristic group of a ketone?
33. How does the oxidation of a ketone differ from that of an aldehyde?
34. What is the characteristic group of an acid?
35. Which of the following acids is unsaturated? How did you decide? $C_{17}H_{33}COOH$; $C_{15}H_{31}COOH$; C_3H_7COOH.
36. What is the difference in formula between oleic acid and stearic acid? How may oleic acid be converted to stearic acid?
37. What are three uses of benzoic acid?

38. What is meant by a diprotic acid? Name two.
39. Name three organic salts used in medicine and indicate a way in which each is used.
40. What is the characteristic group of an ester?
41. What is saponification? What is a soap?
42. Give the medicinal uses of two esters.
43. What is the characteristic group of an ether?
44. Why is diethyl ether called sulfuric acid ether?
45. What use is made of methyl salicylate?
46. What is meant by "enteric-coated" tablets?
47. What is ketosis? Name the three compounds called ketone bodies.

chapter fourteen   IMPORTANT ORGANIC COMPOUNDS

Ideas for preview or review

   Anions are negatively charged particles, and cations are positively charged particles.
   Acids are proton donors, and bases are proton acceptors.
   Salts are compounds with a cation other than hydrogen and an anion other than hydroxyl.
   Many organic compounds containing nitrogen can form salts.
   pH is a way of expressing the concentration of hydronium ion. At ph 7.0 the concentrations of hydronium ions and hydroxyl ions are equal.
   Dialysis is the passage of dissolved material through a membrane.
   Enzymes, organic catalysts, alter the rates of chemical reactions.

Other elements can combine with carbon in much the same way as oxygen. The most important of these are sulfur, phosphorous, and nitrogen. Some of the metals, such as iron, zinc, cobalt, magnesium, copper, and manganese, are essential constituents of important organic compounds. Many of the metals, trace amounts of which are vital for life, are involved in enzyme systems. Cobalt, for example, is a constituent of vitamin B_{12}.

NITROGEN COMPOUNDS

Amines

The amines can be considered as relatives of ammonia. If one of the hydrogen atoms in ammonia is replaced by a carbon-hydrogen group, we have a primary amine.

Typical of the primary amines is the group —NH_2. This is the amino group. We find this group in another class of compounds, called *amino acids*.

If 2 hydrogen atoms in ammonia are replaced by carbon-hydrogen groups, a secondary amine is formed.

Observe that the group characteristic of a secondary amine is —NH. When all 3 hydrogen atoms are replaced, we have a tertiary amine.

The methylamines have obnoxious odors, reminiscent of decaying fish. They are formed as protein decomposes and also as a result of intestinal putrefaction. Ptomaines are amines formed from the bacterial decomposition of protein foods. The names of two of them, putrescine and cadaverine, may suggest their odors to you. It is highly improbable that any cases of "ptomaine poisoning" actually occur. The extremely foul odor of the ptomaines would deter anyone from eating food in this state of decay. What is commonly called ptomaine poisoning is doubtless due to the presence in food of pathogenic microorganisms.

The amines have been introduced as related to ammonia to emphasize the similarity between these compounds and ammonia. Like ammonia they are basic. This indicates that the amino group is a basic group. An amine can react with acids in much the same fashion as ammonia.

$$NH_3 + HCl \rightarrow NH_4Cl$$
Ammonium chloride

$$CH_3NH_2 + HCl \rightarrow CH_3NH_3Cl$$
Methyl- amine   Methyl- ammonium chloride

The secondary and tertiary amines can react with acids also.

$$(CH_3)_2NH + HCl \rightarrow (CH_3)_2NH_2Cl$$
Dimethyl- amine   Dimethylammo- nium chloride

$$(CH_3)_3N + HCl \rightarrow (CH_3)_3NHCl$$
Trimethyl- amine   Trimethyl- ammonium chloride

A compound similar to the salt ammonium chloride, NH_4Cl, is $(CH_3)_4NCl$. This is a highly ionized compound called tetramethylammonium chloride. It belongs to the class called *quaternary* ammonium salts. Both the prefix *tetra* and the word *quaternary* remind us that there are 4 carbon groups in this type of compound.

hexamethylenetetramine

If an aqueous solution of formaldehyde and ammonia is evaporated, hexamethylenetetramine will be formed as a white solid.

$$6HCHO + 4NH_3 \rightarrow (CH_2)_6N_4 + 6H_2O$$

It is also known as *urotropine* and as *methenamine*. This compound is used medicinally as a urinary antiseptic. The drug is absorbed intact from the intestinal tract and is rapidly excreted in the urine. However, if the urine is acidic, formaldehyde, which is the actual antiseptic, is formed in the urine, according to the reverse of the reaction of formation above. To make sure that the urine is acidic, sodium dihydrogen phosphate is administered between doses of the hexamethylenetetramine. The thinking student should now be wondering why the formaldehyde is not set free in the stomach. About 20 per cent of the drug *is* converted to formaldehyde in the stomach. This action is undesirable, for the formaldehyde is irritating to the stomach mucosa and, of course, reduces the amount of drug available in the urinary tract. The hexamethylenetetramine is therefore best administered at the time of the lowest stomach acidity, i.e., between meals. The real answer to this problem is enteric-coated tablets.

At the same time that the formaldehyde is set free, the ammonia reacts with the acid present to form a salt. This action should suggest a second use for hexamethylenetetramine; it will serve as an agent to reduce hyperacidity of the urine. In such cases, of course, the acid phosphate treatment is omitted.

aniline

Aniline is a primary amine related to benzene. It is important industrially as the starting point of the so-called aniline dyes. Aniline is a basic, poisonous, oily liquid. The attachment of other groups to the molecule reduces its toxicity. Compounds formed from aniline are widely used in medicine.

Important Organic Compounds

amines of physiologic importance

Norepinephrine, formerly called noradrenaline, is a primary amine secreted by the adrenal medulla. This hormone causes the peripheral blood vessels to constrict, resulting in an increase in blood pressure.

Epinephrine, also called adrenaline, is also secreted by the adrenal medulla and is a secondary amine. It is involved in the regulation of blood pressure and in carbohydrate metabolism.

Choline and its ester acetylcholine as well as norepinephrine are involved in the transmission of nerve impulses. Because of this fact, nerves are classified as cholinergic and adrenergic. This information is important in understanding the effects of drugs on the nervous system.

Thiamine, vitamin B_1, as the name implies, is an amine. The word vitamin itself means amines of life. We know today that all the vitamins are not amines. The unabbreviated name for vitamin B_1 would be thioamine. The prefix *thio* in a word means sulfur. Thus, this antiberiberi factor is a sulfur amine.

Cyanocobalamine, vitamin B_{12}, the antipernicious anemia factor, is a complex compound containing the metal cobalt and an amine.

Amides

The group characteristic of amides is
$$-\overset{\overset{\displaystyle O}{\|}}{C}-NH_2.$$
If we compare this structure with the carboxyl group of an acid, $-\overset{\overset{\displaystyle O}{\|}}{C}-OH$, we see that an amino, $-NH_2$, group replaces the $-OH$ group of the acid.

acetamide

Acetamide is a typical amide and has the formula

$$H-\underset{\underset{\displaystyle H}{|}}{\overset{\overset{\displaystyle H}{|}}{C}}-\overset{\overset{\displaystyle O}{\|}}{C}-\underset{}{N}\overset{\overset{\displaystyle H}{|}}{-}H$$

Comparing this formula with that of acetic acid, we can readily see the similarity in structure and the reason for the name.

urea

Urea may be considered as a double amide of carbonic acid.

$$H-O-\overset{\overset{\displaystyle O}{\|}}{C}-O-H \qquad H-\overset{\overset{\displaystyle H}{|}}{N}-\overset{\overset{\displaystyle O}{\|}}{C}-\overset{\overset{\displaystyle H}{|}}{N}-H$$

Carbonic acid   Urea

It is the principle end product of protein catabolism. Amino acids not used for growth or repair are deaminized in the liver and the carbon-hydrogen-oxygen part of the molecule is oxidized in the body to yield energy. Ammonia formed from the amino group reacts with carbon dioxide to form urea which is excreted. Urea, on standing, hydrolyzes to form carbon dioxide and ammonia.

sulfanilamide

Sulfanilamide, which can be prepared from acetamide, is not a true amide. It contains the $-SO_2NH_2$ group in place of the $-CONH_2$ group of the amide. This compound, the first of the sulfonamide drugs, was introduced into medicine in 1937 for the treatment of a wide variety of infections. Its undesirable side effects led to the development of modified compounds, which have proved to be of great value in the control of bacterial growth.

Amino Acids

Amino acids are acids with an amino ($-NH_2$) group or, to think of them the other way around, amines with an acid ($-COOH$) group. We meet them in the study of nutrition, physiology, diet therapy, and proteins and enzymes. Many of the amino acids can be obtained by the hydrolysis of proteins.

The most important are alpha amino acids. The amino group is attached to the first carbon atom in the chain after the carboxyl group, the alpha carbon.

glycine

Glycine is α-aminoacetic acid. We can call it simply aminoacetic acid. This compound is not essential for human nutrition. The body can manufacture it if necessary. The next member of the series the body cannot make. It must be included in the diet and is, therefore, considered as one of the *essential* amino acids. This is a derivative of the 3-carbon propionic acid and has the trivial name alanine. Its chemical name, α-aminopropionic acid, tells us that there are 3 carbon atoms in the compound with an amino group on the alpha-carbon atom.

$$NH_2CH_2COOH \qquad CH_3CH\text{—}COOH$$
$$\phantom{NH_2CH_2COOH \qquad CH_3CH\text{—}}\,|$$
$$\phantom{NH_2CH_2COOH \qquad CH_3CH\text{—}}NH_2$$

Glycine Alanine

Since they contain an acid piece and an alkaline piece, they can form salts with each other by splitting out a molecule of water and can build up into very complex molecules. Remember that protoplasm is protein in nature.

$$NH_2CHCOOH + NH_2CH\text{—}COOH \rightarrow$$
$$\phantom{NH_2CHCOOH + NH_2CH\text{—}}|$$
$$CH_3$$

Glycine Alanine

$$NH_2CH_2CONH\text{—}CH\text{—}COOH + H_2O$$
$$\phantom{NH_2CH_2CONH\text{—}}|$$
$$\phantom{NH_2CH_2CONH\text{—}}CH_3$$

Glycylalanine

p-Aminobenzoic Acid

p-Aminobenzoic acid is a very interesting compound, so well known to biochemists that it is called by its intitials: PABA. Part of the molecule of folic acid, one of the B vitamins, has an arrangement of atoms similar to the arrangement in *p*-aminobenzoic acid. We have mentioned above that alanine is essential for human nutrition but glycine is not. Some organisms can manufacture substances needed for life, whereas others must depend upon outside sources for their supply. Certain types of bacteria can manufacture their own folic acid. To do so they use PABA. Let us compare the formulas of sulfanilamide and PABA:

p-Amino-benzoic acid Sulfanilamide

When sulfanilamide is administered, these bacteria pick it up as they would PABA to use in the synthesis of folic acid. What happens? The vital folic acid cannot be made, and the bacteria cease to grow. Since the patient does not make his own folic acid, the sulfanilamide does not harm him. This is the idea in *chemotherapy:* to find chemicals that will inhibit the growth of microorganisms without causing injury to the patient.

Three of the simple amino acids contain sulfur atoms. They are cysteine, pronounced sist'-e-in, cystine, pronounced sist'-in, and methionine. Two molecules of cysteine can react to form cystine. Cystine is occasionally found in the urine of persons with jaundice or liver involvement. This condition is called cystinuria.

The amino acids are soluble in water and can dialyze through membranes. They can ionize either at the amino group, forming positively charged cations, or at the carboxyl group, forming negatively charged anions. The kind of ion formed depends upon the hydrogen-ion concentration of the solution. At a definite pH, which differs for each acid, the molecule can ionize at both ends at the same time. This pH is called the *isoelectric* point of the acid, and the doubly charged ion is called a *zwitterion*. At the normal pH range of blood (7.35 to 7.45) the amino acids exist as negatively charged anions.

The amino acids have very unpleasant odors. It is told that, when prisoners were liberated from concentration camps at the end of the Second World War, some were in such advanced stages of starvation that they could not digest the food that was then available to them. What was to be done? The protein was hydrolyzed outside the body and fed as amino acids. This was such an unpleasant concoction that many persons refused it. The problem was solved by administering the acids by the intravenous route. Today in hospitals intravenous feedings containing amino acids are called protein hydrolysate.

Nitrogen Heterocycles

In all the cyclic and aromatic compounds thus far considered, all the atoms in the ring have been carbon atoms. If a ring contains two or more different kinds of elements, it is described as *heterocyclic*. Many compounds important in body processes are composed of heterocyclic rings. A few are included here to illustrate some of the important atom groupings found in vitamins, enzymes, hormones, and drugs.

pyrrole

A five-membered ring with 1 nitrogen atom as the heteroatom is called *pyrrole*.

The pyrrole arrangement of atoms is found in the two extremely important compounds: hemoglobin and chlorophyll. In both these substances there are 4 pyrrole groups linked together by carbon bridges, forming an inner ring of 16 atoms. In the center of the heme compound, there is an iron atom; in chlorophyll, the center atom is magnesium. It is interesting to speculate that nature can use the same molecular skeleton for such important compounds in man and in the green plants. A group of enzymes called *cytochromes* control the rate of oxidation in the cells. These enzymes have the same pyrrole-and-iron structure as hemoglobin but are not made of the protein globin. Instead, the cytochromes are composed of a heme piece and different proteins.

imidazole

A five-membered ring could contain two nitrogen atoms. Imidazole is an example.

This structure is found in the amino acid histidine and in a compound that can be formed from histidine, *histamine*. Excess histamine in the body causes reactions like an allergy. Histamine causes the capillaries to expand, producing a drop in blood pressure. It may also contract the bronchioles, producing difficult breathing as in asthma. Increased rate of secretion of gastric juice is another possible result. To minimize the effects of histamine, *antihistamines* are used. Many of these are tertiary amines.

pyridine

The six-membered ring with nitrogen as the heteroatom is pyridine.

The pyridine structure is found in nicotinic acid and in the amide of nicotinic acid, nicotinamide. Both of these substances are vitamins belonging to the B group. The body must have these compounds in order to manufacture certain coenzymes, substances that must be present for enzymes to function properly. One of the sulfonamide drugs developed from sulfanilamide is sulfapyridine.

Two nitrogen atoms in a six-membered ring permit three possible arrangements: ortho, pyridazine; meta, pyrimidine; para, pyrazine. Since these rings contain 2 nitrogen atoms, they are known as *diazines*.

purines

The nuclei of cells contain complex organic acids called *nucleic acids*. On hydrolysis, these acids form phosphoric acid and derivatives of pyrimidine. During metabolism, the nucleic acids form *uric acid*. This is the material which, when deposited in the joints, causes the painful symptoms associated with gout. Uric acid and its sodium salt also cause kidney stones. Uric acid belongs to a class of compounds called *purines*. These include purine itself, uric acid, caffeine, xanthine, and several others. The purines are composed of two rings fused together, a pyrimidine and an imidazole ring.

As mentioned above, the purines are formed from nucleic acids. Foods rich in nucleic acids are meat, fish, peas, and lentils. Patients are sometimes put on a low-purine or a purine-free diet. Such a diet would exclude not only most of the protein-rich foods but also coffee, tea, and cocoa.

Of the so-called purine bases found in ribonucleic acid and deoxyribonucleic acid, only two are purines. The others are pyrimidines. In DNA, the purines adenine and guanine link up with the pyrimidines thymine and cytosine, respectively. In RNA, the purines adenine and guanine link with the pyrimidines uracil and cytosine, so that adenine links with uracil and guanine with cytosine. These bonds between the purines and pyrimidines in the nucleic acids are hydrogen bonds.

barbiturates

These compounds, used as sedatives and hypnotics, are based on barbituric acid.

Important Organic Compounds

This compound, like uric acid, contains no carboxyl group; yet it is capable of forming salts. This, we recall, was also the case with phenol, so-called carbolic acid. Barbituric acid and its derivatives are all synthetic drugs. A German chemist, Adolf von Baeyer, made barbituric acid from urea and a dibasic acid found in apples, malonic acid.

[Chemical reaction diagram: Urea + Malonic acid → Barbituric acid]

By substituting different groups for the 2 hydrogen atoms attached to the carbon atom in the ring, drugs can be made with properties varying from sedative to anesthetic effects, some acting for short periods and others of long duration. With 2 ethyl groups in place of the 2 hydrogen atoms, we have diethyl-barbituric acid (Veronal). Other modifications give us phenobarbital (Luminal), a relatively long-acting drug used as a hypnotic or sedative and to relieve or prevent convulsions; pentobarbital (Nembutal), a short-acting sedative; secobarbital (Seconal), a short-acting hypnotic; thiopental (Pentothal), an anesthetic and hypnotic; and amobarbital (Amytal), a hypnotic and sedative of shorter duration than phenobarbital.

What can we conclude from this recital? By changing a few atoms in the molecule or by changing their arrangement, we can change the potency of the drug, its speed of acting, or the length of time its effects can be felt. These drugs in small doses produce a mild sedation, a relaxation of tensions, and a general reduction of anxiety. They are used in the analysis and treatment of some kinds of mental disease. Some of them have been used by the police in questioning suspects in the hope that the relaxed persons would be prone to tell the truth. They have, for that reason, been called "truth serum." However, it is false to assume that a person will tell the truth under the influence of these drugs. In larger doses, these barbiturates are used as "sleeping pills." Still larger doses give the anesthetic effect. For deep anesthesia or for long operations, the barbiturates are used as preanesthetics. These compounds are frequently administered in the form of their sodium salts.

In recent years, the "tranquilizers" have been replacing the barbiturates where a relaxing effect is desired.

alkaloids

Through the ages, man had discovered that the ingestion of parts of certain plants had astounding effects on men and animals. From the dreamy state of indolence achieved by the "lotus eaters" to the curious behavior of horses that had eaten the loco weed, from the paralyzing and deadly effects of the curare-tipped arrows of the South American Indians to the fantasies produced by the pipe of opium, some plant products can produce a wide variety of reactions. Chemists have attempted to extract the active principles causing these reactions from the remaining plant substances. In this project, they have been largely successful. One group of plant extracts is composed of carbon, hydrogen, and nitrogen or carbon, hydrogen, nitrogen, and oxygen. Their structure is heterocyclic, and they react with acids to form salts. These substances are much more soluble as salts than in their native state. These substances are called *alkaloids* and have been defined as *basic nitrogenous substances of plant origin, small amounts of which cause marked physiologic reactions*. Since the chemist has been able to duplicate some of these compounds and has also been able to alter the structure of the molecules of others to yield an improved compound, perhaps the words "of plant origin" should be omitted from the definition.

Reserpine, the first tranquilizer, contains

several heterocyclic rings and is extracted from the roots of *Rauwolfia* plants, commonly called the snakeroot plant. Reserpine can induce quiet without sleep. Chemists have been able to modify the reserpine structure to form two new drugs, a new sedative and a new hypotensive drug.

Strychnine is obtained from the seeds of the nux vomica tree that grows in the East Indies. It causes muscles to contract by stimulating the nervous system. In small amounts it is used as a tonic and as a stimulant in collapse. Strychnine is also used in cases of overdose of sedative. Since strychnine stimulates muscles, an overdose would cause convulsions, rigidity, and death.

Curare, the dried extract of *Strychnos* vine, is known to the readers of mystery thrillers as the deadly South American arrow poison. It yields the alkaloid *curarine,* which relaxes muscles. It has an opposite effect from the strychnine alkaloid, since it paralyzes motor nerve endings, causing muscle relaxation. It is used as an antispasmodic and as an antidote for strychnine poisoning.

The belladonna plant (deadly nightshade) contains several alkaloids. *Atropine,* the best known, is used in the form of the sulfate externally, as a mild painkiller (anocyne); in the eyes, to dilate the pupils (mydriatic); and internally, as a stimulant for the respiratory and circulatory systems. The name *belladonna* means "fair lady." European ladies used to instill a few drops of belladonna extract into their eyes as part of their cosmetic preparation for a big social affair. As a result, the dilated pupils made their eyes appear large and lustrous. Anyone who has ever had an eye examination in which atropine was used will realize that for several hours after the drops have been used the vision is quite blurred. One suspects that the eyes of the fair ladies, although perhaps beautiful, were blank and partly blind.

Morphine, *codeine*, and *papaverine* are opium alkaloids, obtained from the opium poppy. They find a wide use in medicine. Morphine is named for Morpheus, the god of sleep. It is a powerful sedative and pain reliever. Codeine has effects similar to those of morphine, but they are not so strong and codeine is less likely to cause addiction. Unlike morphine and codeine, papaverine is not habit forming. It relaxes involuntary muscles and is used as an antispasmodic and to relax or dilate blood vessels. It, therefore, is used in the treatment of peripheral vascular disease and to relieve pain associated with impaired circulation. An alcoholic, camphorated solution of opium extract is called *paregoric*. This mixture eases pain, slows down peristalsis, and helps mitigate some of the reactions of the gastrointestinal tract to X rays.

The leaves of coca trees yield *cocaine,* a local anesthetic, which has been largely replaced by other more improved anesthetics. *Physostigmine* is obtained from Calabar beans from an African plant. This alkaloid causes the pupil of the eye to contract, the opposite effect to that of atropine. In a very serious eye disease, glaucoma, the fluid pressure in the eyeball increases, producing pain, degeneration of the optic nerve, and sometimes blindness. Physostigmine brings relief by reducing the pressure. Another alkaloid used in this disease is *pilocarpine,* which is made from the leaves of a shrub that grows in Brazil.

Caffeine, a mild stimulant, vasodilator, and diuretic, is the alkaloid in coffee and tea.

Theobromine, from the cacao tree, from South America, is the alkaloid in chocolate and cocoa. Its physiologic effects are the same as those of caffeine. These effects are very mild in comparison with those of some other alkaloids.

Nicotine, the alkaloid in the tobacco plant, like strychnine, is a spastic poison. In vascular disease, it causes spasm of the peripheral arteries. The most important aspect in the treatment of Buerger's disease is to make certain that the patient gives up smoking for good.

The list of naturally occurring alkaloids is an extensive one. Many are historically as well as chemically interesting: *coniine,* from poison hemlock—the cup that killed Socrates; *quinine,* the antimalarial drug, which helps control a scourge that has plagued man from ancient times even to the present day; *ergotoxine* from a fungus on the rye plant, the ingestion of which caused cattle to abort, which is used today in obstetrics because it stimulates the uterus and facilitates its contraction.

How has the chemist helped the science of medicine in its search for safe yet potent drugs? For the local anesthetic cocaine, the chemist has prepared the synthetic substitute procaine (Novocain). This compound retains the anesthetic effect of cocaine but has removed some of the undesirable side effects such as nausea. Meperidine (Demerol) is a synthetic drug, resembling morphine in its effects but not as powerful and also not as dangerous. Methadone is another synthetic narcotic, which depresses the perception of pain. It is also used as an antispasmodic and for its relaxing effects. To supplement the supplies of the very bitter quinine, quinacrine (Atabrine) was developed. Quinine was obtained from the jungles of South America and from plantations in southern Asia. During the Second World War, the American soldiers in the eastern Theater of the war were given Atabrine tablets once and sometimes twice a day. These synthetic drugs were a lifesaving potion in those parts of the world teeming with malaria-infested mosquitoes. Since that time, more powerful antimalarial drugs have been prepared: chloroquinone, oxychloroquinone, and pentaquinone. These are the most powerful antimalarial drugs ever discovered.

We have mentioned some of the alkaloids. There are many more. These few should give one an idea of their importance in medicine. They are all salt formers and are administered in the form of the salt because of its greater solubility. Nurses work with morphine sulfate and strychnine nitrate, as well as many others. Substances that precipate alkaloids are called *alkaloidal reagents.* These same compounds also precipate proteins. Picric acid forms loose molecular combinations with the alkaloids and proteins. Tannic acid from tea and phosphotungstic acid can also tie up the poisonous alkaloids and reduce their physiologic effects. Potassium permanganate, $KMnO_4$, 1 part to 10,000, is used as an antidote for alkaloid poisoning, particularly with strychnine, quinine, and morphine.

A FEW SULFUR COMPOUNDS

Sulfur, with atomic number 16, is directly below oxygen in the periodic table. Where sulfur has a valence of 2, it can form organic compounds quite similar to oxygen.

The —COH arrangement is characteristic of alcohols. Compounds containing —CSH groups are sulfur alcohols and are called *thiols* or *mercaptans.* The name "thiol" is made up of *thio* for sulfur and *ol* for alcohol. "Thiool" is too cumbersome, so the term is shortened to "thiol." The —SH group is also called *sulfhydryl* for *sulf*ur-*hydro*gen. This is an extremely important group of atoms. —SH groups are found in certain enzyme systems, of which they are vital parts. Very minute amounts of such elements as copper, lead, mercury, and arsenic react with the sulfhydryl groups, poison the enzymes, and inactivate the system. It has been discovered that injection of compounds containing sulfhydryl groups into laboratory animals either immediately before or immediately after exposure to ionizing radiations, such as X rays or gamma rays, reduces the susceptibility of the animals to radiation sickness and permits them to survive much larger doses of radiation than would be otherwise possible. We are, of course, interested in this because it is a possible way of protecting human beings from radiation.

The simplest mercaptan is methylmercaptan CH_3SH. The next is ethyl, CH_3CH_2SH. All of them are highly odoriferous;

in fact, their odors are extremely disagreeable.

The arrangement of atoms typical of the class of compounds called ethers is —C—O—C—. If the oxygen atom is replaced by sulfur, we have a *thioether*, —C—S—C—. They are also called *alkyl sulfides*.

Mustard Gas. The dichloride derivative of diethylsulfide, called *mustard gas*, was used in the First World War as a poison gas. Its molecular skeleton is

$$Cl-C-C-S-C-C-Cl$$

Mustard gas is a vesicant; that is, it causes blisters on contact with the skin. Exposure of the skin to the fumes of this substance causes severe and painful burns, and sometimes the results are fatal.

Nitrogen Mustards. Nitrogen mustards comprise a group of compounds with a skeleton like mustard gas, except that a nitrogen atom replaces the sulfur atom in mustard gas.

$$-C-C-N-C-C-$$

These substances are being studied extensively for their effects on the growth of cancer cells.

COMPOUNDS WITH COLOR

Many dyes, stains, and indicators are organic compounds that depend upon the arrangements of the atoms in the molecules for their color. This group includes *fluorescein*, which is an orange powder. When it is dissolved in an alkaline solution, it has a greenish-red fluorescence. It is used to detect the presence of foreign particles in the eye and lesions in the cornea. Bromine acts on fluorescein to form a red or rosy-colored compound called *eosin*. Red ink usually contains eosin. It is a beautiful dye for silk and is used in biology to stain tissue slides. A related compound containing mercury is *mercurochrome*. The bright red water solution of mercurochrome is used as an antiseptic.

Akin to eosin, fluorescein, and mercurochrome are two compounds that nurses use often: phenolphthalein and phenolsulfonphthalein. Phenolphthalein is colorless, but in alkaline solutions it forms a salt that has a red color. Because of this property, it is used as an indicator. Strangely enough, it is also a laxative.

Phenolsulfonphthalein (P.S.P.) is an indicator that exists in two color forms, yellow and red, depending upon pH. As an indicator, it is called *phenol red*. The P.S.P. test is used to measure the ability of the kidneys to excrete. The cells in the glomeruli of the kidneys have several functions, one of which is to excrete into the urine certain substances present in the plasma. The dye, in its red form, is injected intravenously. At specified time intervals urine samples are collected, and the amount of the dye excreted in each sample is measured. In some disorders of the liver the excretion ability of the tubules is increased, whereas in kidney dysfunction, such as nephritis, it is lowered.

Study Exercises

1. Define an amine, showing its relationship to ammonia.
2. What uses are made of hexamethylenetetramine? Explain the action in each case.
3. What is the characteristic group of amides? Name two amides used in medicine and give the use of each.
4. What are heterocyclic compounds?
5. In what respect are chlorophyll and hemoglobin similar?

Important Organic Compounds

6. What are some of the physiologic effects of histamine?
7. Using the barbiturates as examples, explain how the chemist can modify the properties of compounds.
8. What are alkaloids? Name five alkaloids used in medicine and give a use for each.
9. What are the characteristic groups in amino acids?
10. Explain why sulfanilamide is toxic to certain bacteria?
11. What is protein hydrolysate?
12. What is the sulfhydryl group? Why are compounds containing these groups of such interest?
13. Why are nitrogen mustards important?
14. Give two uses of phenolsulfonphthalein.

chapter fifteen CARBOHYDRATES

Ideas for preview or review

  Hydrolysis is a chemical reaction in which water is one of the reactants.

  Enzyme systems consist of organic catalysts produced by cells. Most enzyme systems are specific. They react only with particular compounds or in particular chemical reactions.

  Energy is the power to do work.

  Potential energy, stored in a compound, may be liberated during a chemical reaction.

  A reducing agent causes another substance to undergo reduction, that is, to lose oxygen, to gain hydrogen, to gain electrons, or to decrease in valence.

  Reducing agents are in turn oxidized.

  The aldehyde group, CHO, is easily oxidized to the carboxyl group, COOH, which is characteristic of organic acids.

  The ketone group can be oxidized to form two molecules with shorter carbon chains than were present in the original ketone.

The name "carbohydrate" suggests that these compounds are hydrates of carbon, but carbon does not form hydrates. The name resulted from the fact that the early compounds studied had a composition in which the ratio of hydrogen to oxygen is the same as that in water. This idea is seen in such formulas as $C_6H_{12}O_6$, which might be written as $C_6(H_2O)_6$, and $C_{12}H_{22}O_{11}$, which might be written as $C_{12}(H_2O)_{11}$. However, today we have genuine carbohydrates in which this ratio does not exist, for example, rhamnose, $C_6H_{12}O_5$. On the other hand, we have many compounds in which the molecular formula shows this ratio but which we would never consider as carbohydrates, e.g., acetic acid, $C_2H_4O_2$, or formaldehyde, CH_2O.

COMPOSITION

The simple carbohydrates resemble the alcohols with several hydroxyl groups, except that an aldehyde or a ketone group occupies the position of one of the hydroxyl groups. This resemblance is shown in the following formulas.

```
   CH₂OH           HC=O            CH₂OH
   HCOH            HCOH            C=O
   HOCH            HOCH            HOCH
   HCOH            HCOH            HCOH
   HCOH            HCOH            HCOH
   CH₂OH           CH₂OH           CH₂OH
   Sorbitol        Glucose         Fructose
   an alcohol      an aldose       a ketose
   C₆H₁₄O₆         C₆H₁₂O₆         C₆H₁₂O₆
```

It must be understood that these formulas for glucose and fructose really amount to the general formulas of the simple sugars. Chemical names ending in *-ose* indicate sugars. The sugars containing 6 carbon atoms are called hexoses; 5 carbon atoms, pentoses; 4 carbon atoms, tetroses; 3 carbon atoms, trioses. Sugars that have an aldehyde group (—C(H)=O) are called aldoses; those containing a ketone group (—C—C(=O)—C—) are called ketoses. The two names can be combined to show the number of carbon atoms and the type of group present. Thus

175

aldohexose indicates a sugar containing 6 carbon atoms and an aldehyde group such as glucose; a ketohexose such as fructose is a 6-carbon sugar with a ketone group.

ISOMERISM

Note that glucose and fructose have the same molecular formulas. They are therefore isomers. Many biologically active compounds possess a special kind of isomerism called *optical isomerism*. To have this property, a compound must have at least one carbon atom to which four different groups or atoms are attached. This makes possible two different arrangements of the atoms in space, so that they are mirror images of each other and cannot be superimposed one upon the other. The situation is comparable to right- and left-hand gloves. With the palm sides facing each other, they appear to match, but because of the opposite arrangements of the thumbs and the pinky fingers, gloves, unlike mittens, cannot be worn on either hand. Consider the compound glyceraldehyde:

```
   HC=O            HC=O
    |               |
   HCOH            HOCH
    |               |
   CH₂OH           CH₂OH
 D(+)-glyceraldehyde   L(−)-glyceraldehyde
```

The middle carbon in each compound has four different groups attached, namely, aldehyde, alcohol, hydrogen, and OH. Such a carbon is an asymmetric carbon atom.

Ordinary light waves vibrate in all planes. When a beam of light is passed through a crystal or a prism that can block out all the waves except those traveling in one plane, the light is plane-polarized. When such a beam of light is passed through a solution of each of the two forms of glyceraldehyde, one solution rotates the light to the right and the other rotates it to the left. The one rotating the light to the right is D-glyceraldehyde and the other is L-glyceraldehyde. Other compounds with structures related to these two forms are D or L compounds. Although the spatial arrangements may be similar, such compounds may or may not rotate the light to right or left. If the light is indeed rotated to the right, the compound is dextrorotatory and this is symbolized by a plus sign (+) and if the light is rotated to the left, the material is levorotatory and designated with a minus sign (−). Since D-glyceraldehyde rotates the light to the right, it is D(+)-glyceraldehyde. The other form is L(−)-glyceraldehyde.

The formulas for both glucose and fructose written above indicate that they are both D forms. Note the positions of the OH groups in each compound. However, D-glucose rotates polarized light to the right and is sometimes called dextrose, while D-fructose rotates polarized to the left, from whence it gets its common name levulose. The correct names for these two sugars are D(+)-glucose and D(−)-fructose.

As we learn more and more about the shapes of things, we realize that the wrong shape in a molecule could hinder or nullify its effect in a biochemical reaction. For example, the hormone epinephrine or adrenaline comes in two forms, the D form which is levorotatory and the L form which is dextrorotatory. Of these two forms of the hormone, the D form is 15 times more powerful than the L form. Again, certain microorganisms can thrive on one form of an optical isomer but cannot utilize the other.

The greater the number of asymmetric carbon atoms in the molecule, the greater is the number of possible isomers. We see that this type of spatial arrangement of atoms in molecules can be important in biological chemical changes and in drugs.

CLASSIFICATION

The carbohydrates are divided into three classes depending upon whether or not they react chemically with water and, if they do, on the number of products formed. This re-

Carbohydrates

action with water is called *hydrolysis*. The simple sugars, the monosaccharides, do not undergo hydrolysis. They dissolve in water but do not react chemically with it.

Monosaccharides

Trioses: $C_3H_6O_3$, e.g., glycealdehyde, important in oxidation in the muscles.

Pentoses: $C_5H_{10}O_5$, e.g., ribose, a constituent of nucleic acid. In rare metabolic disease, pentoses may appear in the urine and need to be distinguished from blood sugar.

Hexoses: $C_6H_{12}O_6$, e.g., glucose, or dextrose (this is the blood sugar); fructose, or levulose; galactose.

Disaccharides

Disaccharides are so called because, upon hydrolysis, they form 2 molecules of the hexoses. They are represented by the formula $C_{12}H_{22}O_{11}$. Sucrose, on hydrolysis, yields glucose and fructose. Lactose, on hydrolysis, yields glucose and galactose. Maltose, on hydrolysis, yields 2 molecules of glucose.

Polysaccharides

Polysaccharides hydrolyze to form many molecules of simple sugars. The formula of a polysaccharide is written $(C_6H_{10}O_5)_n$.

The subscript used in this formula indicates an indefinite number. There is some doubt as to the actual number of molecules of simple sugar contained in these compounds. It is known, however, that the number varies with the different polysaccharides. Sometimes different letters are used, as subscripts, to emphasize this: glycogen $(C_6H_{10}O_5)_x$, on hydrolysis, yields x molecules of glucose; dextrin $(C_6H_{10}O_5)_y$, on hydrolysis, yields y molecules of glucose; starch $(C_6H_{10}O_5)_n$, on final hydrolysis, forms n molecules of glucose; cellulose $(C_6H_{10}O_5)_z$ is not hydrolyzed in the human digestive tract but can be used by cud-chewing animals, such as the cow and the goat.

OCCURRENCE

Glucose, $C_6H_{12}O_6$, also called dextrose and grape sugar, is widely distributed, being found in the blood of animals and in the juices of fruits—grapes, for instance. A careful distinction should be made between glucose and commercial "glucose," which is used so much in candy making. The latter does not occur naturally, nor is it a chemical compound; it is a variable mixture of glucose, dextrin, and maltose made by the incomplete acid hydrolysis of starch, usually cornstarch. National advertising is largely responsible for the confusion in this respect.

Fructose, $C_6H_{12}O_6$, is also called levulose or fruit sugar. It occurs along with dextrose in fruit juices and in honey, of which it makes up about 40 per cent.

Sucrose, $C_{12}H_{22}O_{11}$, called cane sugar because of its occurrence in sugar cane, is found also in the sugar beet and in the sap of the sugar maple. In this form it is one of the purest food products, being about 99.85 per cent pure. Maple sugar contains impurities that impart to the sucrose its characteristic flavor.

Lactose, $C_{12}H_{22}O_{11}$, is the sugar of milk and is distinctly of animal origin. It occurs in the milk of mammals in varying quantities, usually from 4 to 6 per cent. It is for this reason that cow's milk must be fortified with carbohydrate, when it is fed to infants, to make it more nearly like mother's milk, which contains 6 to 8 per cent carbohydrate.

Maltose, $C_{12}H_{22}O_{11}$, is the product of the enzymatic hydrolysis of starch. It is found in foods as the result of the action of such amylases or starch-splitting enzymes as occur in germinated barley or malt. It is also an intermediate formed during the acid hydrolysis of starch.

Starch $(C_6H_{10}O_5)_n$, is the most important food

polysaccharide. It is found chiefly in the roots and seeds of plants, to the extent of 90 per cent in some cases. The starch granules have different shapes depending upon their source, and by microscopic examination the source may be identified. It is obtained from its natural sources by the physical processes of grinding and washing with water.

Glycogen, $(C_6H_{10}O_5)_x$, is produced in the animal body and is stored in the liver. The liver of an adult contains about 1 lb glycogen. In smaller amounts it is found also in scallops, oysters, and even in certain insects. This carbohydrate plays the same role in the animal kingdom as does starch in the vegetable kingdom; i.e., it serves as a reserve food supply.

Dextrin, $(C_6H_{10}O_5)_y$, is formed during the partial hydrolysis of starch, or when starch is heated to 200° to 250°C. Thus it is that the outside of a loaf of bread is turned brown in baking. Toasting bread results in the formation of some dextrin on its surface.

Cellulose, $(C_6H_{10}O_5)_z$, is the most prominent carbohydrate in the vegetable kingdom for it constitutes the cellular framework of plants. It occurs in a nearly pure condition in certain plants; e.g., raw cotton is comparatively pure cellulose. It may be obtained from wood on the removal of the lignin by calcium bisulfite.

FORMATION

The carbohydrates in plants result from the photosynthesis reaction. Carbon dioxide from the air and water in the plants unite to form a hexose in the presence of the chlorophyll and under the influence of the light of the sun.

The usual equation for the reaction of photosynthesis is

$$6CO_2 + 6H_2O + \text{Solar energy} \xrightarrow{\text{Chlorophyll}} \underset{\text{Glucose}}{C_6H_{12}O_6} + 6O_2$$

Chlorophyll is an enzyme in this reaction.

The equation is an oversimplification; the exact mechanism of photosynthesis is unknown at present. The photosynthetic reaction is probably the most important chemical reaction in nature. Not only does it supply food and plant structure, but it is the means of storing up solar energy as chemical energy, which can be released on subsequent oxidation.

$$\underset{\text{Glucose}}{C_6H_{12}O_6} + 6O_2 \rightarrow 6CO_2 + 6H_2O + 688.5 \text{ kcal}$$

Further, it maintains the oxygen-carbon dioxide balance in the atmosphere.

Disaccharides are formed in the plants by the condensation of 2 molecules of hexose with the splitting out of a molecule of water.

$$2C_6H_{12}O_6 \rightarrow C_{12}H_{22}O_{11} + H_2O$$

Attempts to synthesize sucrose by this reaction have been unsuccessful.

Polysaccharides are formed in a similar fashion by uniting many molecules of the hexose.

$$nC_6H_{12}O_6 \rightarrow (C_6H_{10}O_5)_n + nH_2O$$

The production of starch in certain plants can be followed easily; for instance, most of us enjoy peas and corn in the early stages of their development, the sugar stage. Later, when the sugar has been converted into starch, they are not so tasty but more filling. In the ripening of certain fruits there is apparently a change from starch to sugar. The ripening of the banana is an example of this.

PHYSICAL PROPERTIES

In general, the carbohydrates are white. Dextrin with its light brown color is an exception in this respect, although it is possible to have even white dextrin.

The carbohydrates vary in their solubility from very soluble to insoluble. Among the monosaccharides, fructose and galactose are more soluble than glucose. Of the disaccharides lactose is less soluble than sucrose and maltose. Cellulose is noted for its in-

Carbohydrates

solubility. The other polysaccharides form colloidal suspensions recognized by the opalescent appearance of the mixtures. Dextrin pastes are very sticky, which accounts for the use of dextrin in the mucilage on postage stamps.

The sweet taste of carbohydrates varies. Fructose is the sweetest sugar, being twice as sweet as glucose and many times sweeter than galactose. The sweetness of sucrose is familiar to everyone, for this sugar is the usual sweetening agent. Lactose hardly reminds us of a sugar from the taste aspect. If it were sweet, milk would be cloying to infants and others on a milk diet. Since a substance must be in solution to affect the taste buds, the polysaccharides must be tasteless.

CHEMICAL PROPERTIES

Combustion

All carbohydrates are combustible, and they all form carbon dioxide and water.

$$C_6H_{12}O_6 + 6O_2 \rightarrow 6CO_2 + 6H_2O + \text{energy}$$

This reaction should be recognized as the reverse of photosynthesis, and it emphasizes once more that the energy evolved originated in the sun.

acid hydrolysis

If a sucrose solution is boiled in the presence of dilute hydrochloric acid as a catalyst, sucrose will be converted into glucose and fructose through hydrolysis.

$$\underset{\text{Sucrose}}{C_{12}H_{22}O_{11}} + H_2O \xrightarrow{\text{HCl}} \underset{\text{Glucose}}{C_6H_{12}O_6} + \underset{\text{Fructose}}{C_6H_{12}O_6}$$

This hydrolysis is also called the *inversion* of sugar, and the resulting mixture of glucose and fructose is called *invert sugar*.

Starch may be hydrolyzed in the presence of acid to form glucose.

$$\underset{\text{Starch}}{(C_6H_{10}O_5)_n} + nH_2O \rightarrow \underset{\text{Glucose}}{nC_6H_{12}O_6}$$

enzymatic hydrolysis

The disaccharides and the digestible polysaccharides are hydrolyzed in the digestive tract through the action of specific enzymes. There is no digestion of monosaccharides since they do not hydrolyze, and there are no enzymes for their digestion.

The enzymes are named for the sugars the reaction of which they catalyze and the name ends in the suffix *-ase*.

$$\underset{\text{Sucrose}}{C_{12}H_{22}O_{11}} + H_2O \xrightarrow{\text{Sucrase or invertase}} \underset{\text{Glucose}}{C_6H_{12}O_6} + \underset{\text{Fructose}}{C_6H_{12}O_6}$$

$$\underset{\text{Lactose}}{C_{12}H_{22}O_{11}} + H_2O \xrightarrow{\text{Lactase}} \underset{\text{Glucose}}{C_6H_{12}O_6} + \underset{\text{Galactose}}{C_6H_{12}O_6}$$

$$\underset{\text{Maltose}}{C_{12}H_{22}O_{11}} + H_2O \xrightarrow{\text{Maltase}} \underset{\text{Glucose}}{2C_6H_{12}O_6}$$

$$\underset{\text{Glycogen}}{(C_6H_{10}O_5)_x} + xH_2O \xrightarrow{\text{Glycogenase}} \underset{\text{Glucose}}{xC_6H_{12}O_6}$$

$$\underset{\text{Starch}}{(C_6H_{10}O_5)_n} + H_2O \xrightarrow{\text{Amylase}} \underset{\text{Dextrins}}{(C_6H_{10}O_5)_x} + \underset{\text{Maltose}}{C_{12}H_{22}O_{11}}$$

Dextrins and maltose are subsequently hydrolyzed to glucose.

The final products of carbohydrate digestion are the monosaccharides: glucose, fructose, and galactose.

reducing action

Compounds, such as aldehydes, are good reductants and can bring about a change in valence in other substances. Ketones with hydroxyl groups on adjacent carbon atoms are also reductants. This property can be used to test for the presence of aldoses and ketoses. All the monosaccharides are reductants. If a disaccharide or polysaccharide has no free aldehyde or ketone group, it will not react with agents such as Benedict's solution or Clinitest. Fehling's solution, Benedict's solution, Folin's reagent, and Clinitest tablets are all composed of Cu^{++} in some form. All these mixtures are blue. Cu^{++} can be reduced to cuprous hydroxide, which is yellow, or cuprous oxide, which is an orange red. Depending upon the amounts of material present, various colors from green through yellow, orange, and red may be produced. By use of a color chart an estimate of the amount of reducing agent present in a sample may be obtained.

$$2Cu^{++} \rightarrow Cu_2O$$
$$\text{Blue} \quad\quad \text{Red}$$

$$C_6H_{12}O_6 \rightarrow C_6H_{12}O_7$$
$$\text{Glucose} \quad\quad \text{Gluconic acid}$$

The aldohexose glucose is the principal sugar in the blood and body fluids. There may be small amounts of fructose and galactose and, in some cases, pentoses. In diabetes, the amount of glucose in the blood increases and appears in the urine (glycosuria). The presence of glucose in the urine may also be associated with the consumption of a large amount of carbohydrates at one time or with emotional states, severe mental strain, convalescence from severe fevers, or poor posture.

Benedict's test, Clinitest, and Dreypak are commonly used to test for the presence of reducing sugar in urine. Folin's reagent is used to measure the amount of glucose in the blood and spinal fluid. In meningitis, the sugar in spinal fluid decreases.

Sugar, to many lay persons, means sucrose. The nurse should be certain that the patient who must test his urine sample at home knows the difference between glucose and sucrose. If a patient suspected that the reagent used to test the urine samples was too old to give correct results, could he check it by dissolving some table sugar in water and testing it in the same way as he would his urine sample? Since glucose and sucrose have different properties, this procedure would be useless (see Table 15–1).

fermentation

Some monosaccharides, in the presence of enzymes from yeast, undergo a chemical change to form ethyl alcohol and carbon dioxide.

$$C_6H_{12}O_6 \xrightarrow{\text{Zymase}} 2C_2H_5OH + 2CO_2$$
$$\text{Glucose} \quad\quad\quad \text{Ethanol} \quad \text{Carbon dioxide}$$

This reaction is sometimes called simple fermentation or alcoholic fermentation. Pentoses do not undergo simple fermentation nor does galactose. Yeast also contains enzymes able to catalyze the hydrolysis of starch to maltose, of sucrose to glucose and fructose, and of maltose to glucose but no enzymes able to hydrolyze lactose. Lactose can be fermented with enzymes from lactic acid bacteria, forming lactic acid.

The fermentation reaction can be used to differentiate the various sugars that might be present in urine. For example, the urine sample of a nursing mother gives a positive reaction with Benedict's test but does not ferment with yeast. What would you suspect the sugar in the urine to be? Could it be lactose?

iodine reaction

Table 15–1 also shows the action of the various carbohydrates with iodine. The exact nature of this reaction is not known. It is believed to be a question of adsorption by the polysaccharide. The product, formed with starch, goes under the rather loose name of iodostarch. The color produced with cold starch is dissipated upon heating, but it returns as the temperature is reduced.

By a combination of the tests recorded in Table 15–1, it is possible to identify many of the carbohydrates qualitatively: sucrose is the only sugar that does not reduce Benedict's solution; lactose is the only disaccharide that does not ferment; and the polysaccharides may be distinguished by the color of the iodine test. Iodine is a test for

Carbohydrates

Table 15-1 PROPERTIES OF CARBOHYDRATES

Carbohydrate	Reduce?	Ferment?	Color with Iodine
Glucose	Yes	Yes	None
Fructose	Yes	Yes	None
Galactose	Yes	No	None
Sucrose	No	Yes	None
Lactose	Yes	No	None
Maltose	Yes	Yes	None
Starch	No	No	Blue
Dextrin	No*	No	Wine
Glycogen	No	No	Red
Cellulose	No	No	None

*One form of dextrin reduces.

starch, but this reaction may be considered from another point of view, namely, that starch is a test for iodine. Starch is sometimes used as an antidote for iodine.

STRUCTURE

Although many of the properties of carbohydrates can be explained in terms of the open-chain structures, others cannot. Experimental evidence indicates that the carbohydrates can and do exist in ring forms in equilibrium with the open-chain forms.

Similarly, sucrose formed by the loss of a molecule of water from the aldehyde group of glucose and the ketone group of fructose has a double ring structure. It gives a negative Benedict test because both of these groups are tied up. Maltose, formed from 2 molecules of glucose, is bound by the aldehyde group of one glucose molecule and an alcohol group of the other thus giving a positive Benedict test.

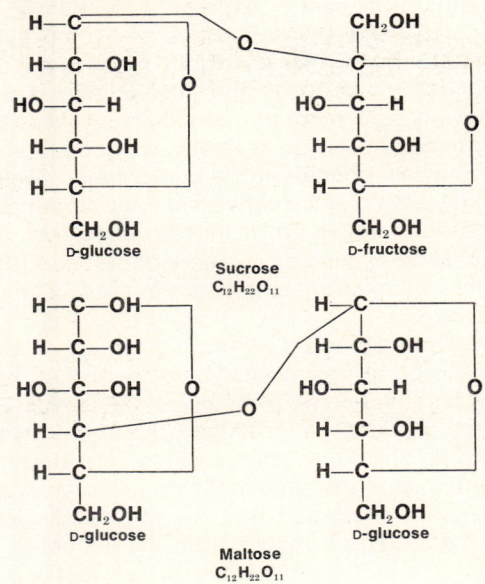

METABOLISM

Anabolism

The monosaccharides absorbed through the intestinal wall are transported to the liver, where part of the glucose and all the fructose and galactose are converted into glycogen.

$$n C_6H_{12}O_6 \rightarrow (C_6H_{10}O_5)_n + n H_2O$$
Glucose → Glycogen
Fructose
Galactose

Note here that, even though molecules of the three monosaccharides enter the liver, only glucose comes out. The glycogen is stored in the liver, where it acts as a reserve supply of glucose. The name *glycogen* means "former of glucose"; as glucose is needed to maintain the sugar level in the blood, the glycogen is hydrolyzed to glucose.

$$(C_6H_{10}O_5)_n + n H_2O \rightarrow n C_6H_{12}O_6$$

The amount of glucose in the blood is remarkably constant over a period of time, about 0.1 per cent or, expressed another way, 100 mg per cent, which means 100 mg per 100 ml blood. After a carbohydrate meal the glucose content of the blood increases, causing a temporary condition of *hyperglycemia*. There are several ways of counteracting this change in the blood sugar level: (*a*) glucose may be converted to glycogen as described above, (*b*) it may produce fat, (*c*) it may be eliminated by the kidneys, and (*d*) it may be oxidized to produce energy (catabolism).

The method by which glucose becomes fat is not entirely clear. The glycerol part of the fat could come from glyceric aldehyde, a form of which is produced during muscle contraction. Fatty acids in food fats always contain an even number of carbon atoms. It is therefore presumed that the fatty-acid part of the stored fat results from the condensation of a sufficient number of molecules of some substance formed in the body containing 2 carbon atoms, e.g., acetaldehyde, CH_3CHO.

The kidneys have the ability to hold back glucose up to a concentration of about 160 mg per 100 ml (renal threshold). In hyperglycemia the concentration of glucose in the blood may well exceed this figure, and the excess spills into the urine. This condition is called *alimentary glycosuria* and is of no significance, since it is merely an attempt to return the blood sugar to its normal level.

Catabolism

The equation for the oxidation of glucose is

$$C_6H_{12}O_6 + 6O_2 \rightarrow 6CO_2 + 6H_2O + 688{,}500 \text{ cal}$$

This value for the amount of energy evolved is the number of calories formed per mole of glucose. Dividing it by 180, the molecular weight of glucose, we arrive at about 3,875 cal per Gm glucose, or 3.87 kcal per Gm. The reaction is not so simple as the above equation indicates. The glucose is first converted to glycogen, and then there follows a long succession of changes, resulting in the formation of pyruvic acid and then lactic acid, the product of muscular activity. About 80 per cent of the lactic acid is reconverted to glycogen, and the rest is oxidized to carbon dioxide and water.

The conversion of pyruvic acid, obtained either directly from glucose or by the dehydrogenation of lactic acid, to carbon dioxide and water takes place by means of the *tricarboxylic acid cycle,* also called the *citric acid cycle* and the *Krebs cycle.* This is the series of reactions by which any foodstuff can be completely oxidized in the body. Foodstuffs other than carbohydrates must be converted first to pyruvic acid or some member of the cycle before it can be oxidized to carbon dioxide and water. The overall equation for the cycle is

$$CH_3COCOOH + 3H_2O + 5[O] \rightarrow 3CO_2 + 5H_2O + \text{energy}$$

These reactions are controlled by several very complicated enzyme systems. The enzymes are themselves regulated by hormones. We shall mention two, which have antagonistic effects: insulin, which tends to reduce the concentration of blood glucose, and glucagon, which tends to increase it.

When glucose is oxidized in the body, the

Carbohydrates

volume of oxygen needed for the oxidation is exactly equal to the volume of carbon dioxide formed from the oxidation. The volume of carbon dioxide formed divided by the volume of oxygen used is called the respiratory quotient (RQ).

$$RQ = \frac{6CO_2}{6O_2} = 1$$

Study Exercises

1. Classify the following carbohydrates as monosaccharides, disaccharides, or polysaccharides: glucose, glycogen, maltose, sucrose, fructose, dextrin, galactose, and lactose.
2. Show by means of an equation how carbohydrate is formed by photosynthesis.
3. Why are certain sugars reducing sugars while others are not?
4. Why is it that sucrose does not reduce either Benedict's solution or Clinitest?
5. Write the equation for the oxidation of glucose.
6. Name two carbohydrates synthesized in the human body.
7. Write word equations for the hydrolysis of sucrose, lactose, maltose, glycogen, and starch. Include the appropriate enzyme in each case.
8. Identify each of the following carbohydrates, which might under certain conditions be found in a urine sample: (a) it gives a positive Benedict's test, gives no color with iodine, and does not undergo yeast fermentation; (b) it gives a positive Benedict's test, gives no color with iodine, and undergoes yeast fermentation.
9. Explain the following occurrence: tincture of iodine, which is brown, was spilled on a white starched uniform, and the stain turned blue.
10. What are the end products of carbohydrate digestion?
11. What is the principal sugar in the general circulation?
12. How is the blood sugar level maintained?
13. What is meant by renal threshold?
14. What would be the effect on the concentration of blood glucose if the liver lost its ability to synthesize or store glycogen?

chapter sixteen   LIPIDS

Ideas for preview or review

The fats or lipids are a much maligned and misunderstood class of chemical compounds. Food-faddist and misleading advertisements sometimes leave the general public confused. The nurse will want to know the scientific facts and will want to help patients clarify their thinking in this area. Recent research results have led to questioning many of our previous ideas about fat metabolism. As a result, one must keep one's mind open but not be gullible.

Many diagnostic findings are like detective work; they provide clues. From the clues, the situation in the body must be determined. Take, for example, measurements of oxygen consumption and carbon dioxide production by a patient in a given period of time. We know that carbon dioxide is formed in the body from oxidation. By comparing the volumes of oxygen and carbon dioxide, the respiratory quotient for the patient can be determined. A diabetic patient has a RQ of 0.7. This is the RQ for fat, and it tells the doctor that the patient is oxidizing fat exclusively. From measurements on patients, the physician can determine the type of oxidation proceeding in the body.

Lipids are compounds that are insoluble in water but soluble in the usual fat solvents, such as ether, chloroform, and carbon tetrachloride. Lipids are widely distributed in nature in the tissues of both plants and animals.

The production of fats in plants has long been the subject of investigation. There seems to be no evidence that plants can synthesize fats from carbon dioxide and water. There is evidence, however, that fats may be produced in plants from carbohydrates. For instance, it is observed that as the quantity of fat increases in a plant, the carbohydrate content decreases. Animals, on the other hand, derive their fat from their diet. Any food, carbohydrate, protein, or fat, taken in excess of body needs can be stored as fat. It is common knowledge that corn is fed to cattle to fatten them and that pigs are turned loose in the peanut fields of Virginia to fatten on the peanuts left in the soil. Many of us know only too well that an excess of carbohydrates in our diet adds to the waistline.

It is interesting to compare the fats from various animal sources in the light of the source animal's insulating coat and natural habitat. The fat in the animal body must be at least in a semisolid state to allow free action of the muscles. An animal with a thick, protective fur could store fat with a higher melting point than one with little or no such protection. An animal living in a very cold climate must store fat with a lower melting point than one in a warmer climate. A little thought will produce illustrations of both.

CLASSIFICATION

The lipids are usually classified as follows:

I. Simple lipids: esters of fatty acids and various alcohols

a. Fats and oils, glyceryl esters, e.g., butterfat, olive oil
   b. Waxes, esters of high molecular weight, monohydric alcohols, e.g., beeswax, cerumen (wax in the ear)
II. Compound lipids: esters containing groups other than those from an alcohol and a fatty acid
   a. Phospholipids: fats that contain phosphoric acid and a nitrogenous base, e.g., lecithin in egg yolk, nerves, blood, bile, and brain, cephalins in brain
   b. Glycolipids: compounds composed of a fatty acid and a carbohydrate (derivative of a polyhydric alcohol), containing nitrogen but no phosphorus; cerebrosides, e.g., kerosin and phrenosin in nerves
III. Derived lipids: compounds derived from groups I and II by hydrolysis and that are soluble in fat solvents
   a. Sterols: high-molecular-weight cyclic monohydric alcohols, e.g., cholesterol, ergosterol, calciferol
   b. Fatty acids of high molecular weight, e.g., stearic acid

The simple lipids contain carbon, hydrogen, and oxygen. These are concentrated forms of stored energy. As a good housewife reduces material to be stored to compact bundles, so the body reduces ingested carbohydrates, proteins, and fats in excess of present need to be changed and stored in compact form. The simple sugar glucose has the formula $C_6H_{12}O_6$, in which the ratio of carbon to oxygen is 1:1 and that of hydrogen to oxygen is 2:1. Stearin, also called glyceryl tristearate, a common fat in the body, has the formula $C_3H_5(C_{18}H_{35}O_2)_3$. The molecular formula would be $C_{57}H_{110}O_6$. Here the ratio of carbon to oxygen is approximately 9:1 and the ratio to hydrogen to oxygen, approximately 18:1. Bear these facts in mind when studying the respiratory quotients of carbohydrates and fats. The RQ for fat is 0.7.

FORMATION

An alcohol ($-\overset{|}{\underset{|}{C}}-OH$) can react with an acid ($-\overset{O}{\overset{\|}{C}}-O-H$) by splitting out a molecule of water to form a compound known as an ester. The process of making esters is called, naturally enough, *esterification*.

$$C_2H_5OH + CH_3COOH \xrightarrow{H_2SO_4} CH_3COOC_2H_5 + H_2O$$
Ethanol   Acetic acid   Ethyl acetate
(alcohol)   (acid)   (ester)

The simple lipids are esters of glycerol and organic acids. Since glycerol has 3 hydroxyl groups, it can react with 3 molecules of acid to form its ester.

$$C_3H_5(OH)_3 + 3C_3H_7COOH \rightarrow$$
Glycerol   Butyric acid

$$(C_3H_7COO)_3C_3H_5 + 3H_2O$$
Glyceryl tributyrate or butyrin (butterfat)

The glyceryl esters are called *glycerides*.

If all 3 acid molecules are alike, the ester is called a *simple* ester; but if they are different, it is a *mixed* ester. Most of the fats and oils are made of long-chain acids such as palmitic acid, $C_{15}H_{31}COOH$; stearic acid, $C_{17}H_{35}COOH$; oleic acid, $C_{17}H_{33}COOH$; linoleic acid, $C_{17}H_{31}COOH$.

Palmitic and stearic acids are called *saturated acids* because each carbon atom is attached to the adjacent carbon atom by 1 valence bond.

$$\begin{array}{cccc} H & H & H & H \\ | & | & | & | \\ -C & -C & -C & -C- \\ | & | & | & | \\ H & H & H & H \end{array}$$

Oleic acid is short 2 hydrogen atoms, indicating 2 valence bonds between 2 adjacent carbon atoms. This condition is called *unsaturation*.

$$\begin{array}{cccc} H & H & H & H \\ \| & | & | & | \\ -C & -C=C & -C- \\ | & & & | \\ H & & & H \end{array}$$

In linoleic acid, with 4 hydrogen atoms

Lipids

Table 16-1 COMMON FATTY ACIDS AND GLYCERIDES

Name	Formula	Melting Point	Structure
Palmitic acid	$C_{15}H_{31}COOH$	65°C	Saturated
Stearic acid	$C_{17}H_{35}COOH$	71°C	Saturated
Oleic acid	$C_{17}H_{33}COOH$	−4°C	1 double bond
Linoleic acid	$C_{17}H_{31}COOH$	−5°C	2 double bonds
Linolenic acid	$C_{17}H_{29}COOH$	−10°C	3 double bonds
Arachidonic acid	$C_{19}H_{31}COOH$	−49°C	4 double bonds
Glyceryl palmitate (tripalmitin)	$(C_{15}H_{31}COO)_3C_3H_5$	44°C	Saturated
Glyceryl stearate (tristearin)	$(C_{17}H_{35}COO)_3C_3H_5$	64°C	Saturated
Glyceryl oleate (triolein)	$(C_{17}H_{33}COO)_3C_3H_5$	−12°C	3 double bonds
Glyceryl linoleate (trilinolein)	$(C_{17}H_{31}COO)_3C_3H_5$	Liquid	6 double bonds
Glyceryl linolenate (trilinolenin)	$(C_{17}H_{29}COO)_3C_3H_5$	Liquid	9 double bonds
Glyceryl arachidonate (triarachidonin)	$(C_{19}H_{31}COO)_3C_3H_5$	Liquid	12 double bonds

short, there are 2 sets of double valence bonds in the compound.

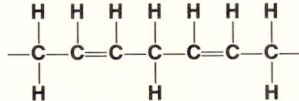

The terms saturated and unsaturated refer to the valence bonds in the compound and have no reference to saturated and unsaturated solutions.

Table 16-1 shows the common glycerides found in fats and oils together with their melting points. From this table it is evident that the solid glycerides are saturated compounds and the liquid glycerides are unsaturated compounds. Whether a given substance is a fat or an oil is determined by its composition. Tallow contains a high percentage (75 per cent) of the saturated glycerides and a low percentage (25 per cent) of the unsaturated; it is a solid fat. Cottonseed oil contains a low percentage (25 per cent) of saturated glycerides and a high percentage (75 per cent) of the unsaturated; it is a liquid oil. The consistency of the various fats, then, depends upon the relative amount of saturated and unsaturated glycerides present in them. Butter is peculiar in that it contains a number of the glyceryl esters of the lower members of the fatty-acid series, chiefly glyceryl butyrate, which gives to the butter its characteristic flavor. It is an interesting fact that the glycerides that occur naturally are those having an even number of carbon atoms in the acid group contained in them.

PROPERTIES

Physical

Fats and oils are insoluble in water. Their specific gravity is less than 1. Since they are lighter than an equal volume of water, the fats and oils collect on the surface of a water-fat mixture. However, fat and oil can be broken down into such minute particles that the fat droplets remain suspended in a water mixture, forming an emulsion. This is the case with "homogenized" milk. The cream does not float to the top.

Fats and oils have a shiny appearance and a slippery or greasy feeling.

Their state depends on the temperature, solid fats changing to liquids with increased temperature as indicated in Table 16-1.

When pure, the fats and oils are colorless or white.

Chemical

oxidation

The simple lipids are combustible, yielding carbon dioxide, water, and energy.

$$C_{57}H_{104}O_6 + 80O_2 \rightarrow 57CO_2 + 52H_2O + \text{energy}$$
Glyceryl oleate

Notice the great number of oxygen molecules needed to oxidize 1 molecule of fat. We can see why carbohydrates and proteins yield 4 kcal per Gm and fats yield 9 kcal per Gm.

From these data we can calculate the respiratory quotient for fat:

$$RQ = \frac{57CO_2}{80O_2} = 0.7$$

hydrolysis

Upon being heated with water, the fats hydrolyze somewhat, forming glycerol and the corresponding free acid; for example,

$$(C_{17}H_{35}COO)_3C_3H_5 + 3HOH \rightarrow$$
$$C_3H_5(OH)_3 + 3C_{17}H_{35}COOH$$

In deep-fat frying, water from the food may react with the glycerides. The formation of the free acids can be easily detected by their odors. The rancidity of butter is due to small amounts of butyric acid, formed from glyceryl butyrate. Rancidity, in general, may be caused by hydrolysis or by an oxygenation of the fat; sometimes both take place. The action is hastened by heat, light, and moisture. Fats stored at room temperature become rancid more rapidly than those stored under refrigeration.

Overheating or burning of fat causes the formation of *acrolein*. This is an unsaturated aldehyde called acrylic aldehyde and is made from glycerol by the removal of 2 molecules of water.

$$C_3H_5(OH)_3 \rightarrow C_2H_3CHO + 2H_2O$$

Acrolein has a disagreeable, penetrating odor and affects the eyes, producing a copious flow of tears. It is also irritating to the stomach mucosa. Much of the indigestion caused by fried foods is due to acrolein, formed by improper cooking methods and absorbed by the food.

enzymatic hydrolysis

All body fluids are very dilute water solutions. The lipases, enzymes that catalyze the hydrolysis of fats in the body, are in dilute water solutions. Since fat does not mix with or dissolve in water, ingested fat must be broken up into tiny droplets (emulsified), which can be distributed through the water, in order for sufficient surface contact between the fat and the lipases to occur. Fat that is in the emulsified state when it is eaten can be hydrolyzed in the stomach through the action of gastric lipase. Egg yolk contains emulsified fat. Because the fat in this food exists as minute drops, we do not recognize the shiny appearance of fat nor does the fat separate from the other ingredients in egg yolk. Since the gastric contents are acidic, the products of hydrolysis in the stomach are glycerol and acids. Glycerol is readily soluble in water, but the long-chain fatty acids are not and they form colloidal mixtures.

Unemulsified fats must first be broken down into droplets before they can be digested. This is accomplished in the small intestines through the action of bile salts from the liver. The bile salts are poured into the small intestines, where they reduce the surface tension, permitting the fat to become dispersed as droplets mixed in with pancreatic juice and intestinal juice. The bile salts are not enzymes. In chemical structure, they resemble cholesterol. Enzymes specific for hydrolysis of fat in the intestines are pancreatic lipase and intestinal lipase. Since the intestinal contents are alkaline, the products of digestion here are glycerol and soaps. Some investigators claim that the intestinal fluids are not sufficiently alkaline to form soaps and that the products of fat hydrolysis here are also glycerol and colloidal fatty acids. The truth of the matter is that we don't know exactly how this process works as yet. The problem is how, if some of the products are colloidal, they can pass through the intestinal membranes. Colloidal particles are normally unable to diffuse through membranes. We do know that the bile not only aids in the emulsification of fat but also plays an important role in the absorption of fat products from the intestines. Through the action of the bile salts, the fatty acids produced are broken down into extremely small microdroplets, called micelles. These, along with glycerol, pass through the intestinal wall where some of them reunite to form droplets of neutral fat. It is the emulsified fat in milk that gives that food its "milky appearance." When the droplets are absorbed into the lacteals, those structures assume a milky appearance, hence the name lacteals, which means *of milk.* The lacteals of the lymphatic system unite with other lymphatic vessels, finally to join the thoracic duct from whence the microdroplets of fat enter the general circulation. This is an interesting result. Ingested fat, after digestion, becomes little droplets of fat again. All the carbohydrates and proteins, with the exception of the simple sugars, are chemically changed in the process of digestion, but part of the fat is carried in the blood as emulsified fat. Part of the fat is transported bound to protein, particularly to plasma albumen, and part is bound to cholesterol.

saponification

Soaps are salts of the higher fatty acids and can be made from fats and oils by the addition of lye or alkalis. Not all soaps are soluble or are good cleansing agents. Sodium and potassium soaps are the detergent soaps. Sodium salts make solid soaps and potassium salts make liquid or soft soaps. Green soap, used in the hospital, is a mixture of soft soap and alcohol. When the detergent soaps are added to hard water containing calcium or magnesium ions, the insoluble calcium and magnesium soaps are formed.

In obstructive jaundice, the bile salts are not present in the intestines to aid in the emulsification and absorption of fat. The fat is excreted in the feces causing a condition called steatorrhea or fatty stool. This condition is often accompanied by low serum calcium, because the calcium reacts with the fat forming the calcium soap and is thus excreted along with the fat.

hydrogenation

A comparison of the formula of a solid glyceride will show that the difference is a few atoms of hydrogen. This, of course, results from the fact that the solid glycerides are saturated compounds, whereas the others are unsaturated. If we could add hydrogen to the unsaturated liquid glyceride we could produce a solid. This addition is carried out in the process of *hydrogenation,* which has already been discussed in the case of the unsaturated hydrocarbons. The following equation illustrates the hydrogenation of a liquid glyceride.

$$(C_{17}H_{33}COO)_3C_3H_5 + 3H_2 \xrightarrow{Ni} (C_{17}H_{35}COO)_3C_3H_5$$
Liquid triolein — Solid tristearin

Hydrogenation is an extremely important process because it enables us to convert plentiful vegetable oils into scarce solid fats. The consistency of the solid is governed by the extent to which the hydrogenation is carried out, i.e., the amount of unsaturated compound left unchanged. On the other hand, hydrogenation may be completed and unsaturated oils added to reduce the consistency to that desired.

Hydrogenation is also important in the soap industry. Soaps produced from unsaturated glycerides are not so satisfactory as those from the saturated compounds. It therefore pays to hydrogenate the soap stock first in order to produce a better product.

WAXES

Waxes are mixtures in which an ester of a high-molecular-weight monohydric alcohol and a high-molecular-weight fatty acid predominate. For instance, beeswax is largely myricyl palmitate ($C_{15}H_{31}COOC_{30}H_{61}$). Cholesterol esters are also waxes.

PHOSPHOLIPIDS

These are complex glyceryl esters in which phosphoric acid takes the place of one fatty acid and a nitrogenous base has reacted with a hydrogen atom of the phosphoric acid. The lecithins and cephalins are typical.

CH$_2$OOCR
CH OOCR
CH$_2$O—P(=O)(OH)—OCH$_2$CH$_2$N(CH$_3$)$_3$
OH
A lecithin

HOCH$_2$CH$_2$N(CH$_3$)$_3$
OH
Trimethylhydroxy-
ethyl-ammonium hydroxide
choline

The cephalins are similar to the lecithins in composition except for the base they contain. Instead of choline, most cephalins contain cholamine, β-aminoethyl alcohol.

CH$_2$OOCR
CH OOCR
CH$_2$O—P(=O)(OH)—OCH$_2$CH$_2$NH$_2$
OH
A cephalin

$\overset{\alpha}{\text{HOCH}_2}\overset{\beta}{\text{CH}_2}\text{NH}_2$
β-Aminoethyl alcohol
cholamine

Lecithins are important for the transport of fats in the body, incidentally providing phosphoric acid for the building of new cells. They are very good emulsifying agents and find wide application industrially. Cobra venom contains an enzyme that is able to remove 1 fatty-acid molecule from lecithin to form lysolecithin. This compound is able to cause hemolysis of the blood and accounts for at least part of the toxicity of the venom.

Cephalins play an important part in the clotting of blood, for, it is believed, they are formed in the disintegration of blood platelets.

GLYCOLIPIDS

Glycolipids are also called *cerebrosides* because of their occurrence in the brain and nerve tissues. They are also called *galactolipids* because upon hydrolysis they yield galactose (a 6-carbon sugar). They contain nitrogen but no phosphorus. In addition to the sugar, hydrolysis produces fatty acids and sphingosine. This latter compound is a complex alcohol containing an amino group (NH$_2$). Kerasin and phrenosin are the most common glycolipids.

STEROLS

The most common of the sterols is cholesterol, which is related to many substances of biologic interest, e.g., hormones and vitamins. The formula of cholesterol is given here to establish the sterol nucleus. This nucleus, which will be encountered repeatedly, is a derivative of phenanthrene and cyclopentane.

Sterol nucleus

Cholesterol

Cholesterol occurs in the tissues of the nervous system and in the blood. Gallstones are composed mainly of cholesterol.

Cholesterol occurs naturally in blood, and with advancing years its concentration seems to increase; plaques of cholesterol can be deposited on the walls of blood vessels. These deposits have at least four effects: (a) they decrease the free-flowing diameter of the blood vessel, (b) they decrease the elasticity of the arterial walls, (c) they introduce a jagged lining to the walls and interfere with the rate of blood flow by friction, (d) they may cause clots in blood vessels. In addition, they may become dislodged and cause stoppages in vital organs (such as the heart).

Remember that cholesterol is found normally in the blood and cells. The range usually given is 120 to 260 mg per 100 ml serum. Because an increase or decrease of cholesterol may be indicative and helpful in the diagnosis of certain disorders such as damage to the liver, thyroid malfunction, and anemia, laboratory tests can be performed, one to determine cholesterol level and another to determine that of cholesterol esters. Both tests are time-consuming: the first takes 2 days; the second, 1 week. In most cases, the results of these tests are not too conclusive, since the role of cholesterol is not completely understood.

Much research is being done to determine the relationship between fat in the diet and cholesterol. Some researchers have found evidence to indicate that diets rich in saturated fats (stearin or beef suet, palmitin) contribute to high blood cholesterol levels and that persons in whose diet the fat content is principally unsaturated fats (olein or olive oil) are less prone to this condition. The evidence is not conclusive, but with the increasing numbers of older persons in our population, this is a very important problem.

METABOLISM OF FAT

After a meal, the percentage of fat in the blood increases. To return this percentage to normal several avenues are available.

Anabolism

Several lipids that are essential for the proper functioning of the body are synthesized from fats. Obviously, the synthesis of these substances uses up only a small amount of the fat present in the blood. Fat is an essential constituent of protoplasm and cell membranes. In the transport of substances through cell membranes, fat-soluble substances dissolve in the membranes themselves and enter the cell with ease. Fats are also found in bone marrow and in nerves. A larger part of the fat is stored under the skin and around many of the organs. It is called *adipose tissue* or *depot fat* and serves several purposes.

Through the use of "tagged" atoms, radioactive isotopes whose course may be followed easily, an interesting fact regarding depot fat has been discovered. It was once thought that stored fat was static; i.e., it remained there until it was used up. It is now known that if an animal is fed an adequate diet containing these tagged atoms, no increase in weight will occur but fat containing these atoms will be deposited. Conversely, when tagged atoms are not included in the subsequent diet, fats containing them gradually disappear from the deposits even though the amount of fat remains constant. It is evident, then, that depot fat is constantly changing.

It is interesting to note further how the nature of depot fat in an animal is affected by the diet. If the fat is synthesized from carbohydrates, the depot fat will be characteristic of the animal; i.e., a high-carbohydrate diet produces low-melting-point fat in a hog but high-melting-point fat in a steer. On the other hand, a diet high in oils such as soybean oil produces a soft, flabby fat with a very low melting point. High-fat diets are not commonly eaten by human beings, but such a diet has a profound effect on the desirability, and hence the market value, of food animals.

In the catabolism of fats the two parts of the molecule go their separate ways, as it were. A possible series of compounds showing the oxidation of glycerol to carbon dioxide would be as follows:

$$\begin{array}{ccc} CH_2OH & CH_2OH & CH_3 \\ | & | & | \\ CHOH & \to CHOH \to & C=O \to CO_2 + H_2O \\ | & | & | \\ CH_2OH & CHO & COOH \\ \text{Glycerol} & \text{Glyceric} & \text{Pyruvic} \\ & \text{aldehyde} & \text{acid} \end{array}$$

Glycerol may also be converted into glucose by the following series of reactions:

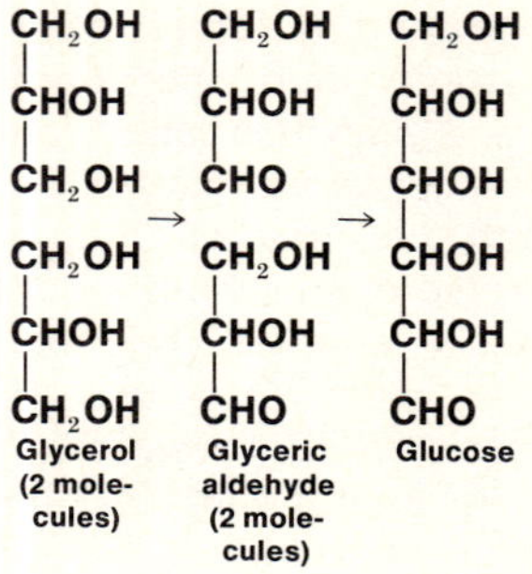

$$
\begin{array}{ccc}
CH_2OH & CH_2OH & CH_2OH \\
CHOH & CHOH & CHOH \\
CH_2OH & CHO & CHOH \\
CH_2OH & CH_2OH & CHOH \\
CHOH & CHOH & CHOH \\
CH_2OH & CHO & CHO \\
\text{Glycerol} & \text{Glyceric} & \text{Glucose} \\
\text{(2 mole-} & \text{aldehyde} & \\
\text{cules)} & \text{(2 mole-} & \\
 & \text{cules)} &
\end{array}
$$

(Glycerol → Glyceric aldehyde → Glucose)

Catabolism

There is no direct proof that the fatty acids can produce energy in the muscles. However, it is believed that they can, because in the absence of carbohydrate, the muscles continue to contract. Since it is the carbohydrates and fats of the diet that usually provide the energy, it is to be presumed that in the absence of carbohydrates energy is derived from the fats.

The physiologic chemist Knoop, in 1904, first explained a way that fatty acids could be oxidized in the tissues. His explanation is called *Knoop's theory of β-oxidation*. According to this theory, the fatty acids are oxidized 2 carbon atoms at a time. For instance, starting with stearic acid, if 2 carbon atoms are removed, palmitic acid results. This may then be oxidized, splitting off 2 carbon atoms, etc. The following series of compounds shows how carbon dioxide and water may result in the last stages of normal catabolism. The earlier stages would follow the same pattern.

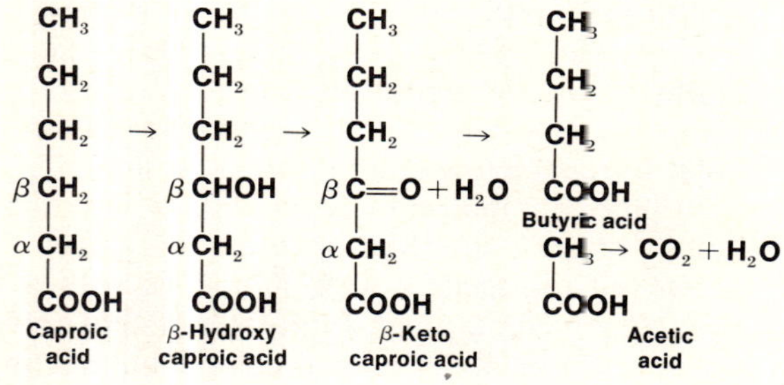

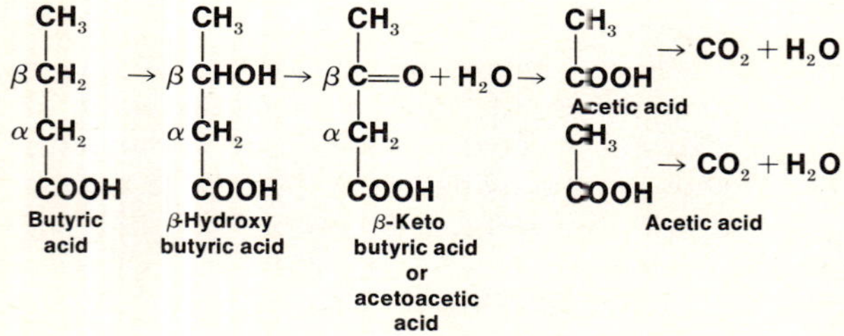

$$
\begin{array}{ccccc}
CH_3 & CH_3 & CH_3 & CH_3 & \\
CH_2 & CH_2 & CH_2 & CH_2 & \\
CH_2 \to & CH_2 \to & CH_2 \to & COOH & \\
\beta\ CH_2 & \beta\ CHOH & \beta\ C{=}O + H_2O & \text{Butyric acid} & \\
\alpha\ CH_2 & \alpha\ CH_2 & \alpha\ CH_2 & CH_3 \to CO_2 + H_2O & \\
COOH & COOH & COOH & COOH & \\
\text{Caproic} & \beta\text{-Hydroxy} & \beta\text{-Keto} & \text{Acetic} & \\
\text{acid} & \text{caproic acid} & \text{caproic acid} & \text{acid} &
\end{array}
$$

$$
\begin{array}{ccccc}
CH_3 & CH_3 & CH_3 & CH_3 \to CO_2 + H_2O \\
\beta\ CH_2 \to & \beta\ CHOH \to & \beta\ C{=}O + H_2O \to & COOH \\
\alpha\ CH_2 & \alpha\ CH_2 & \alpha\ CH_2 & \text{Acetic acid} \\
COOH & COOH & COOH & CH_3 \to CO_2 + H_2O \\
\text{Butyric} & \beta\text{-Hydroxy} & \beta\text{-Keto} & COOH \\
\text{acid} & \text{butyric acid} & \text{butyric acid} & \text{Acetic acid} \\
 & & \text{or} & \\
 & & \text{acetoacetic} & \\
 & & \text{acid} &
\end{array}
$$

If oxidation of glucose is not taking place because of starvation or diabetes mellitus, β-hydroxybutyric acid, acetoacetic acid, and acetone appear in the blood to appear later in the urine. Acetone also appears in the breath. The presence of these acids in the blood reduces the pH of the blood, and acidosis, accompanied by coma, may result.

This variation in the oxidation of the fatty acids is as follows:

$$\beta\ \underset{\substack{|\\ \text{COOH}\\ \beta\text{-Hydroxy}\\ \text{butyric}\\ \text{acid}}}{\overset{\substack{\text{CH}_3\\|}}{\text{CHOH}}} \underset{\alpha\ \text{CH}_2}{\overset{}{\longrightarrow}} \beta\ \underset{\substack{|\\ \text{COOH}\\ \beta\text{-Keto}\\ \text{butyric acid}\\ \text{or}\\ \text{acetoacetic}\\ \text{acid}\\ \text{or}\\ \text{diacetic acid}}}{\overset{\substack{\text{CH}_3\\|}}{\text{C}=\text{O}}} \underset{\alpha\ \text{CH}_2}{\overset{}{\longrightarrow}} \underset{\substack{\text{CH}_3\\ \text{Acetone}}}{\overset{\substack{\text{CH}_3\\|}}{\text{C}=\text{O}}} + \text{CO}_2$$

Fortunately, the use of insulin makes possible the normal oxidation of glucose in the body and prevents the development of this acidosis. There is no reason today why a diabetic person who is properly instructed and cooperative should fear the loss of his life from this disease.

Since it can be digested in the digestive tract, insulin must be administered by needle. This is a disadvantage. Today, some patients can use oral substitutes along with careful dietary control. One such oral medication is tolbutamide. Unlike insulin, this is not a protein, but for some persons it helps keep the disease in check.

Study Exercises

1. Define simple, compound, and derived lipids and give two examples of each.
2. What are the differences between saturated and unsaturated fats?
3. How can unsaturated fats be changed to saturated fats?
4. Why is fat considered to be a concentrated energy food?
5. Since the body is able to produce bile salts to emulsify fats and lipases to catalyze their hydrolysis, why do many persons consider fried foods hard to digest?
6. Trace the path and changes that occur in a portion of ingested butter (unemulsified fat) from the mouth to the blood stream.
7. What are four possible effects of the deposition of patches of cholesterol in the walls of the arteries?
8. Why does fat yield more energy per gram than carbohydrate and protein?
9. What is acrolein? Why is it formed by all fats?
10. Mention three physiologic uses of lecithins.
11. Explain briefly the catabolism of fats.
12. Why are ketone substances sometimes formed when fat is oxidized in the body?

chapter seventeen PROTEINS

Ideas for preview or review
   Solutions are homogeneous mixtures of particles the size of average molecules or smaller.
   Larger particles or large molecules form colloidal mixtures.
   Generally speaking, colloidal particles do not pass through intact membranes.
   Mineral acids, such as nitric, hydrochloric, and sulfuric, are strong electrolytes, which exist mainly in the form of charged ions.
   Organic acids, which are compounds containing the carboxyl group (—COOH), exist principally in the form of molecules and form relatively few charged ions.
   Neutral particles can acquire a negative charge either by the loss of protons or the addition of electrons.
   Edema is the accumulation of fluid in the tissues.

Proteins are the giants in the world of molecules: giants not only in size, but also in the important role they play in body processes. They are called *macromolecules* and form not only organs and tissues but enzymes, many hormones, antibodies, viruses, and genes as well. Some persons think of proteins as meat, fish, eggs, milk, cheese, beans, etc. These are foods *rich in proteins.* Proteins have names like albumin, globulin, casein, and gelatin. If the source of albumin is milk, it is called lactalbumin; if eggs, ovalbumin; if blood, serum albumin.

COMPOSITION

Proteins contain very few kinds of atoms. All of them contain carbon, hydrogen, oxygen, and nitrogen. Most of them contain sulfur, and phosphorus occurs in a few proteins. Iodine is a common constituent of the proteins of marine animals. The percentage composition of a few simple proteins is shown in Table 17–1. The molecular weight of a protein is hard to determine with accuracy. At least we may get an idea of the minimum molecular weight from the percentage composition. For instance, if we assume that a molecule of a given protein contains only 1 atom of sulfur and the percentage of sulfur in it is 1.00 per cent, the molecular weight must be at least 3,200. There is no reason to believe that the molecule contains only a single atom of sulfur; probably it contains many of them. Lactoglobin, from milk, has a molecular weight of about 42,000, and its formula is probably $C_{1,864}H_{3,012}O_{576}N_{468}S_{21}$.

In addition to being tremendously large, some protein molecules are extremely fragile. It is interesting to wonder how the atoms in such a great molecule would be arranged in space. Some kinds of proteins have atoms arranged in chains. These are tough and fibrous, as in hair and silk; others have atoms arranged in coils like the thread on a screw. Still others have double and triple coils or spirals; these are easily shattered. It is well to remember that the biologic activity as well as the physical and chemical properties of the proteins depend upon the arrangement of the atoms in space as well as upon the kinds of atoms of which the molecule is composed. Changes in the structure of proteins in the lens of the eye gives rise to cataracts. These changes can be brought about by injury, ex-

Table 17-1 COMPOSITION OF SOME PROTEINS*

Protein	C	H	N	S	O
Egg albumin....	52.75	7.00	15.51	1.62	23.02
Serum albumin..	52.93	7.05	15.89	1.82	22.31
Gelatin	50.11	6.56	17.81	0.23	25.26
Gliadin	52.72	6.86	17.66	1.03	21.73
Zein	55.23	7.26	16.13	0.60	20.78
Average	52.75	6.95	16.60	1.06	22.62

*All values are expressed as per cent.

posure to radiation, glandular malfunction, or aging.

Proteins are composed of varying arrangements of small molecules called *amino acids*. These are compounds containing a proton donor group (—C(=O)—O—H, carboxyl group), which gives to the compound acidic properties, and a proton acceptor group (—N(H)—H, amino group), which imparts basic properties. These compounds are both acidic and basic at the same time. Such compounds are called *amphoteric*.

CLASSIFICATION

Proteins can be classified according to the products formed on hydrolysis. Hydrolysis is a chemical reaction with water, yielding two or more products.

I. Simple proteins produce only amino acids on hydrolysis.
   a. Albumins: relatively simple proteins, which form colloidal solutions with water, e.g., ovalbumin, lactalbumin, serum albumin.
   b. Globulins: larger than albumins and insoluble in pure water. Globulins are found in blood serum and are further classified as alpha globulins, beta globulins, and gamma globulins. Gamma globulins are particularly important because they contain antibodies.
   c. Fibrinogen: necessary for the clotting of blood.
   d. Protamines: the simplest proteins, found in the sperm of certain fish. Protamine insulin and protamine zinc insulin are used in the treatment of diabetes.
   e. Other classes of simple proteins include such substances as keratin from hair, elastin from elastic tissue, and collagen from connective tissue and bone.

II. Conjugated proteins hydrolyze to form amino acids plus other kinds of compounds.
   a. Phosphorproteins yield amino acids and phosphoric acid on hydrolysis, e.g., casein from milk, vitellin from egg yolk.
   b. Nucleoproteins form nucleic acids and amino acids. As the name implies, nucleoproteins are found in the nuclei of cells. Two very important nucleic acids are ribonucleic acid (RNA), which contains ribose, a pentose (5-carbon sugar), and is found principally in the cytoplasm of the cell, and deoxyribonucleic acid (DNA) (it has 1 less oxygen atom than RNA), which is found in the chromosomes.
   c. Chromoproteins yield amino acids and a colored compound when hydrolyzed, e.g., hemoglobins in red blood cells.
   d. Glycoproteins form amino acids and carbohydrate, e.g., mucin in saliva, mucoid from tendons.

III. Derived proteins are formed from proteins by the action of various agents: acids, heat, etc., e.g., proteans, metaproteins, coagulated proteins, proteoses, peptones, polypeptides, and peptides.

This method of classification is based upon the answer to the question: What compounds are formed when the protein is hydrolyzed?

Another method of classification results from the question: Does the protein supply all the amino acids necessary for the animal to live and grow? If the answer is yes, the protein is called a complete protein, such as casein from milk; if no, it is an incomplete protein, such as gelatin or zein from corn. This classification is used in nutrition work.

FORMATION

The important amino acids involved in the formation of proteins are those in which the amino group (—NH$_2$) and the acid group (—COOH) are both on the same carbon atom. These are called alpha amino acids and their skeleton would be

$$\begin{array}{c} H \quad\quad O \\ | \quad\quad \| \\ N-C-C-O-H \\ | \quad\quad | \\ H \end{array}$$

Carbon has 2 more valences, 1 of which carries a hydrogen atom and the other, hydrogen or a side chain.

The simplest amino acid, glycine, carries 2 hydrogen atoms on the alpha carbon atom:

$$\begin{array}{c} H \quad H \quad O \\ | \quad | \quad \| \\ N-C-C-O-H \\ | \quad | \\ H \quad H \end{array}$$

This would be the general formula for all the other amino acids.

$$\begin{array}{c} H \quad H \quad O \\ | \quad | \quad \| \\ N-C-C-O-H \\ | \quad | \\ H \quad \text{Side chain} \end{array}$$

If the side chain is —CH$_3$, the acid is alanine (amino-propionic acid).

$$\begin{array}{c} H \quad H \quad O \\ | \quad | \quad \| \\ N-C-C-O-H \\ | \quad | \\ H \quad HCH \\ \quad\quad | \\ \quad\quad H \end{array}$$

A side chain —CHOHCH$_3$ gives threonine (α-amino-β-hydroxy-butyric acid).

$$\begin{array}{c} H \quad H \quad O \\ | \quad | \quad \| \\ N-C-C-O-H \\ | \quad | \\ H \quad HC-OH \\ \quad\quad | \\ \quad HCH \\ \quad\quad | \\ \quad\quad H \end{array}$$

If we call the side chain R, we could write the formula for any alpha amino acid as

$$\begin{array}{c} H \quad H \quad O \\ | \quad | \quad \| \\ N-C-C-O-H \\ | \quad | \\ H \quad R \end{array}$$

A solution of an amino acid is essentially neutral,[1] but it will react to form a salt with 1 molecule of either an acid or a base. For instance, with an acid the reaction is

$$\begin{array}{c} \text{RCHNH}_2 \\ | \\ \text{COOH} \end{array} + \text{HCl} \rightarrow \begin{array}{c} \text{RCHNH}_3\text{Cl} \\ | \\ \text{COOH} \end{array}$$

But, with a base, such as sodium hydroxide, a sodium salt is formed.

$$\begin{array}{c} \text{RCHNH}_2 \\ | \\ \text{COOH} \end{array} + \text{NaOH} \rightarrow \begin{array}{c} \text{RCHNH}_2 \\ | \\ \text{COONa} \end{array} + \text{H}_2\text{O}$$

If this is true, there should be no reason why 1 molecule of an amino acid could not react with another molecule of the same or with 1 molecule of a different amino acid.[2] For instance, using 2 molecules of glycine,

---

[1] Three of the amino acids have several basic groups and 1 acid group, making their solution basic. Further, three others have several acid groups and 1 basic group, making their solutions acidic.

[2] It should be understood that such a reaction takes place only in the tissues. That is, amino acids do not react with each other directly and can only be made to do so indirectly with great difficulty.

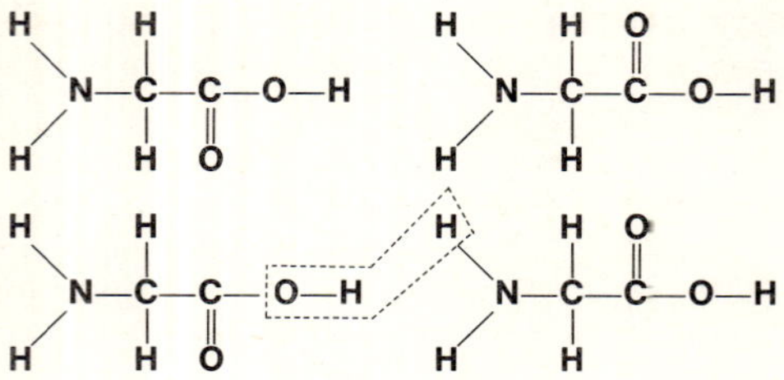

Splitting out a molecule of water,

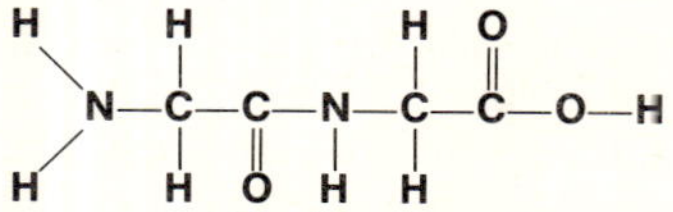

Glycyl glycine

The link between the two amino acids is called a *peptide link*. It is

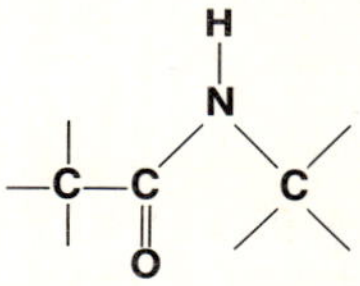

and can be detected by the biuret test. The above reaction can be written semistructurally as follows:

$$NH_2CH_2\overset{O}{\overset{\|}{C}}—OH + H\,NHCH_2COOH \longrightarrow$$

$$NH_2CH_2\overset{O}{\overset{\|}{C}}—NHCH_2COOH + H_2O$$
Glycyl glycine

Glycyl glycine is a dipeptide because it has been formed from 2 molecules of amino acid. Three molecules would produce a tripeptide and many molecules a polypeptide. Regardless of the number of molecules of amino acid used, the product formed will always have the same characteristics; i.e., it will be acidic on one end and react with a base and be basic on the other end and react with an acid; or it can react with another molecule of an amino acid. Note that each time 2 molecules react, a molecule of water is obtained.

In this fashion, more amino acids can be added, forming a larger molecule called *polypeptide*. Still larger molecules have different properties and are called *peptones*. The addition of more molecules of amino acids would make a *proteose,* and on the addition of still more we would recognize the properties of *proteins.*

From this discussion it must be evident that a protein is composed of a great many molecules of amino acids linked together as in the formation of glycyl glycine above. There are some 20 to 25 amino acids, but it is not necessary for us to know their formulas or even their names to be able to discuss proteins. When we consider that these amino acids may react in all sorts of numbers of molecules and in all sorts of combinations, we obtain some idea of the enormous number of proteins that could exist. Needless to say, our discussion must be a general one.

PROPERTIES

Since protoplasm is composed largely of proteins, we can translate much of the knowledge about proteins as directly applicable to protoplasm.

Physical

Proteins, being such large molecules, cannot form true solutions. Some form colloidal solutions, and some others, such as those in skin and hair, are insoluble in water. The physical properties of many proteins are, therefore, the properties of colloidal mixtures.

a. Many are opalescent; i.e., they have a slightly milky appearance.

b. They cannot pass through membranes; i.e., they cannot dialyze. Proteins cannot be absorbed through the intestinal wall, nor can they escape from the cells in the body. This is a very important property of blood proteins. Since they are retained by membranes, they remain within the closed circulatory system and help maintain the osmotic pressure of the blood. Consider these questions. Fluids pass out of the arterial capillaries into tissues; why is the direction of flow reversed in the venous capillaries? Albumin is a normal constituent of blood; why is it considered abnormal in the urine?

Chemical

formation of ions

No matter how many amino acids units are joined to form peptides, polypeptides, peptones, proteoses, or proteins, there is still a free amino group (basic) and a free carboxyl group (acid).

$$H_2N-CHR-CO-NH-\text{(Rest of the molecule)}-CHR-COOH$$

If we write R for part of the molecule, we can simplify the formula thus:

$$H_2N-CHR-COOH$$

The carboxyl group can give up a proton and become negatively charged

$$[H_2N-CHR-COO]^- + H^+$$

Anion

or the amino group can pick up a proton and become positively charged.

$$H^+ + H_2N-CHR-COOH \rightarrow [H_3N-CHR-COOH]^+$$

Cation

By this mechanism blood proteins can help to regulate the amount of hydrogen ion in the blood. If hydrogen ions tend to increase (acidosis), the proteins can pick them up and become positively charged. Hydrogen ions are thereby removed from the circulating fluid. If hydrogen ions tend to decrease (alkalosis), the proteins can liberate hydrogen ions and become themselves negatively charged.

action of ions of heavy metals

By heavy metals we mean such metals as lead, silver, and mercury, which have a high atomic weight. Just as we may think of nitric acid as hydrogen nitrate, so may we think of a protein as hydrogen proteinate. And as nitric acid reacts with salts to form nitrates, so we may think of proteins as forming proteinates. The proteinates of the heavy metals are insoluble. The reaction with silver ions is

$$R-\underset{NH_2}{CH}-COOH + Ag^+ + H_2O \rightarrow R-\underset{NH_2}{CH}-COOAg \downarrow + H_3O^+$$

Silver proteinate

Mercuric ions would act in a similar fashion, though we must bear in mind the difference in the valence of the mercuric ion, Hg^{++}. The disinfecting action of silver ions and of mercuric ions results from the formation of these insoluble proteinates with the protein of bacteria. It must be remembered, however, that this reaction is not specific for the protein in bacteria. That is, if the bacteria are lodged in a protein mass such as sputum, the disinfecting solution may react with the protein of the sputum and effectively cover up or seal the bacteria within the mass. A more penetrating disinfectant, such as chlorine water, would be more effective.

Silver, mercuric, and lead salts are poisonous when taken internally because of the reaction with the proteins of the body. This same type of reaction provides the action of the antidote for silver and mercury poisoning. Since the function of an antidote is to render the poison insoluble so as to reduce its absorpton, egg albumin will provide the necessary protein for the formation of the insoluble proteinate in the stomach. This solid can then be removed by means of an emetic or with the aid of a pump.

Lead poisoning is more subtle than the others because lead is a cumulative poison. Small quantities ingested are only partly excreted, so there is a slow accumulation until the fatal dose is reached.

Poisoning by mercuric chloride, or corrosive sublimate, has been serious in the past for a number of reasons: (a) it was a common household disinfectant; (b) it was dispensed in the usual tablet form; and (c) its solution is colorless and easily confused with water. Today, accidental poisoning from mercuric chloride is rare because (a) it is contained in a blue bottle the shape of which—round on one side and square on the other—is different from all others in common use; (b) it has largely been replaced by other more effective and safer disinfectants; (c) the tablets are now molded in the shape of a coffin and spaced on a string; and (d) the tablets are dyed blue to color the solution and prevent confusion with water.

the effect of heat (denaturation)

The action of heat on proteins varies with the temperature. The coagulation of the egg in a custard is brought about at a relatively low temperature and results in a gel-like structure. If the custard is cooked at a high temperature, the egg will be harder and there will be a definite separation of the solids and liquids.

If an egg is added to hot water, it will be immediately coagulated and a compact product will result. Milk proteins coagulate if the temperature is too high. A double boiler is recommended for heating milk, since the temperature of the steam is lower than that of a free flame. Since milk proteins are soluble in cold water, milk glasses should be rinsed in cold water rather than in hot to avoid the coagulation of the protein on the glass. Blood stains should be removed with cold water instead of hot for the same reason.

The use of heat in sterilization depends upon the coagulation of the protein of the bacteria. Sterilization of dressings, etc., is best accomplished in the autoclave because of the higher temperature.

Albumin in the urine may be detected by heating it. The urine is usually heated in a test tube by warming the upper part. The formation of a precipitate, or coagulation in the urine, may indicate the presence of albumin or of phosphates. Glacial acetic acid added to the precipitate will dissolve it if it is caused by phosphates.

action of salts

It has been stated that the soluble proteins form colloidal solutions. That being the case, they may be salted out of solution. When it is desirable to obtain a protein for study, it may be salted out with ammonium sulfate. Of course, albumin is separated out when it is heated, but in that case it is insoluble; however, if the protein is salted out, it is unchanged and lends itself well to experimentation.

action of acids

Organic Acids. The low hydronium-ion concentration produced by an organic acid such as acetic acid seems to have different effects on different proteins. A small amount of vinegar added to the water in which fish is boiled makes the fish much firmer. On the other hand, in boiling tough meats, the presence of a small amount of vinegar catalyzes the hydrolysis of the connective tissue to make the meat more tender. This is comparable to the action of the commercial meat tenderizers.

Mineral Acids. Nitric acid coagulates soluble proteins. Once more we may use this action to detect albumin in the urine. The test tube containing the urine is held aslant, and the concentrated nitric acid is poured down the side of the tube to form a heavy layer. The action takes place as the two liquids diffuse, resulting in the formation of a fluffy white ring. This test is known as *Heller's test*.

Tannic Acid. This acid is chosen as a representative of those acids which form insoluble salts with proteins. In these cases the protein acts as a base, forming a salt such as protein tannate. Tannic acid has been used in the treatment of second-degree burns. The acid is sprayed on the burned area to form the insoluble protein tannate. The result is an impervious coating that excludes the air and protects the area from infection. Incidentally, the acid has a mild disinfecting action by reacting with the protein of any bacteria present. Silver nitrate solution, a more powerful germicide, may be sprayed on the area after the initial spraying with tannic acid. As the burn heals, the artificial eschar that was formed gradually shrinks and finally drops off. Such acids as tannic acid are used to harden the protein in hides during the tanning process. Tannic acid is extracted from tea if it is boiled.

effect of alcohol

Ethyl alcohol coagulates proteins. For this reason 70 per cent alcohol can be used as a disinfectant, the protein of the bacteria being affected. The effective concentration of alcohol as an antiseptic is 70 per cent by *weight*. Alcohol in concentrations varying only slightly from this value is less effective. In fact, 70 per cent alcohol by *volume* is almost useless as a germicide. High concentrations of alcohol may dehydrate bacteria without killing them. The use of 50 per cent alcohol is recommended for back rubs. The alcohol hardens the protein of the skin and reduces sweating. Both of these actions tend to prevent the decubitus ulcers that may otherwise develop on bedridden patients.

hydrolysis

The products of protein hydrolysis served as the basis for one method of classification

$$\text{Simple protein} + H_2O \rightarrow \underset{\text{Final product}}{\text{amino acids}}$$

$$\text{Conjugated protein} + H_2O \rightarrow \text{amino acid} + \text{other compounds}$$

Enzymatic Hydrolysis (Digestion). Enzymes that catalyze the hydrolysis of proteins are called *proteases*. These catalysts regulate a stepwise breakdown of the large protein molecule by splitting off amino acids. As the protein molecule becomes smaller, it no longer has the properties of true protein and is called *derived* protein.

$$\text{Protein} + xH_2O \xrightarrow{\text{Protease}} \text{proteose} + x \text{ amino acids}$$
$$\text{Proteose} + xH_2O \xrightarrow{\text{Protease}} \text{peptone} + x \text{ amino acids}$$
$$\text{Peptone} + xH_2O \xrightarrow{\text{Protease}} \text{polypeptides} + x \text{ amino acids}$$
$$\text{Polypeptides} + xH_2O \xrightarrow{\text{Protease}} \text{peptides} + x \text{ amino acids}$$
$$\text{Peptides} + xH_2O \xrightarrow{\text{Protease}} x \text{ amino acids}$$

Amino acids are the only products formed as a result of the digestion of simple proteins.

In the digestion of a conjugated protein such as the glycoprotein mucin, carbohydrases (enzymes that catalyze the hydrolysis of carbohydrates) would be involved in addition to the proteases. The final products of the digestion of mucin would be amino acids and simple sugars.

oxidation

Proteins can react with oxygen, as anyone knows who has burned a steak. In a calorimeter, an instrument for measuring the heat evolved, a protein such as albumin evolves 5.80 kcal per Gm. This results from the formation of carbon dioxide, water, and nitrogen. In the body, however, nitrogen and part of the carbon and hydrogen are split off before the remaining part of the molecule is oxidized. The actual amount of heat produced by the oxidation of protein in the body is about 4.35 kcal per Gm, which is usually rounded off to 4 kcal per Gm.

Sulfur and phosphorus found in protein can be oxidized to sulfates and phosphates.

When protein is oxidized in the body, the volume of carbon dioxide produced is less than the volume of oxygen used in the oxidation. The respiratory quotient (the volume of oxygen used divided into the volume of carbon dioxide produced) for protein is given as 0.8.

Color Tests for Proteins

A number of tests are used qualitatively to detect proteins. It must be understood, however, that no one reaction is characteristic of all proteins, nor will a given protein necessarily respond to all tests. Instead, each test indicates a particular type of linkage or the presence of specific groups in the amino acids present in the proteins. However, from a practical standpoint, by applying several of the tests a compound can be shown to belong to this class.

biuret reaction

When a protein is treated with a small amount of sodium hydroxide solution and a dilute solution of cupric sulfate is added drop by drop, a reddish-violet to violet-blue color is produced. This reaction is characteristic of all compounds that have 2 amino groups attached to different atoms. The name of the test arises from the fact that biuret, $NH_2CONHCONH_2$, gives the test. If protein is completely hydrolyzed to amino acids, this test will be negative.

millon's reaction

In this test the protein is warmed with a small amount of Millon's reagent, a solution of mercuric nitrate in dilute nitric acid. A yellow or white color changing to red on heating indicates the presence of a protein. This test is given by those proteins which yield tyrosine on hydrolysis and by other compounds that contain the $—C_6H_4OH$ group.

xanthoproteic reaction

Concentrated nitric acid produces a yellow color when it acts on any protein material containing an amino acid with a benzene ring in it. The color is turned to orange if the solution is made basic. Almost everyone who has worked in a chemical laboratory has at one time or another observed this test on his fingers when he has gotten concentrated nitric acid on them. In such cases the color is more pronounced after using soap because of the slight alkalinity of the soap solution.

hopkins-cole test

To a mixture of a protein and a glyoxylic acid (CHOCOOH) solution, concentrated sulfuric acid is added carefully down the side of the tube to form a layer. If tryptophan is present in the protein, a violet ring will appear between the two layers.

Some Biologically Important Proteins

albumin

Albumin is a relatively small molecule, as proteins go, and forms colloidal mixtures in water and dilute salt solutions. It is coagulated by heat. Serum albumin is of vital importance in maintaining the osmotic pressure of the blood. The concentration of albumin in serum varies, but the normal range is usually described as 4 to 5.2 Gm per 100 ml serum.

globulins

Globulins are much larger molecules than albumins and form colloidal mixtures in dilute salt solutions. Serum globulins are not as effective as albumin in regulating the osmotic pressure of the blood. Gamma globulin is involved in the transport of antibodies, and its concentration increases when the body is invaded by certain disease organisms. The concentration of serum globulins is usually 1.3 to 2.7 Gm per 100 ml serum.

In some abnormal conditions the smaller albumin molecules leak through body membranes, causing a decrease in the osmotic pressure of the blood, and edema results. A laboratory test, the A/G ratio, can be used as a diagnostic tool. This is the relative concentration of albumin and globulin in the serum. Low A/G ratios are found in malnutrition, in certain disorders of the liver, in nephritis, and in nephrosis.

The enzymes urease, trypsin, chymotrypsin, amylopsis, and steapsin are globulins, as is also fibrinogen.

myosin

Myosin constitutes about 80 per cent of muscle protein and is coagulated upon death. The phenomenon known as *rigor mortis* is produced by this coagulation.

collagen

Connective tissue and cartilage consist largely of collagen. This word has the same root as "colloid," and the suffix added to this root makes the name mean "glue forming." Long boiling at 100°C converts this protein into gelatin, which is softer and soluble. The rendering tender of tough pieces of meat, such as corned beef, by long boiling is a process of converting this protein into gelatin. It is a well-known fact that this corned beef can be held together as pressed corn beef by wetting it with the stock obtained in the cooking. Tannin, chrome alum, and other reagents change collagen to a harder substance that resists bacterial decomposition. In the tanning industry, the hides are freed from hair by a treatment with lime, after which the skin is changed to leather by exposure to some one of these tanning agents.

Disorders such as rheumatic fever, rheumatoid arthritis, and serum sickness produce changes in the collagenous connective tissue such that the collagen may disintegrate, harden, or die. The processes causing these reactions are unknown today.

keratin

Keratin is the protein of the skin, nails, and hair and is remarkable for its insolubility.

fibrinogen

Fibrinogen occurs in the blood plasma of all vertebrates to the extent of about 0.4 per cent. Upon exposure to air by a cut or wound, it is converted by enzymatic action into *fibrin,* which precipitates and so stops the flow of blood. Fibrin is a derived protein.

hemoglobin

Hemoglobin is a conjugated protein of the red blood cells that forms, upon hydrolysis, the simpler protein *globin* and an iron compound, *hemin.* It is rather interesting to compare hemoglobin with chlorophyll, inasmuch as they have somewhat similar roles to play in the two kingdoms, plant and animal. They seem to have similar structures, chlorophyll being a compound of magnesium and hemoglobin being one of iron.

In the blood, which is slightly alkaline, hemoglobin exists in the form of the negatively charged ion formed by the loss of hydrogen. Hemoglobin ions react promptly in an oxygen-rich atmosphere to form the loose compound *oxyhemoglobin.* In this way most of the oxygen is transported from the lungs to the tissues, where in the decreased oxygen concentration the oxyhemoglobin dissociates, forming oxygen and hemoglobin. This process can be repeated over and over.

Metabolism of Proteins

During our study of the digestion of proteins we learned that they are hydrolyzed to form the constituent amino acids. In considering the metabolism of proteins, we are concerned with the fate of each amino acid. The limits of our discussion will not permit the consideration of each amino acid, but we must confine ourselves to general reactions only.

anabolism

The use of isotopic nitrogen (N^{15}) in artificially prepared amino acids has enabled investigators to trace the course of the acids and the amino group in the body. Such studies have forced us to alter some of our ideas regarding amino acid metabolism. For instance, it is now known that the amino acids in the blood are in a dynamic equilibrium with cell proteins, because after feeding amino acids containing tagged atoms, those atoms are found incorporated in the proteins of the body.

The maintenance of nitrogen equilibrium[1] in the body was originally interpreted to mean that the body could not accumulate or store protein. We now know that if the food protein is increased, there is a definite lag in the increase of nitrogen excretion and the body shows a new nitrogen equilibrium at a higher level. Therefore, there must have been a storage of new protein in the body. Conversely, following such a high protein intake, the body resists the effect of a protein-deficient diet over a relatively long period.

[1] In nitrogen equilibrium the total nitrogen of the food is equal to the total nitrogen excreted. The body can maintain this equilibrium, provided the intake is not below a certain minimum and the food proteins are adequate. If the nitrogen excreted is greater than the nitrogen of the food, the equilibrium gives way to a negative balance. Such a balance follows starvation and wasting diseases. On the other hand, a positive balance results when the nitrogen excreted is less than the nitrogen of the food; this occurs during growth or recovery from starvation or wasting disease, during convalescence, and following surgery.

The protein is stored chiefly in the liver, kidney, and intestinal tissues, and this protein is the first to be used up when the protein intake is greatly reduced. Under these conditions the loss of body protein is reduced by the addition of carbohydrate or small amounts of fat to the diet. These foodstuffs are therefore called *protein sparers*. This protein-sparing effect demonstrates still further that proteins can be used for storage as well as replacement of tissues.

catabolism

Amino acids not used in the production of body proteins undergo a process of *oxidative deamination* in the liver. Just how this is accomplished may only be postulated, but the net result is

$$R-\underset{\underset{NH_2}{|}}{CH}-COOH \xrightarrow{[O]} NH_3 + R-\underset{\underset{O}{\|}}{C}-COOH$$

Amino acid → Keto acid

or specifically,

$$CH_3-\underset{\underset{NH_2}{|}}{CH}-COOH \xrightarrow{[O]} NH_3 + CH_3-\underset{\underset{O}{\|}}{C}-COOH$$

Alanine → Pyruvic acid

These products then go their separate ways.

The ammonia is converted to urea in the liver.

$$2NH_3 + CO_2 \rightarrow NH_2CONH_2 + H_2O$$

and the urea, being very soluble, is readily transported to the kidneys, where it is eliminated in the urine.

The keto acids resulting from the deamination of the amino acids are oxidized to carbon dioxide and water and yield energy at the same time. If these acids are not needed for immediate energy, they may be stored for future use in the form of body fat.

Proteins and Heredity

We have mentioned that protoplasm is protein in nature. Some very exciting research is being carried on these days, using tagged atoms, X rays, electron microscopes, etc., in an effort to find the answer to such questions as: How do the chromosomes influence the dividing cells of an organism to ensure the development of hereditary traits? Are there chemical directors to dictate the development of such traits as hair and eye color and body build, as well as kinds of cells—liver cells, muscle cells, etc.? Does the fertilized egg contain a master plan for the finished organism?

Questions like these lead us to a study of the nucleoproteins and of their very important constituents, deoxyribonucleic acid (DNA) and ribonucleic acid (RNA).

The sperm cell is composed almost entirely of DNA and since the sperm cell is as important as the egg in contributing hereditary material, DNA must in some way contain the answer to at least some of these questions. The genes in the chromosomes are specific parts of the DNA chain. One theory of heredity is the *one-gene–one-enzyme* theory according to which one gene determines the synthesis of amino acids in the proper sequence to form one enzyme. Enzymes are very important proteins which control the rate of both anabolic and catabolic reactions in the body. If an enzyme is absent or defective, a necessary reaction in the body may not occur or it may be abnormal. Enzymes therefore control growth.

As we know, enzymes are chains of specific amino acids joined by peptide links in the proper sequence. These then control the synthesis of other substances in the cell.

The molecule of DNA is composed of a double coiled strand; the outer aspect of each strand is composed of alternate ribose-phosphate units. Attached to the ribose units on the inner aspect of each strand is a purine or a pyrimidine base unit. Hydrogen bonds hold a purine from one strand to a pyrimidine of the other.

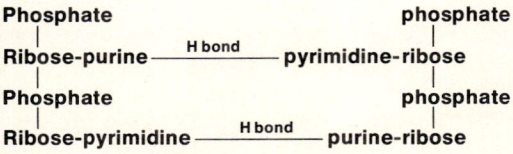

The purines in DNA are guanine and adenine, and the pyrimidines are thymine and cytosine. Therefore a linkage might be

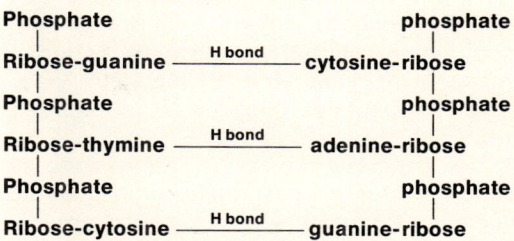

In cell division, the hydrogen bonds break and each strand of DNA can duplicate itself so that each daughter cell contains all the genetic material in the proper order. The order of the bases is important because it is this sequence in DNA in the nucleus that determines the arrangement of these units in the messenger RNA that is synthesized from the DNA pattern or template.

Guanine always couples with cytosine, and adenine couples with either thymine or uracil. In RNA there is no thymine. Using the right-hand strand of DNA above as a pattern for a strand of RNA we would get

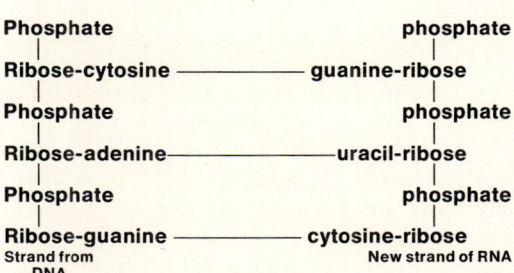

The messenger RNA travels from the nucleus to the ribosomes in the cytoplasm where it becomes attached. When this occurs, amino acids from the cytoplasm are brought to the ribosome by a smaller molecule called transfer RNA. The arrangement of any three consecutive bases on a strand of messenger RNA is thought to be the *code* that determines the specific amino acid and the place where it belongs in the protein chain. Another code of three bases (a triplet)

determines when the molecule is the right size and the synthesis of that particular protein is stopped.

Using the codes on the messenger RNA as the directions, the ribosomes add the right amino acid in the right place by forming peptide links and building up peptides, polypeptides, etc., up to the desired protein.

As each cell divides, it passes on to its daughter cells an exact replica of the genetic nuclear information which is used as a model for manufacturing messenger RNA which determines, in turn, what proteins will be manufactured by the cell. These proteins in turn determine the cell's activities, and the characteristics of the developing organism are thus determined.

Some diseases are thought to be caused by missing or defective enzymes. Such a condition is PKU. The initials stand for phenylketonuria. It is characterized by the presence in the urine of abnormal phenylketones, which accumulate in the blood and then spill over in the urine, because an enzyme needed for the normal metabolism of an essential amino acid, phenylalanine, is missing.

Scientists wonder if, in the future, such defects could be corrected if the defective or missing gene could be repaired, replaced, or compensated for in some way.

Study Exercises

1. Define simple, conjugated, and derived proteins and give three examples of each.
2. Name five proteins found in the body.
3. What is a peptide link? How is it formed? What is a dipeptide? What is a polypeptide?
4. How do proteins help maintain the osmotic pressure of the blood?
5. Why would one expect that serum albumin would be retained within intact capillaries and would be absent from a normal urine sample?
6. Explain how proteins can act as either anions or cations.
7. Why are the salts of heavy metals poisonous?
8. What would be an antidote for mercuric nitrate poisoning?
9. Why must the metal-antidote mixture be removed from the stomach by pump or gastric lavage (washing out the stomach)?
10. Why are silver nitrate, mercuric chloride, and ethyl alcohol effective disinfectants?
11. How does heat destroy microorganisms?
12. Why should blood stains be rinsed first in cold water?
13. Give two simple methods for testing for the presence of albumin in urine.
14. Some burn ointments and treatments make use of tannic acid. Why?
15. What are the end products of digestion of simple proteins?
16. A sample of albumin and protease was incubated in the laboratory to bring about the digestion of the protein. How could one test for the completeness of protein digestion?
17. Why is 70 per cent ethyl alcohol a more effective disinfectant than the more concentrated 95 per cent solution?
18. What is meant by protein sparer?
19. What does deamination mean?
20. How is urea formed in the liver?
21. Mr. Brown is on a high-protein diet. This diet yields more calories than he needs for normal growth and development. Would one expect him to gain weight? Give reasons for your answer.
22. What is the source of urea in the urine?
23. What are the products of the oxidation of proteins in the body?

Chapter eighteen METABOLIC REGULATORS

Ideas for preview or review

Basal metabolism or basal metabolic rate (B.M.R.) is a measure of the rate of oxidation in the body when the body is at rest, the stomach is empty, the mind is at ease, and the environmental temperature is comfortable.

Oxidation reactions generate heat.

Heat production in the body is more closely related to body surface than to body weight.

$$\underset{\text{Glucose}}{C_6H_{12}O_6} + 6O_2 \rightarrow 6CO_2 + 6H_2O + \underset{\text{Heat}}{688 \text{ kcal}}$$

By knowing the amount of oxygen used in the oxidation of glucose, we can calculate the amount of heat generated.

One method of measuring the basal metabolic rate involves determining the amount of oxygen consumed by a patient in a given period of time. From this figure, the amount of oxygen used in an hour can be calculated.

Under basal conditions, the consumption of 1 liter of oxygen is equivalent to 4.8 kcal.

Under basal conditions, a normal adult man generates about 40 kcal per hr per sq meter of body surface. An adult woman produces about 36 kcal per hr per sq meter.

A man who produces 55 kcal per sq meter per hr is generating 15 kcal above average.

$$\frac{15}{40} = \frac{3}{8} \quad \text{or} \quad 37.5 \text{ per cent}$$

This measurement is reported as +37.5.

An adult woman who produces 28 kcal per hr per sq meter of body surface is generating 8 kcal per hr below normal.

$$\frac{8}{36} = \frac{2}{9} \quad \text{or} \quad 22 \text{ per cent}$$

This is recorded as −22.

Readings are considered normal within the range of +20 and −20.

B.M.R. measurements are of considerable clinical significance, particularly in disorders involving the thyroid glands.

The presence or absence of minute amounts of certain substances in the body means the difference between normal function and disease or even death. These metabolic regulators include the hormones, proenzymes, enzymes, coenzymes, activators, vitamins, antigens, and antibodies. Their actions are related and interdependent. We know that some of the essential vitamins are part of enzyme systems. For example, vitamins B_1, B_2, and B_6 are coenzymes. Hormones function as regulators of enzyme systems, some acting to initiate or accelerate enzyme action, others to inhibit or prevent enzymes from functioning. The hormone insulin, from the pancreas, causes a drop in glucose concentration in the blood by transporting glucose from the blood across the cell membranes into the tissues, and by accelerating the enzymatic conversion of glucose to glycogen in the liver. A hormone from the pituitary (diabetogenic hormone) facilitates the breakdown of glycogen to glucose, causing an increase in blood sugar levels. We see here a system of checks and balances, giving under normal functioning a beautifully balanced chemical system in which cells live within very narrow limits of concentration changes.

When most of these substances were discovered, little was known of their chemical structure. They were detected through clinical manifestations of malfunction or absence. So we have the early method of naming the vitamins after letters of the alphabet and the rather crude distinction between hormones and enzymes; namely, hormones are substances secreted by ductless glands and poured into the general circulation to exhibit their effects at some distance from their point of origin, and enzymes are secreted through ducts to the site of their activity or are produced by the cells in which they function. Hormones are messengers, and enzymes are catalysts.

Recent research has vastly improved our knowledge in this field. Since these substances exist in such small quantities, the problem of isolating them in a pure form was a difficult one. Today this is no longer a difficulty. There are several reasons for this improved state of affairs. A method of analysis called *chromatography* enables an investigator to separate very small amounts of material quickly and quantitatively. Spectrophotometers and ultracentrifuges are now mass produced so that even a small laboratory can afford to have them. These instruments greatly facilitate research work. The availability at reduced prices of isotopes has stimulated the use of these substances to label molecules or atoms in molecules. Labeled molecules can be followed during metabolic changes, and their fates can be determined. The electron microscope shows details of structure heretofore unavailable, and new theories, like the spiral or helix theory of protein structure, have made X-ray diagrams of these large molecules intelligible. Induced genetic changes in cells have shed light on enzyme chemistry. These and other techniques have revolutionized our knowledge of the vital metabolic regulators.

ENZYMES

Enzymes are globular proteins that act as catalysts. They increase or decrease the rate of specific chemical reactions. By this we mean that one kind of enzyme will speed up one chemical reaction or one kind of chemical reaction, but not another, although the second reaction may be very similar to the first. For example, the enzyme sucrase in the intestinal juice will catalyze the hydrolysis of sucrose, $C_{12}H_{22}O_{11}$, to glucose and fructose but does not affect the hydrolysis of maltose or lactose, both isomers of sucrose (compounds having the same molecular composition, $C_{12}H_{22}O_{11}$, but different arrangements of atoms). Catalysts alter the rate of chemical reactions but can be recovered unchanged at the end of the reaction. Actually some of the enzyme molecules are broken or worn out in the course of the reaction. This is not surprising, since globular proteins are very fragile. However, a little enzyme goes a long way. One milligram of sucrase, 0.001 Gm, will catalyze the hydrolysis of about 200 Gm sucrose. This is 1 part of enzyme in 200,000 parts of sucrose. The material with which or upon which an enzyme works is called the *substrate*. In the example above, sucrose is the substrate. The present method of naming enzymes generally adds the suffix *-ase* to the root name of the substrate.

Substrate	*Enzyme*
Sucrose	Sucrase
Maltose	Maltase
Urea	Urease
Glycogen	Glycogenase
Cholesterol esters	Cholesterinase
Lecithin	Lecithinase

Some of the earliest known enzymes still go by their old names: pepsin, trypsin, amylopsin, etc.

Proenzymes

Many enzymes exist in an inactive form until the proper conditions for their function are present. These inactive forms are called proenzymes and are named with the suffix *-ogen*, e.g., pepsinogen, trypsinogen. Pepsinogen is a protein with a molecular weight of 42,000. In the presence of hydrochloric acid, it breaks up, forming pepsin, with a weight of 38,000, and a protein piece. This extra piece of protein had tied up the enzyme property until conditions were right for its activity, namely, a high concentration of hydrogen ion.

Coenzymes

In many cases, the presence of another substance is necessary for the enzyme to function. These extra substances are called coenzymes. They are usually organic medium-sized molecules. Being medium sized, they do not evidence the properties of colloids. (The enzymes, being proteins, are also colloids.) Some coenzymes are very simple; HCl is a coenzyme for pepsin. Some coenzymes in oxidation systems contain high-energy phosphate groups; coenzyme I and coenzyme II are examples. (Sometimes metals such as magnesium, iron, zinc, copper, or manganese are needed for enzyme function. These metals are called *activators*.) Niacin, also called nicotinic acid, the antipellagra vitamin, is an essential part of coenzyme I. Folic acid, another vitamin, is a coenzyme.

ENZYME ACTION

We can see from this brief discussion that enzymes do not work alone. We refer more correctly to enzyme "systems." Several theories have been advanced to account for enzyme action. One states that enzymes may start chain reactions. Another theory claims that enzymes supply a surface upon which chemical action can occur. An enzyme may have a surface shape that allows it to fit upon the shape of the substrate. Such an action would be comparable to the fitting of a particular key in a particular lock. Some enzymes act as master keys and fit more than one substrate. These enzymes can catalyze more than one reaction. If an enzyme is a hydrolase (catalyst for a hydrolysis reaction), a molecule of water and a molecule of the substrate are brought into such a relationship that the hydrolysis reaction can occur. The action of enzymes is influenced by changes in hydrogen-ion concentration; i.e., each enzyme is capable of catalytic action only within a certain range of pH values. Enzyme action is affected by temperature changes. High temperatures destroy enzyme activity. Finally, enzyme action, like catalysis, is reversible. The enzymes do not change the equilibrium point of a reaction, but the systems, under enzyme influence, reach equilibrium rapidly. This may well explain many of the facts of metabolism. Anabolism and catabolism are in many cases reverse chemical reactions, and the shift in the equilibrium here is caused by the same conditions that affect the equilibrium of any reversible reaction, e.g., the relative concentrations of the reactants and the products. For instance, the production of glycogen $(C_6H_{10}O_5)_x$ from glucose $(C_6H_{12}O_6)$ is an example of anabolism or synthesis, and the reverse, glycogen forming glucose, illustrates catabolism. These two actions may well be due to the same enzyme system, but the direction may be due to the relative concentrations of glucose, glycogen, and water.

CLASSIFICATION OF ENZYMES

Enzymes can be classified according to the type of chemical reaction with which they are involved.

1. Oxidoreductases. As the name implies, these enzymes catalyze oxidation-reduction reactions. Here we find oxidases, hydrogenases, and dehydrogenases. If the reactions involve molecular oxygen, they are called oxidases. Although removal of hydrogen is also an oxidation reaction, enzymes concerned with this change are called dehydrogenases. Since oxidation also means loss and gain of electrons, enzymes involved in electron transfer come in this category. The final conversion of food to carbon dioxide and water in the body occurs through two series of reactions, the Krebs citric acid cycle and the electron transfer system. Both sets of reactions are catalyzed by oxidoreductases.

2. Transferases. These enzymes move groups of atoms. Transmethylases move methyl groups; transaminases move amino groups from one molecule to another.

3. Hydrolases. These enzymes are involved in hydrolysis reaction, the chemical reaction of a substance with water with the formation of simpler products. In this class we find the digestive enzymes. Those concerned with the hydrolysis of carbohydrates are the carbohydrases, one of which is sucrase. We have also lipases, peptidases, esterases, phosphatases, etc.

4. Lyases. These enzymes remove groups from compounds. Examples are decarboxylases which split off carboxyl groups, and carbonic anhydrase which breaks up the carbonic acid molecule to form carbon dioxide and water.

5. Isomerases. Enzymes move atoms or groups around in the same molecule thus forming isomers. An example is the formation of glucose-6-phosphate to glucose-1-phosphate. In this reaction, the phosphate group moves from one terminal carbon atom in the molecule to the other.

6. Ligases or synthetases. Through the action of these enzymes, two molecules are combined and at the same time a high-energy compound such as adenosinetriphosphate (ATP) is decomposed.

Of prime importance to an understanding of body function is an understanding of the hydrolases and the oxidoreductases.

enzymatic hydrolysis

Substances such as glucose and fructose are able, when dissolved, to dialyze through the walls of the intestines. For that reason they do not require digestion, and we find no enzymes for them.

Other foods must undergo chemical changes before they can dialyze through the intestinal walls. To facilitate these chemical changes, the food particles are broken up into small bits by chewing and by the churning action of the stomach. They are intimately mixed with the digestive fluids containing catalysts, the enzymes, for the required reactions. A diagram of the digestive tract is given in Fig. 18–1.

salivary digestion

Composition of Saliva. The composition of saliva varies, depending upon the stimuli that arouse the reflex secretion of the saliva. Under normal conditions saliva is secreted in sufficient quantity to keep the mouth moist, but in emotional states it is stopped. The secretion is increased by (*a*) the tasting and smelling of food, (*b*) the chewing of foods or even the movement of the jaws or the presence of a solid substance in the mouth, (*c*) the sight or thought of good-tasting food. Saliva collected while chewing paraffin contains about 99.5 per cent water. The remaining part contains *mucin,* a glycoprotein that confers upon saliva its stringy consistency and gives saliva its lubricating quality, salivary amylase (ptyalin), maltase, and such inorganic ions as chloride, carbonate, phosphate, sulfate, sodium, potassium, calcium, and magnesium. A normal person secretes about 1,500 ml saliva daily. It is an almost neutral solution.

Action of Saliva. Salivary amylase catalyzes the hydrolysis of starch. The actual reaction must be very complicated, but it may be represented in more or less sketch fashion as follows:

Metabolic Regulators

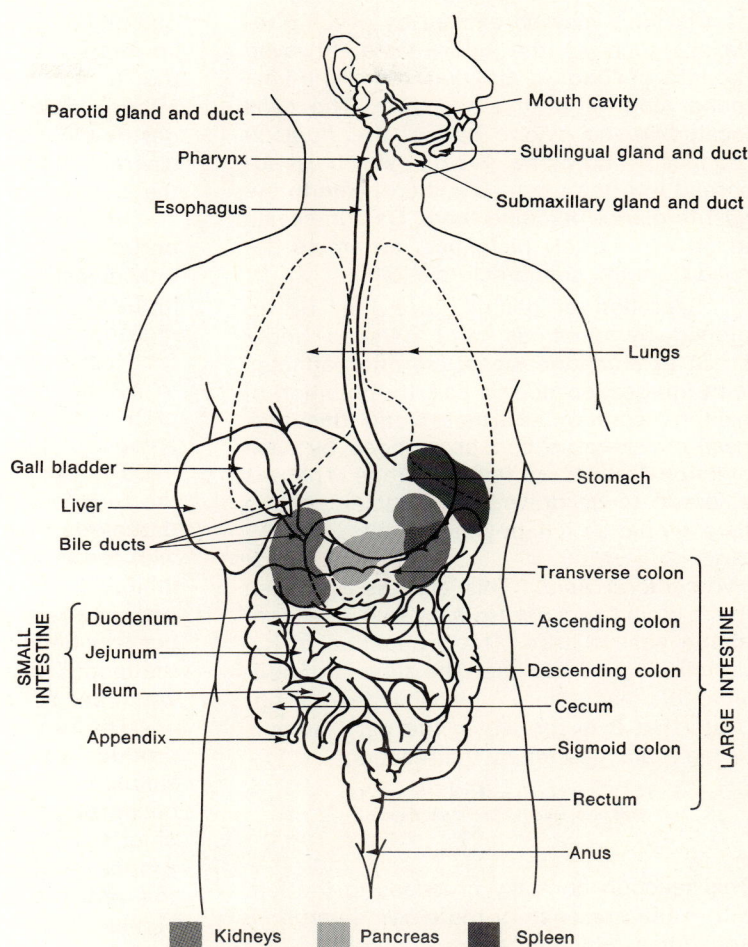

Fig. 18-1 Diagram of the digestive tract. *(From Harold S. Diehl and Ruth E. Boynton, "Personal Health and Community Hygiene," 2d ed., McGraw-Hill Book Company, New York, 1951.)*

$$(C_6H_{10}O_5)_n \rightarrow (C_6H_{10}O_5)_x \rightarrow C_{12}H_{22}O_{11}$$
$$\text{Starch} \quad\quad \text{Dextrins} \quad\quad \text{Maltose}$$

The enzyme maltase catalyzes the hydrolysis of maltose to glucose.

$$C_{12}H_{22}O_{11} \xrightarrow{\text{Maltase}} C_6H_{12}O_6$$

Any glycogen in food is also hydrolyzed by salivary amylase into glucose. Cellulose is not hydrolyzed by the salivary enzymes. Unruptured grains, which are covered with a thin coating of cellulose, are unaffected.

Practically speaking, very little digestion occurs in the mouth with the exception of carbohydrate food particles that cling to the teeth.

gastric digestion

Gastric Juice. Studies made on gastric juice show that it consists of hydrochloric acid and the enzymes gastric protease (pepsin) and gastric lipase. Secretion of gastric juice may be excited by psychic reflexes, such as the odor, taste, or even thought of food. In addition, there is a hormonal control of the secretion. The cells located in the pyloric part of the stomach produce a hormone *gastrin,* which is absorbed into the blood and then taken to the gastric glands, exciting them. Gastrin seems to be very largely histamine, an amine derived from the amino acid histidine.

Secretion of gastric juice may be inhibited by emotions, and persons under strain or emotional upset are often afflicted with indigestion due in part to inhibition of gastric secretion. Further along this line, gastric peristalsis can be stopped by such emotional states as fear and rage. There is a lesson to be drawn here; namely, eating time should be a happy time.

Hydrochloric Acid. This acid as produced in the stomach varies from 0.2 to 0.5 per cent in the gastric juice. The production of hydrochloric acid has been the subject of considerable discussion. A plausible theory suggests that it is formed in the parietal cells of the fundic glands by the reaction

$$\underset{\text{Present in blood}}{H_2CO_3} + NaCl \rightleftharpoons \underset{\substack{\text{Excreted} \\ \text{into} \\ \text{stomach}}}{HCl} + \underset{\substack{\text{Resorbed} \\ \text{into blood}}}{NaHCO_3}$$

This reaction normally proceeds to the left, but in the presence of the enzyme *carbonic anhydrase,* which catalyzes the reaction

$$H_2CO_3 \rightleftharpoons CO_2 + H_2O$$

production of hydrochloric acid is favored, provided the latter is removed from the reacting system. At any rate, it is very evident that hydrochloric acid is present in the stomach in concentrations that are toxic to living cells and so must be produced on the spot, as it were. Further, it is known that the pH of the blood increases soon after eating, due no doubt to the reabsorption of sodium bicarbonate as shown above. This is reflected in the increase of the pH of the urine at this time, an effort on the part of the body to maintain a constant pH of the blood. The quantity of hydrochloric acid secreted varies so markedly under pathologic conditions that it may be a valuable aid in diagnosis. Hyperacidity is generally present in cases of gastric ulcer and certain neurotic conditions, whereas hypoacidity occurs in carcinoma of the stomach and in certain types of anemia.

Hydrochloric acid has a number of important functions in gastric digestion. The only digestion of carbohydrates taking place in the stomach is the continuation of salivary digestion in the less acidic portions of the food mass. The pH of saliva varies from 6.6 to 7.2, but ptyalin seems to function best at pH 6.0. In the stomach, therefore, a small amount of acid may lower the pH of the food mass to provide the optimum conditions for the functioning of ptyalin. However, as hydrochloric acid becomes thoroughly mixed with the food mass, the action of ptyalin is inhibited and finally stopped. The pH of gastric juice is about 1.0 when freshly excreted but may become 3.0 or 4.0 as it is mixed with the food. The action of ptyalin is inhibited at about pH 5.0 and stopped at about pH 3.0.

The secreting cells of the gastric glands produce a proenzyme of pepsin called *pepsinogen.* The substance is activated and becomes pepsin in the presence of hydrochloric acid. The acid, further, provides the proper medium for the maximum activity of this enzyme, which will be discussed later. Finally, the hydrochloric acid provides a sufficiently high hydronium-ion concentration in the gastric juice to inhibit the growth or kill many of the bacteria that enter the stomach with the food.

Action of Pepsin. This enzyme is formed from pepsinogen through activation by hydrochloric acid. It converts proteins into proteoses and peptones, the first stage of their degradation. Pepsin is most active at about pH 1.5. Its activity is inhibited at pH 3.0 and stopped at pH 4.0. On the other end of the scale, a pH lower than the optimum reduces the activity, and long exposure may even destroy pepsin.

Action of Gastric Lipase. The hydrolysis of fats in the stomach is very slight. In order to increase the speed of reaction, the area of the interface between the fat and the reactant must be increased. This is accomplished by emulsification in the intestine. Emulsified fat such as butterfat in milk is partially digested in the stomach through the action of gastric lipase, but the great bulk of the fat digestion takes place in the small intestine.

intestinal digestion

After the action in the stomach has gone on for 2 to 7 hr, the food mass plus the gastric juice, now called *chyme,* moves on into the intestine through the pylorus. The cause of the relaxing and contracting of the pyloric sphincter is not completely understood, but it seems that the acidity of the chyme has the greatest effect; that is, the high acidity of the chyme in contact with the mucosa on the gastric side of the pylorus causes the sphincter to relax, and when enough chyme has passed through to bring acid in contact with the lining of the intestine, it contracts. Thus it is that the chyme slowly but at intervals moves into the intestine.

Pancreatic Juice. This fluid, which contains a high percentage of water, also contains the enzymes *amylopsin* (pancreatic amylase), *steapsin* (pancreatic lipase), *lactase, maltase,* and *carboxypolypeptidase,* which attacks polypeptides on the acidic end of the molecule. In addition, the juice contains *trypsin* and *chymotrypsin,* which are secreted as *trypsinogen* and *chymotrypsinogen.* Trypsinogen is activated by the *enterokinase* of the intestinal juice, and the chymotrypsinogen is activated by trypsin. Trypsin and chymotrypsin are proteases, converting proteins to polypeptides. In addition, trypsin accelerates blood clotting, and chymotrypsin clots milk. Salts of weak acids present, such as carbonates and phosphates, hydrolyze to produce a pH of 9.0. The activation of the pancreas to produce this juice seems to be caused by the hydrochloric acid as it comes in contact with the cells of the duodenum. These cells produce a hormone, *secretin*, which upon being absorbed into the blood is conducted to the pancreas and excites its cells.

Bile. Strictly speaking, bile is not a digestive fluid because it does not include any enzymes. However, since it plays such an important part in digestion, it is included in our discussion. The bile components of interest to us are bile salts, bile pigments, and cholesterol.

Bile is produced in the liver and concentrated in the gallbladder, where part of it is stored through the absorption of water from the bile. The secretion of bile by the liver is continuous and seems to be stimulated by the presence of bile salts in the intestine; that is, the bile salts are absorbed and conveyed to the liver, they stimulate the liver, and then they are included in the bile again— a continuous process. Evidence for this is found in the relatively small amount of bile salts or products of their decomposition found in the feces. The flow of bile from the gallbladder into the duodenum is caused by the contraction of the gallbladder, which has been activated by the hormone *cholecystokinin*. This hormone seems to be formed when fat in the chyme arrives in the intestine. An adult normally produces about 500 ml bile per day.

The bile salts are the sodium salts of glycocholic and taurocholic acids. These salts have a tremendous ability to lower the surface tension of water and thus act as emulsifying agents.

The bile pigments obviously provide the color. The chief compounds are *bilirubin* (reddish) and *biliverdin* (greenish). The latter results from the oxidation of the former. The normal color of feces and of urine is caused by compounds derived from bilirubin. This pigment results from the catabolism of hemoglobin and is extracted from the blood by the liver. Black and blue spots show the formation of bilirubin from the hemoglobin of injured blood cells. When bile is not discharged owing to an obstruction in the duct or to a liver disorder, these pigments accumulate in the blood and the skin has the yellow color characteristic of jaundice.

Cholesterol is eliminated from the body in the bile, and when the bile becomes concentrated in the gallbladder, cholesterol may crystallize to form gallstones.

Intestinal Juice. This juice contains the enzymes *sucrase, maltase,* and *lactase* to hydrolyze the corresponding sugars. There are also present a number of *peptidases,* which are capable of bringing about the final stages of the hydrolysis of proteins to amino acids (erepsin effect). These peptidases include *aminopolypeptidase,* which attacks polypeptides at the amino group, and *dipeptidase,* which hydrolyzes dipeptides to amino acids. The substance *enterokinase* is also present in the intestinal juice. This juice is secreted by the glands of the intestinal mucosa and has a pH of about 8.0.

Enzyme Action on Chyme. While the three liquids involved in intestinal digestion have been considered separately, it must be remembered that they act simultaneously. Not only do they provide the necessary enzymes for the final conversion of carbohydrates, fats, and proteins to compounds that may be absorbed, but they establish the proper conditions for the optimum activity of the enzymes. Further, the bile through its emulsifying action increases the surface area of fats subjected to the lipases. It must be evident, then, that absence of bile would greatly retard the digestion of fats and, in general, interfere with digestion, because fats tend to form a coating over particles of carbohydrate and protein of the chyme and thus prevent access of the water-soluble enzymes into the interior of those food particles.

The action of the various enzymes is best shown in Table 18–1, which summarizes digestion.

absorption

Absorption of foods takes place in the small intestines, which are well equipped for this purpose. Their great length (about 23 ft), their great surface due to the villi in the walls, and the length of time chyme remains in the intestines are factors aiding absorption here.

A few substances are otherwise absorbed. For instance, nitroglycerin is usually administered by placing tablets under the tongue. It is absorbed more efficiently by the sublingual rather than the intestinal route. Alcohol and various essential oils, e.g., oil of garlic containing allyl sulfide, $(C_3H_5)_2S$, are absorbed from the stomach.

The mechanism of absorption has not been adequately explained. It is quite evident that part of it may be due to dialysis, to pressure and concentration differences, but part is due to transport mechanisms in the cell walls of the intestinal lining. Transportation is in turn under the control of other enzyme systems.

We have described in some detail the hydrolytic enzymes involved in digestion, but hydrolytic enzymes in other parts of the body are equally important. For example, the conversion of glycogen to glucose in the liver and in the muscles is catalyzed by the enzyme *glycogenase.*

ENZYMATIC OXIDATION-REDUCTION

The absorbed food can be used for growth and repair and for the manufacture of necessary substances such as enzymes and hormones. It can be stored for future use in the form of glycogen and fat or it can be oxidized to provide energy to do work such as contraction of muscle and active transport of substances across cell membranes. In addition, the oxidation of food supplies heat to the body.

Whether the food was originally carbohydrate, fat, or protein, the final step in its oxidation is the same. Amino acids to be used for energy are deaminized in the liver

Table 18-1 ACTION OF VARIOUS ENZYMES

Foodstuff	Digested in	Secretion	Enzyme	Substrate	Products Formed
Carbohydrates	Mouth	Saliva	Ptyalin	Starch	Maltose
				Glycogen	Glucose
			Maltase	Maltose	Glucose
	Stomach	Gastric juice	No carbohydrases		
	Intestine	Pancreatic juice	Lactase	Lactose	Glucose and galactose
			Maltase	Maltose	Glucose
			Amylopsin	Starch	Maltose
		Intestinal juice	Sucrase	Sucrose	Glucose and fructose
			Maltase	Maltose	Glucose
			Lactase	Lactose	Glucose and galactose
Lipids	Mouth	Saliva	No lipases		
	Stomach	Gastric juice	Gastric lipase	Emulsified fats (cream)	Glycerol and fatty acids
	Intestine	Pancreatic juice	Steapsin	Fats	Glycerol, fatty acids, and soaps
Proteins	Mouth	Saliva	No proteases		
	Stomach	Gastric juice	Pepsin	Proteins	Proteoses and peptones
	Intestine	Pancreatic juice	Trypsin and chymotrypsin	Proteins, proteoses, and peptones	Polypeptides
			Carboxypolypeptidase	Polypeptides	Peptides and amino acids
		Intestinal juice	Aminopolypeptidase	Polypeptides	Peptides and amino acids
			Dipeptidase	Dipeptides	Amino acids

where the amino groups are used to make urea and the remaining carbon, oxygen, and hydrogen fragments can be oxidized along with carbohydrate, glycerol, and fatty acids, which are degraded to 2- or 4-carbon chains.

Oxidation of these substances outside the body requires high temperatures, and once the reactions are started, they proceed rapidly with the evolution of light and heat. In the body, the reactions occur at much lower temperatures and the rates of reaction are regulated and controlled. In addition, much of the energy is stored in compounds from which it is readily available. When the body needs energy, it needs it right away and the energy must be instantly available. The body makes this possible by storing the energy in adenosine triphosphate (ATP), a so-called high-energy compound. These molecules can rapidly change to adenosine diphosphate (ADP) and at the same time yield energy. Subsequently, the ATP can be regenerated through further oxidations.

This happy state of affairs is made possible through the action of enzymes. The final "common pathway" in the oxidation of food in the body is the Krebs citric acid cycle, also called the tricarboxylic acid cycle.[1] Working along with the Krebs cycle is an electron transport system. The overall picture is the removal of carbon dioxide molecules from the food in the Krebs cycle and the addition of hydrogen from the food to atmospheric oxygen from respiration, with the formation of water in the electron transport system. The system is called a cycle because after a 2-, 3-, or 4-carbon compound has gone through a series of chemical changes in the cycle with the formation of 3 molecules of carbon dioxide, the starting compounds of the cycle are reformed and are ready to pick up another compound and repeat the process.

Each step in both of these systems is under enzymatic control. The reactions take place in the mitochondria of cells. These mitochondria are sometimes referred to as the powerhouses of the cells because it is in the mitochondria that the oxidation-reduction reactions produce energy.

The enzymes for the Krebs cycle are arranged in order on the outer membrane of the mitochondria and the enzymes for the hydrogen and electron transfer, flavoproteins, cytochromes, and other enzymes are arranged in order on the inner membrane. Together, they both oxidize the food and generate ATP from ADP and inorganic phosphate. Since the end reaction in the electron chain is the reaction with atmospheric oxygen, the enzymes are referred to as the respiratory chain.

An overview of the Krebs cycle is as follows:

1. A 3-carbon acid such as pyruvic acid reacts with coenzyme A to form acetyl CoA, at the same time losing a molecule of carbon dioxide.
2. This compound reacts with oxaloacetic acid to form citric acid, and coenzyme A is liberated, free to unite with the next molecule that enters the cycle.
3. The atoms in the citric acid molecule are rearranged through first the addition of a molecule of water and then its removal. Two hydrogen atoms are removed and oxalosuccinic acid is formed.
4. Oxalosuccinic acid loses a molecule of carbon dioxide and α-ketoglutaric acid is formed.
5. α-Ketoglutaric acid loses 2 hydroden atoms and a molecule of carbon dioxide, and we have succinic acid.
6. Succinic acid loses 2 hydrogen atoms to form fumaric acid.
7. Fumaric acid adds water to form malic acid.
8. Malic acid loses 2 hydrogen atoms and oxaloacetic acid is formed. Here we are back at step 2 ready to repeat the cycle.

HORMONES

The ductless or endocrine glands whose secretions are delivered into the blood stream are the manufacturers of the substances that control and correlate the activities of the body. The important hormone-secreting tissues are the thyroid, parathyroid pancreas, adrenal, sex, and pituitary glands, the pylorus, and the intestinal mucosa.

Thyroid

This gland is located on either side of the trachea in the neck, connected by an isthmus across the front. The cells are filled with a colloidal substance, *thyroglobulin,* which upon hydrolysis yields the active principle, *thyroxine.* The formula of the latter shows it to be an iodine derivative of a substance related to the amino acid tyrosine.

[1] Some of the acids in the system have three carboxyl groups; hence the name tricarboxylic acid cycle (TCA cycle).

$$HO-\bigcirc-CH_2-CH(NH_2)-COOH$$
Tyrosine

Metabolic Regulators

In more detail, the Krebs cycle is as follows:

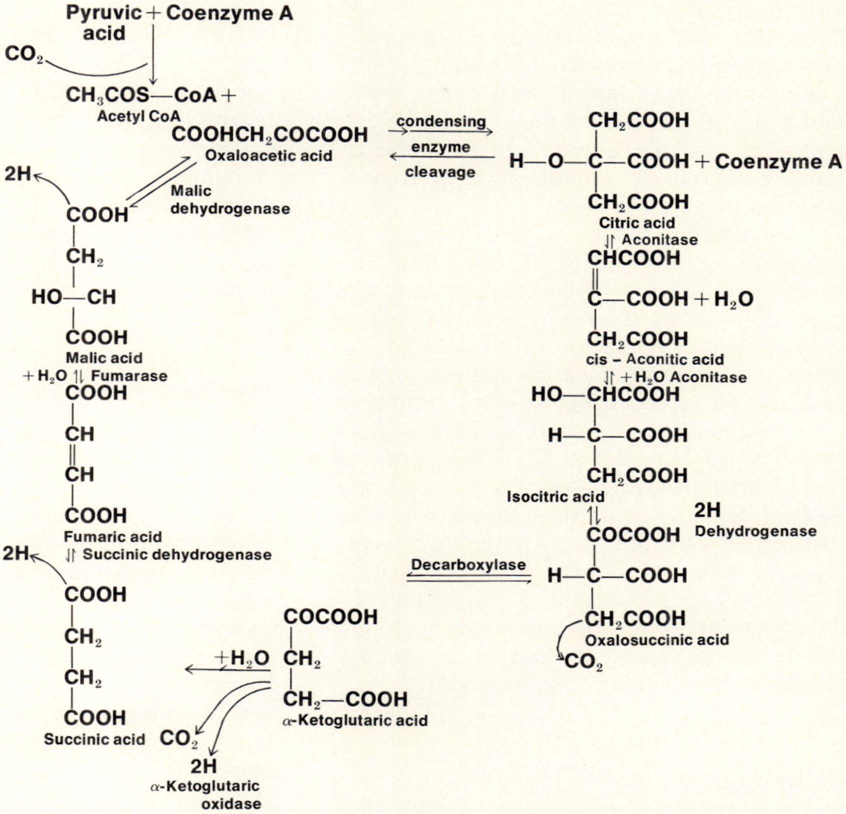

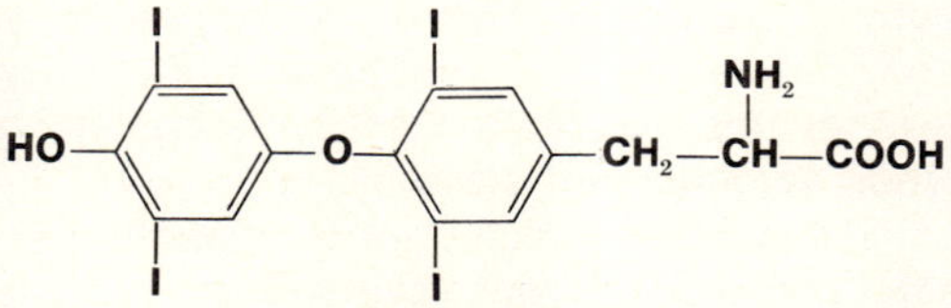

$$\text{Pyruvic acid} + \text{Coenzyme A} \rightarrow CH_3COS\text{—}CoA + CO_2$$

Acetyl CoA then condenses with oxaloacetic acid ($COOHCH_2COCOOH$) via the condensing enzyme (cleavage releases Coenzyme A) to form citric acid:

$$\begin{array}{c} CH_2COOH \\ H\text{—}O\text{—}C\text{—}COOH \\ CH_2COOH \end{array}$$

Citric acid → (Aconitase) → cis-Aconitic acid:

$$\begin{array}{c} CH_2COOH \\ CHCOOH \\ \| \\ C\text{—}COOH + H_2O \\ CH_2COOH \end{array}$$

cis-Aconitic acid + H_2O (Aconitase) → Isocitric acid:

$$\begin{array}{c} HO\text{—}CHCOOH \\ H\text{—}C\text{—}COOH \\ CH_2COOH \end{array}$$

Isocitric acid → (Dehydrogenase, 2H) → Oxalosuccinic acid:

$$\begin{array}{c} COCOOH \\ H\text{—}C\text{—}COOH \\ CH_2COOH \end{array}$$

→ (Decarboxylase, CO_2) → α-Ketoglutaric acid:

$$\begin{array}{c} COCOOH \\ CH_2 \\ CH_2\text{—}COOH \end{array}$$

α-Ketoglutaric acid → (α-Ketoglutaric oxidase, 2H, CO_2, +H_2O) → Succinic acid:

$$\begin{array}{c} COOH \\ CH_2 \\ CH_2 \\ COOH \end{array}$$

Succinic acid → (Succinic dehydrogenase, 2H) → Fumaric acid:

$$\begin{array}{c} COOH \\ CH \\ \| \\ CH \\ COOH \end{array}$$

Fumaric acid + H_2O (Fumarase) → Malic acid:

$$\begin{array}{c} COOH \\ HO\text{—}CH \\ CH_2 \\ COOH \end{array}$$

Malic acid → (Malic dehydrogenase, 2H) → Oxaloacetic acid.

Thyroxine:

HO—(ring with 2 I)—O—(ring with 2 I)—CH_2—$CH(NH_2)$—COOH

Practically all the iodine ingested is found in the thyroid, and the importance of iodine in the diet can not be overemphasized.

The various diseases involving the thyroid show that it regulates the rate of oxidation in the body and indirectly influences the activity of the heart and the nervous system.

The measurement of the basal metabolic rate shows the effect of an excessive activity of the thyroid (hyperthyroidism) or the lack of thyroid secretion (hypothyroidism).

Excessive activity of the thyroid results in exophthalmic goiter, or Graves' disease. The symptoms include a high metabolic rate, a high heart rate, elevated temperature, nervousness, a derangement of reproductive functions in women, and a protusion of the eyeballs. The appetite may be excessive, but in spite of it there may be loss of weight due to the increased metabolism.

A lack of thyroid secretion results in three pathologic conditions: goiter, resulting from the efforts of the gland to make up the deficiency; cretinism; and myxedema.

Cretinism is congenital and is characterized by poor growth, marked obesity, flabbiness, and a complete arrest of mental development. Children so afflicted are called *cretins* and are dwarfed, with heavy heads and abdomens and weak muscles. Cures that approach the miraculous are obtained by feeding thyroid to cretins at fairly regular intervals throughout life.

Myxedema results when the thyroid atrophies or is removed in adult life. The symptoms include puffy swelling or edema of parts of the body, especially of the hands and face; general depression of all bodily functions, accompanied by a low metabolic rate; and usually a depressed mental condition.

The thyroid gland secretes another hormone called thyrocalcitonin. This hormone is a polypeptide which helps to regulate the level of calcium ion in the plasma. It lowers calcium levels probably by inhibiting the release of calcium from the bones. Thyrocalcitonin has a rapid action and together with the hormone from the parathyroid glands affords an accurate method of controlling the level of calcium in the plasma.

Parathyroids

These glands are small and are attached to the thyroid. Removal of the parathyroids results in death in a few days, and probably many of the deaths resulting from thyroid extirpation in the early days were caused by the removal of, or at least the injury to, the parathyroids. The active substance of the parathyroid has been isolated and seems to be a protein. Its function seems to be the maintenance of the proper calcium level in the blood. During hypofunctioning or after removal of the glands, blood calcium decreases and tetany usually appears when the calcium falls below 7.5 mg per 100 ml. Injections of the parathyroid hormone restore the calcium level of the blood, the calcium coming from the bones rather than resulting from increased intestinal absorption of calcium. Hyperfunctioning of the parathyroids results from abnormal growth of the glands or from a tumor in one or more glands. This is accompanied by blood calcium concentration as much as 25 per cent or more above normal, decalcification of the bones, and calcium deposition in the kidneys. The decalcification leads to weak bones and spontaneous fractures. The treatment for this condition is surgical, and removal of the tumors offers an excellent chance for recovery if renal damage has not been too extensive.

Pancreas

Although the pancreas produces an external secretion, pancreatic juice, certain of its beta cells, the islands of Langerhans, are so located that the product of their activity must be delivered to the blood. This secretion not only regulates the proper metabolism of glucose but decreases the production of carbohydrate from noncarbohydrate sources. Failure of the secretion causes typical diabetes. The secretion, *insulin,* is a protein with a molecular weight of about 41,000. Much interest has centered on the relationship of the structure of insulin to its action. However, no particular type of chemical configuration has been found to account for the activity of insulin. Hydrolysis with acid or enzymes results in complete loss of activity. It is for this reason that insulin is ineffective when administered orally.

If insulin is injected into a diabetic person, the blood sugar level drops and the storage of glycogen increases; if the disease has progressed to the point of producing acetone bodies, this production is reduced. The action is rapid, and as a result, there can be a variation in the amount of blood sugar hour by hour. To prolong the effect of the insulin, it is combined with protamine, a simple protein obtained from the sperm of rainbow trout. This combination forms a

Metabolic Regulators

salt. The salt appears to give off insulin at a comparatively slow rate, approaching the normal rate of discharge from the pancreas, and affords steadier control of the disease even with fewer injections a day. Still another form contains zinc and is known as protamine zinc insulin. The choice of the form of insulin used is dictated by the condition of the patient.

The obvious difficulties attending the injection of properly graded doses of insulin several times a day have led to attempts to find some substitute that could be taken orally. This search has produced the amide tolbutamide. Its name would lead us to expect certain groups in the molecule: *tol* for toluene, *but* for butane (4 carbon atoms), and *amide*.

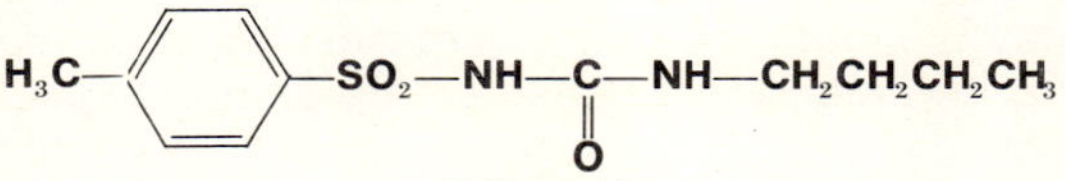

Tolbutamide

This compound is marketed as Orinase. It is not a protein and is not hydrolyzed in the digestive tract. Its chemical structure closely resembles sulfadiazine, but it cannot act as an antibacterial agent. It appears to stimulate the synthesis and release of insulin by the beta cells of the pancreas. Unfortunately, the use of this compound is limited to adults with relatively mild forms of the disease. In addition, the patient must have beta cells; when these cells are lacking, insulin cannot be produced. However, we can expect continued research in this field.

Another type of cell in the pancreas, called alpha cell, secretes a hormone with an effect antagonistic to that of insulin. This hormone is *glucagon*. It stimulates the hydrolysis of glycogen and increases the concentration of glucose in the blood.

Adrenals

These glands, called both *adrenal* (on the kidney) and *suprarenal* (above the kidney), are made up of two separate and very different portions. The inner portion, called the *medulla,* secretes hormones called epinephrine or adrenaline and norepinephrine or noradrenaline.

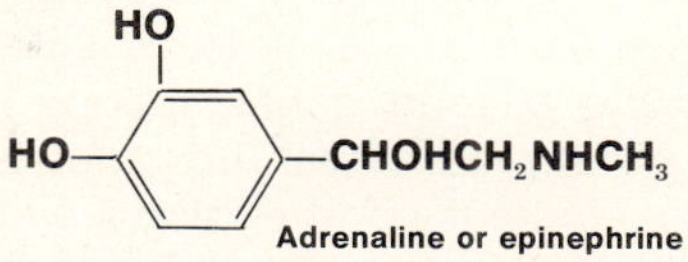

Adrenaline or epinephrine

The outer portion, called the *cortex,* secretes a mixture of hormones, known as *cortin,* which resemble the sex hormones in structure.

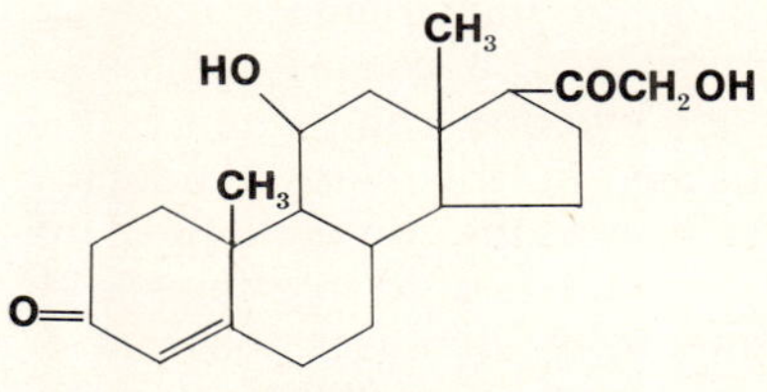

Corticosterone

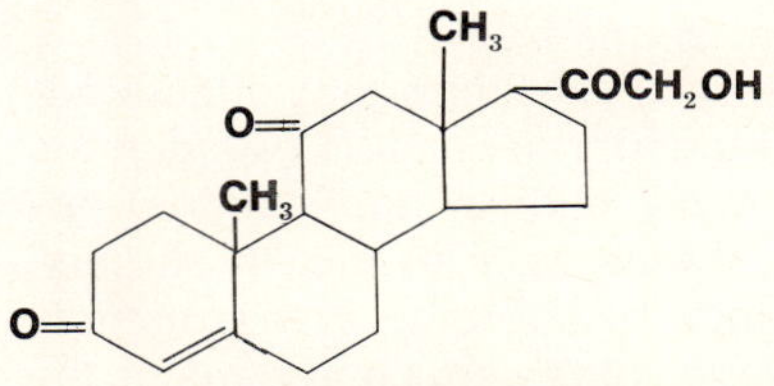

Dehydrocorticosterone

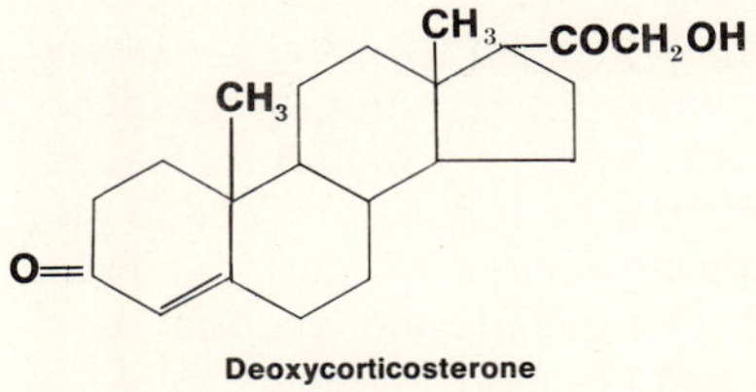

Deoxycorticosterone

Adrenaline, secreted by the medulla, has some most interesting effects, which may be generally described as preparing the body for "flight or fight." It increases the blood pressure by constricting the smaller arterioles. The expression "white with anger" correctly describes the condition arising from the constriction of the blood vessels of the skin caused by the adrenaline secreted. But opposed to this, there is a dilation of the blood vessels of the skeletal muscles. It increases the oxygen consumption; .e., the metabolic rate increases. It promotes *glycogenolysis,* increasing the blood sugar and the amount of lactic acid in the muscles in preparation for the anticipated emergency. A secondary effect of this is emotional glycosuria. It is a powerful cardiac stimulant, steadying the rate and making the heart beat more powerfully. Adrenaline also influences the contractions of nonstriated muscle in general. For instance, it causes dilation of the pupil of the eye and contraction of the papillary muscles in the skin, producing "goose flesh." It stops the contractions of the gastrointestinal tract and so interferes with digestion. Adrenaline increases the rate of the clotting of blood, a desirable effect in this emergency if it involves any bloodletting. In short, the medulla may be considered the "minute man" of the body.

Because of these important effects adrenaline, or epinephrine, as it is called in the U.S.P., has many clinical applications. Applied locally it acts as a styptic. In surgery, its injection into the operative area constricts the small arteries of the area and reduces the loss of blood. Adrenaline is often used in solutions of local anesthetics. Because of the constriction of the blood vessels locally, it limits the absorption of the local anesthetic and thus prolongs the duration of the anesthesia. Consequently, the amount of anesthetic needed may be reduced. Injection into the heart muscle serves to start the heart when it has stopped during an operation. Further, it is used to start the heart action of a newborn baby, if necessary. Adrenaline is a powerful bronchodilating agent, and when a solution of the hydrochloride is nebulized and inhaled, the sufferer from bronchial asthma is greatly relieved. Because of vaso-constriction and shrinkage of the nasal mucosa by adrenaline, it is very useful in reducing nasal congestion associated with a cold, hay fever, etc. Clinically, ephedrine is a keen competitor for adrenaline. A comparison of the formula of ephedrine with adrenaline will show the similarities and differences.

$$\text{C}_6\text{H}_5-\text{CHOHCH}(\text{NHOH})\text{CH}_3$$

Ephedrine

A deficiency of cortin produces Addison's disease, which is characterized by bronze patches on the skin, an increased rate of excretion of sodium in the urine, and a great increase in the potassium content of the blood serum. The blood solids increase as the water content decreases, resulting in a lowered blood pressure. The carbohydrate metabolism is disturbed resulting in a lowered blood sugar level. Injections of cortin tend to restore these conditions to normal. In short, cortin seems to control the sodium-potassium equilibrium in the body and also seems to oppose the action of insulin. The similarity of the cortex hormones to the sex hormones, particularly the androgens (maleness producers), has been pointed out. Hyperactivity of the cortex, which may result from a tumor, produces an overdevelopment of the male characteristics.

Cortisol has an anti-inflammatory effect. It inhibits the action of leukocytes, the development of edema, and the formation of fibrin. It is used in the treatment of diseases such as rheumatoid arthritis to reduce inflammation in the joints.

Allergies result from the introduction of a foreign protein into a susceptible person who becomes sensitized to the protein. Cortisone and cortisol have an antiallergic effect. They inhibit the formation of antibodies and reduce hypersensitivity reactions.

Organ transplants involve the introduction into the patient of foreign tissue. It can be seen that compounds like cortisone and

Metabolic Regulators

cortisol can help to reduce the tendency of the body to reject the transplant.

Sex Glands

These glands, also called the *gonads,* are the reproductive glands: testes and ovaries. Their internal secretions are responsible for secondary sexual characteristics as well as for the full development of the external genital organs. The secretions further favor oxidative metabolism in all animals; it is common knowledge that castration of young animals is practiced for the purpose of making them put on weight rapidly and produce a higher quality of meat. The two male sex hormones are *testosterone* and *androsterone* (derived from the Greek word for male and from the ketosteroid structure). Their similarity to each other and to cholesterol is shown by their formulas.

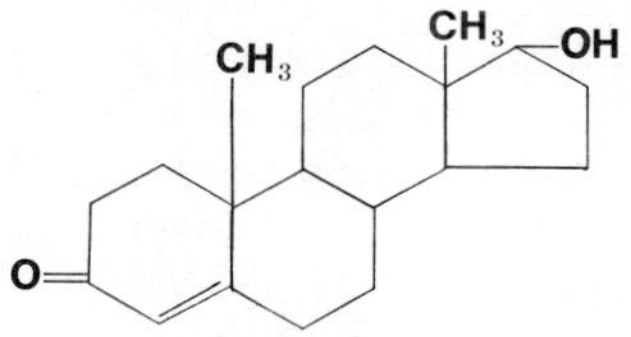

Testosterone

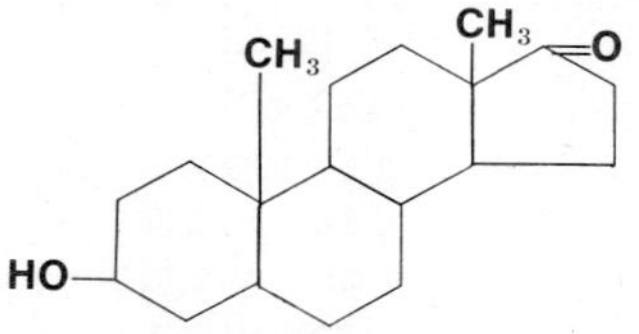

Androsterone

Testosterone, of which the only natural source seems to be the testes, is about ten times more potent in producing secondary sex characteristics, as indicated by the growth response of a capon's comb, than is androsterone. Androsterone is found in the urine, and it is possible that it has been produced from testosterone in the body before excretion in the urine.

The ovaries furnish two internal secretions, one formed by the interstitial cells in the body of the ovary and the second by the corpus luteum. As each follicle develops up to the time it extrudes an ovum from the surface of the ovary, epithelial tissue within the follicle is also developing until, at the time of ovulation, it is a gland of considerable internal secreting capacity. After ovulation, there develops in the ruptured follicle a new glandular structure, the corpus luteum, so called because of its yellow color.

The follicle liquid contains at least two hormones, *theelin* (from the Greek word for female), which is also called *estrone,* and *estradiol.* The name estrone refers to the ability of such hormones to produce the periodic occurrence of *estrus* (the period during which the female is capable of mating) in lower animals and to the presence of a ketone group in its structure. Estradiol is about six times as potent as estrone and is believed to be the primary hormone from which other estrogenic compounds found in the urine are derived by metabolic changes in the body. A third estrogen, which, like estrone, is found in the urine of pregnant women, is *estriol.* It has about one-tenth the potency of estrone. A comparison of the structural formulas of these estrogens shows their similarity and their resemblance to the androgens.

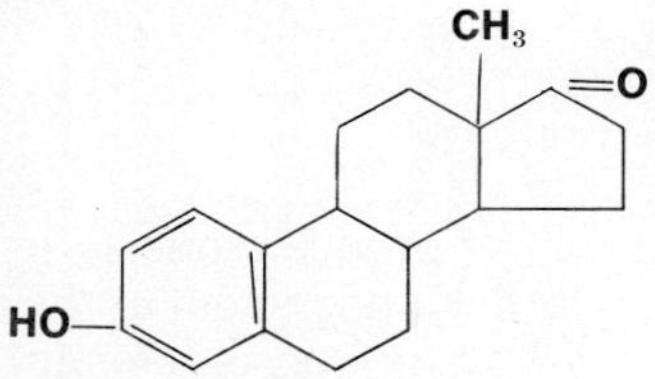

Estrone

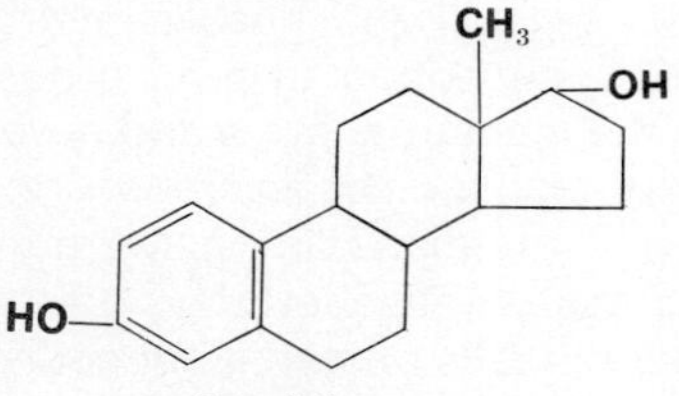

Estradiol

Estriol

These estrogens are responsible for the development of the secondary sexual characteristics that occur at puberty and for the changes in the uterus prior to ovulation, which occurs in the middle of the menstrual cycle.

The hormone produced in the corpus luteum is *progesterone,* which becomes reduced to *pregnanediol* in the body and is excreted in the urine during the latter half of the menstrual cycle. The formulas show their relation.

Progesterone

Pregnanediol

The function of progesterone is to prepare the uterus for the attachment of the fertilized ovum. If fertilization does not occur, the production of both follicular and corpus luteum hormones falls off and a new menstrual cycle begins. If conception occurs, progesterone inhibits the menstrual cycle and develops the mammary glands. The excretion of progesterone and pregnanediol in the urine gradually increases until a few days prior to labor, whereupon it falls off and labor soon follows. The part played by the anterior pituitary in the several phases of the reproductive process will be discussed later.

Pituitary

This gland is located at the base of the brain. It is probably the most important gland of the whole endocrine system because its secretions affect the activity of other glands in addition to having some physiologic effects in their own right. For this reason it is sometimes referred to as the "master gland." The secretions of the two lobes are very different and will, therefore, be considered separately.

anterior lobe

Among the several hormones secreted by this lobe, there is one that controls the growth of young animals. A deficiency in childhood produces dwarfism, a condition that is not to be confused with cretinism, because in dwarfism a person is normal in every respect except size. An excess of this growth hormone produced during childhood leads more or less obviously to giantism. Hyperactivity of the gland usually results from a tumor, and if it occurs in an adult, growth is not symmetrical but is observed chiefly in the bones of the head and the extremities (*acromegaly*). *Prolactin,* a protein, is secreted by the anterior lobe and initiates lactation after the mammary gland has been properly conditioned by the action of the ovarian hormone.

The gonadotropic hormones regulate the development of sex organs and the production of sex hormones. The *follicle-stimulating hormone* (FSH) contains a polysaccharide group and is therefore destroyed by digestive enzymes. This hormone stimulates the development of follicles in the ovary to the point of ovulation. The *luteinizing hormone* (LH) is inactivated by pepsin and trypsin. This hormone causes the production of the corpus luteum, i.e., the continuation of the action initiated by the FSH, and stimulates the production of progesterone. Corresponding effects are produced in the male; i.e., the former hormone is necessary for the proper functioning of cells producing spermatozoa, and the latter one stimulates the production of male sex hormones.

It is interesting to note that during the early stages of pregnancy the placenta produces a substance similar to these gonadotropic hormones. It appears in the urine about a month after conception and may be detected biologically. In Friedman's modification of the *Aschheim-Zondek test,* popularly called the "bunny test," 20 ml urine is injected into the blood of an adult female rabbit that has been kept in isolation for 3 weeks. Two days later, if the urine contained the hormone, the ovaries will show definite luteinization and hemorrhage of some follicles.

The *thyrotropic hormone* stimulates the growth of the thyroid gland and increases the production of thyroid hormone. In fact, it may well be that hyperthyoidism or exophthalmic goiter could be caused, at least in part, by an increase in the secretion of this hormone.

The *adrenocorticotropic hormone* (ACTH), a polypeptide, affects the adrenal gland in such a way as to increase the production of the hormones of the adrenal cortex. Those like cortisol and cortisone (glucocorticoids) help to regulate carbohydrate metabolism and to preserve carbohydrate stores, especially in stress situations. They also have antiinflammatory and antiallergic effects. Mineralocorticoids like aldosterene regulate the reabsorption of sodium and the excretion of potassium in the kidney tubules. The adrenal cortex when stimulated by ACTH also secretes the moderately active sex hormones.

The pituitary exerts a number of definite effects upon metabolism. One factor increases the blood sugar concentration, probably by increasing the rate of hydrolysis of glycogen in the liver. Under normal conditions this situation is counterbalanced by the insulin produced by the pancreas. The balance is upset in diabetes. Another factor increases the catabolism of fatty acids, which results in the formation and excretion of ketone bodies.

posterior lobe

An extract of the posterior lobe contains a number of principles. Injected or absorbed through the nasal mucosa, this extract is used to check polyuria. It is, therefore, described as an *antidiuretic.* Polyuria is accompanied by a consuming thirst and the ingestion of large quantities of water and is a condition associated with *diabetes insipidus.* Apparently, this hormone is necessary for the normal reabsorption of water by the kidney tubules. The extract is used to produce a slow but persistent increase in the blood pressure, which suggests its use in combating shock. This effect on the blood pressure probably results from the general contracting effect the extract has on involuntary muscles, in this case those in arterial walls. To illustrate further, injections cause a marked increase of milk flow, i.e., not through an increase in production but rather through an increase in the rate of flow due to the contraction of the involuntary muscles in the mammary gland, thus squeezing out the milk. Finally, since injections of the pituitary extract cause violent contractions of the muscles of the pregnant uterus, they are used to hasten labor. The antidiuretic hormone is also called ADH or vasopressin. The hormone affecting ejection of milk from the breast and contraction of uterine muscles is named oxytocin.

Relaxin

The reproductive organs—the ovaries, uterus, and placenta—produce during pregnancy a polypeptide that causes a separation of the symphysis pubis, thus increasing the size of the pelvic cavity and facilitating the birth process. This event is called "pelvic relaxation," and the hormone principally responsible is named *relaxin* (Table 25).

Table 18-2 HORMONES

Hormone	Secreting Organ	Chief Biologic Action
Thyroxin	Thyroid gland	Increases tissue oxygen use
Parathyroid hormone	Parathyroid glands	Regulates calcium and phosphate metabolism
insulin	Pancreas beta cells	Decreases glucose, increases glycogen, spares protein
Glucagon	Pancreas alpha cells	Increases glucose, decreases glycogen
Adrenaline	Adrenal medulla	Vasoconstrictor, promotes conversion of glycogen to glucose
Testosterone	Testes	Stimulates male characteristics, favors protein synthesis
Androsterone	Testes	Mild stimulation of male characteristics, protein anabolic
Theelin (estrone)	Ovaries	Estrogenic, favors protein synthesis
Estradiol	Ovaries	Estrogenic
Progesterone	Corpus luteum	Prepares uterus for pregnancy
Corticosterone*	Adrenal cortex	Favors glycogen production and protein catabolism
Aldosterone*	Adrenal cortex	Retains sodium
Deoxycorticosterone (DOC)*	Adrenal cortex	Retains sodium, excretes potassium
Hydroxyandrostenedione*	Adrenal cortex	Mild stimulation of male characteristics
Cortisol*	Adrenal cortex	Favors glycogen production, relieves inflammation, favors protein catabolism, affects thyroid
Prolactin	Anterior pituitary	Stimulates milk production
Growth hormone	Anterior pituitary	Stimulates bone growth and protein synthesis
Follicle-stimulating hormone (FSH)	Anterior pituitary	Stimulates ovarian follicles
Luteinizing hormone (LH)	Anterior pituitary	Induces production of progesterone, causes production of corpus luteum; in the male, stimulates gonadal hormones
Thyrotropic hormone (TSH)	Anterior pituitary	Stimulates thyroid and thyroxin production
Adrenocorticotropin (ACTH)	Anterior pituitary	Stimulates adrenal cortex secretion
Vasopressin (Pitressin)	Poterior pituitary	Antidiuretic, peripheral vasoconstrictor
Oxytocin (Pitocin)	Posterior pituitary	Stimulates uterine muscle contraction
Gastrin	Pylorus	Stimulates secretion of gastric juice
Secretin	Intestinal mucosa	Stimulates secretion of pancreatic juice and bile
Cholecystokinin	Intestinal mucosa	Stimulates gallbladder contraction
Enterogastrone	Intestinal mucosa	Inhibits gastric motility
Relaxin	Reproductive organs during pregnancy	"Pelvic relaxation"

*These hormones collectively are called *cortin*.

VITAMINS

The historical discovery that scurvy and beriberi resulted from the absence of certain substances in the diet led to the detection of other compounds of which minute amounts were necessary for life or for normal growth and development. The first chemical tests of these substances showed that they contained nitrogen and gave reactions characteristic of amines. Since they were necessary for life, they were called "vitamines." Subsequently the final e was dropped, and even though some of them are not related to amines and some are not necessary for life, the term "vitamin" will probably persist in popular usage. Since the nature of these substances was unknown at the time of their discovery, they were designated by letters. This system has led to some confusion, because in the case of several vitamins (B, for instance) what was first thought to be a single substance has since been shown to be a complex of several substances, each having a special function. Since many of the vitamins have been isolated and even synthesized, names descriptive of their chemical structure have come into use. It is common to classify vitamins according to their solubility in water or in fat, and we shall follow that practice.

One of the puzzling things about vitamin research has been the finding that minute amounts of these substances in some cases made the difference between life and death. Another peculiar finding was that certain vitamins were essential for one kind of animal but not for another. Work with laboratory rats led to the discovery of an antisterility factor and an anti-gray-hair factor. When these substances were tried on human beings, the reactions were not the same. From a wealth of research results, plausible theories for these and other enigmatic questions are beginning to emerge.

One theory states that the vitamins may be parts of enzyme systems. Of the substances in the B complex, vitamins B_1, B_2, and B_6 have been definitely identified as coenzymes or as parts of coenzyme systems. In our discussion of enzymes, we pointed out that very small amounts had tremendous biologic effects. The same situation may apply here. We know that, through anabolic processes, organisms can build up very complex compounds. Plants can synthesize most of the substances needed for their growth from very primitive material. Animals are limited in this regard and must ingest more complex material. Animal types differ in what is essential. In man, alanine, CH_3CHNH_2COOH, is an essential alpha amino acid, whereas glycine, CH_2NH_2COOH, is not. Apparently man cannot make alanine from glycine or from anything else. This theory would account for the different vitamin requirements of different organisms. Not all the vitamins need to be supplied in the diet. The notable exceptions are vitamin D, which the body can synthesize in the presence of ultraviolet light, and vitamin K, which is manufactured in the intestines by bacteria that are normally found there.

Water-soluble Vitamins

The water-soluble vitamins are the vitamins of the B complex and ascorbic acid, vitamin C.

b-complex vitamins

Thiamin (Vitamin B_1). The molecule of this vitamin contains a pyrimidine and a thiazole nucleus.

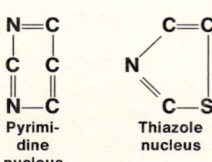

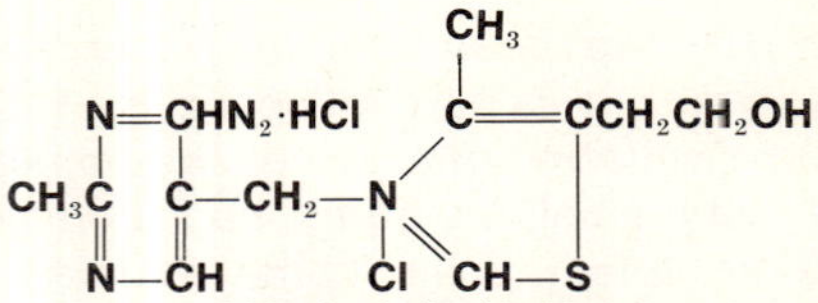

Thiamine hydrochloride

The function of thiamin in the body seems to be metabolic, for when it is deficient, the oxidation of carbohydrates is affected and there is an accumulation of alpha keto acids, e.g., pyruvic acid. Nerve tissue is particularly affected, and thus it is that in *beriberi,* the deficiency disease, paralysis of the legs accompanied by atrophy of the muscles involved occurs and there is an accumulation of pyruvic acid in the blood with consequent excretion of it in the urine.

The international unit (I.U.) of thiamin is equivalent in activity to 0.003 mg crystalline thiamine hydrochloride. The recommended daily intake of this vitamin for adults and growing children is about 2 mg (670 I.U.). During pregnancy and lactation the requirement is higher (2.5 to 3 mg daily).

Riboflavin (Vitamin B_2). Chemically, this vitamin is a compound of ribose, a 5-carbon sugar, and the colored cyclic substance flavin, as shown:

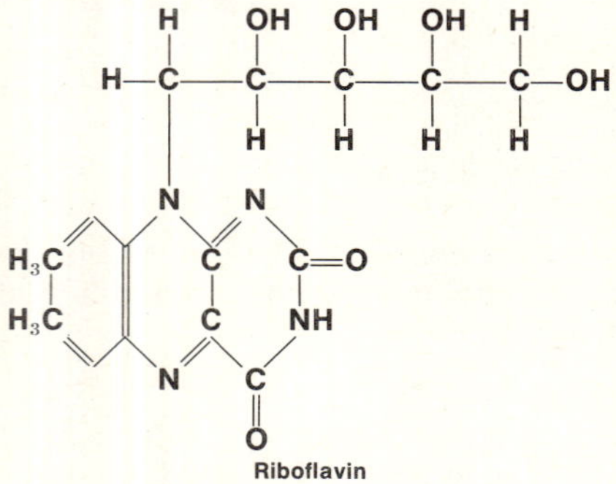

Riboflavin

Riboflavin is apparently an essential constituent of the yellow oxidation enzyme, which is thought to be a phosphoric acid ester of riboflavin in combination with protein. This enzyme is an essential link in the metabolism of carbohydrates and is probably a constituent of all cells. Serious deficiency of this vitamin produces lesions in the angles of the mouth and visual difficulties, which may take the form of photophobia and dimness of vision at a distance or in poor light. Riboflavin is distributed throughout the body, but there is limited storage of the vitamin. Any excess over the requirement of about 2 mg per day is excreted in the urine.

Nicotinic Acid (Niacin). Niacin is the name coined to make the distinction between the vitamin and the alkaloid nicotine quite clear to the public. Both the acid and its amide are effective in counteracting deficiency of this vitamin. The following formulas show their composition:

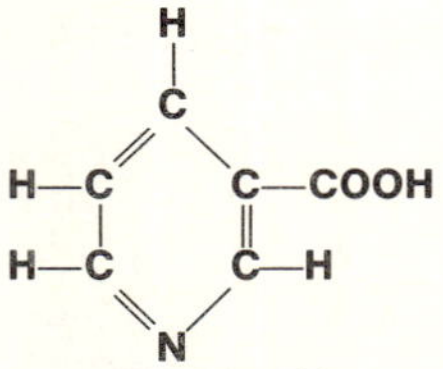

Nicotinic acid (niacin) 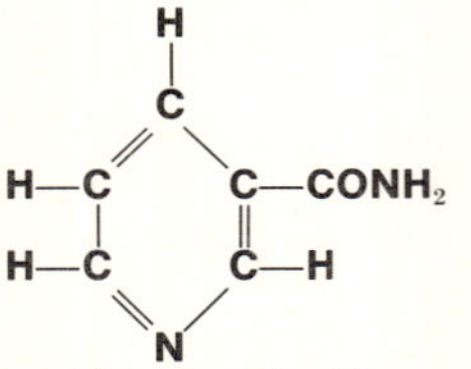   **Nicotinic acid amide (niacinamide)**

A deficiency of this vitamin produces *pellagra,* which has been described as a disease characterized by skin lesions, particularly in those areas exposed to the light, and a bright scarlet tongue and oral mucous membranes, followed by various nervous, gastrointestinal, and mental disorders. Pellagra has long been associated with the consumption of corn, and yet the equally low nicotinic acid content of rice and wheat does not produce the disease. It is evident that the disease is not caused by the simple vitamin deficiency as such. It has been shown that the intestinal flora are capable of converting tryptophan into nicotinic acid. Now, since the protein of corn does not produce tryptophan upon hydrolysis, pellagra results from a corn diet because of the low nicotinic

Metabolic Regulators

acid content *and* the absence of the precursor of the vitamin.

Nicotinic acid seems to function in supplying part of the coenzymes that act during the carbohydrate metabolism. The daily requirement varies from 11 to 20 mg per day, depending upon the caloric intake. Any excess is excreted in the urine as a methyl derivative of the amide.

Pyridoxine (Vitamin B_6). The resemblance between this compound and nicotinic acid is shown by the formula below, both being derivatives of pyridine.

[Structural formulas of Pyridine and Pyridoxine]

So far as man is concerned, no specific symptoms have been attributed to a deficiency of this vitamin. However, it has been used to clear up the last stages of pellagra, for instance, the residual neuromuscular symptoms. The vitamin has been used quite successfully in the treatment of the nausea of pregnancy and of radiation sickness.

[Structural formula of Pantothenic acid]

Pantothenic Acid. This name is derived from the Greek words meaning "from everywhere" on account of its widespread occurrence. There is no evidence that pantothenic acid deficiency has ever been observed in man. However, it is not impossible that some of the symptoms of vitamin-B-complex deficiency could be due to an inadequate amount of pantothenic acid in the diet. The fact that administration of this vitamin prevents gray hair in animals has been exploited, but there is no evidence to show that the graying of human hair is due to a deficiency of this or any other vitamin.

Folic Acid Complex. This name refers to a highly complicated group of substances and alludes to the source of one factor, spinach leaves. It is necessary for the proper formation of blood. Nutritional anemias are due to the absence of the acid from the diet, but pernicious anemia is probably due to an inability to utilize folic acid because of the absence of vitamin B_{12} in the liver.

Vitamin B_{12} (Erythrotin, Cyanocobalamin). This vitamin is the antipernicious-anemia factor from liver extract. Experiments have shown that molecules of vitamin B_{12} and riboflavin contain the same nucleus. This vitamin is essential for the formation of hemoglobin.

Biotin. Absence of this vitamin in the diet produces a condition that has come to be known as *egg-white injury*. When 30 per cent of the protein in the diet is supplied by uncooked egg white, these symptoms may appear: pallor, shedding of scales from the skin, fatigue, muscle pains, and loss of appetite. All the symptoms disappear in the presence of biotin. The egg-white injury is due to the inactivation of biotin by one of the egg albumins, *avidin,* which combines with biotin and makes it unavailable. Egg white or avidin that has been heated 2 min at 100°C cannot inactivate biotin, so even a soft-boiled egg is not dangerous. Since there is an excess of avidin over biotin in an egg, the egg must be cooked in order to utilize the biotin.

[Structural formula of Biotin]

other b-complex factors

Since human beings seem to be unaffected by a deficiency of these factors, they need not be considered by nurses. However, they might be enumerated to indicate their existence: they are inositol; *p*-aminobenzoic acid; choline; vitamins B_3, B_4, B_5 (B_7, B_8, B_9 do not exist), B_{10}, B_{11}, B_{13}, B_{14}, and L; and factors U, W, R, and S.

$$\underset{O}{C}-\underset{OH}{C}=\underset{OH}{C}-\underset{H}{\overset{H}{C}}-\underset{OH}{\overset{O}{C}}-CH_2OH$$
Ascorbic acid

Absorbic Acid

vitamin c

This compound is easily oxidized, which accounts for its easy destruction during the heating and storage of food.

Ascorbic acid functions in the formation of the colloidal intercellular substance that cements the cells together. Its deficiency results in weakened capillary walls and brittle bones and teeth, evidenced by swollen bleeding gums, broken teeth, and hemorrhaging under the skin, so common in *scurvy*.

This vitamin is constantly being destroyed in the body, probably by oxidation. Obviously, the daily intake must be equal to the amount oxidized to avoid deficiency. The daily requirement can be determined accurately by the saturation test. In this test a subject is saturated with the vitamin and then held on a diet free of the vitamin for a number of days, after which the amount needed to produce saturation again is determined. This amount divided by the number of days equals the average daily amount metabolized. Saturation is indicated by the rapid excretion in the urine of the bulk of the vitamin administered. An adult requires 75 to 100 mg daily. Under certain circumstances the rate of destruction is greater, and the daily requirement increases. A person suffering from certain infectious diseases, notably tuberculosis, requires up to 100 per cent increase. An increase is also needed during pregnancy and lactation.

Fat-soluble Vitamins

vitamin a

Vitamin A is also called opthalamin and the anti-infection vitamin. It has the following formula:

Vitamin A ($C_{20}H_{29}OH$)

The body is able to derive this vitamin from the yellow plant pigment carotene. The action seems to take place in the liver and is one of hydrolysis.

$$C_{40}H_{56} + 2H_2O \rightarrow 2C_{20}H_{29}OH$$
β-Carotene   Vitamin A

Vitamin A is fairly stable when heated and is not appreciably affected by boiling water. Also, since it is practically insoluble in water, ordinary boiling procedures do not affect its potency. Canning and freezing do not seem to reduce the vitamin content, but drying and dehydrating cause a considerable loss of the vitamin, probably through oxidation.

The vitamin itself is more rapidly absorbed from the intestinal tract than is carotene. Bile is not necessary for the absorption of the vitamin but is required for the absorption of carotene. Mineral oil, a common intestinal lubricant, has been shown to reduce the absorption of carotene, perhaps because they are both hydrocarbons.

One of the first clinical symptoms of vitamin A deficiency is night blindness, the inability to see well in dim light or to adjust quickly to sudden changes of light. In the retina of the eye the cells that are stimulated by light are rods and cones. The cones respond chiefly to colored light and are essential for normal color vision. The rods are stimulated by dim light and contain a pigment known as *visual purple,* a conjugated protein in which vitamin A supplies the prosthetic (nonprotein) group. Visual purple is bleached to visual yellow by bright light, and when vitamin A is deficient, the retina is unable to resynthesize visual purple rapidly, resulting in a hazardous condition. We have all experienced this normal delay in the ability to see in a dim light when we entered a darkened movie theater from the brightly lighted foyer. Normally, the eye resynthesizes the visual purple in 1 sec, but in vitamin A deficiency 2 or 3 sec may be required. A little arithmetic will point up another hazard. A car traveling 40 miles per hr will cover about 60 ft per sec. Thus, the normal person is blinded for about 60 ft after being exposed to a bright oncoming headlight, and the vitamin-A-deficient person is blinded for 120 to 180 ft. In that difference in blind-driving distance anything can happen.

Vitamin A deficiency causes a hardening of the cells of certain tissues of the body, including the lacrimal glands, the conjunctiva, the cornea, the salivary glands, and the respiratory tract among others. The failure of tear secretion (the antiseptic of the eyes) makes it possible for bacteria to infect the eye, causing the cornea to become cloudy and hardened. This condition, called *xerophthalmia,* may lead to permanent blindness.

The changes in the respiratory tract account for the fact that vitamin-A-deficient persons are more likely to have colds, sinus disease, pneumonia, and other respiratory diseases. This does not mean that colds can be prevented by taking vitamin A.

The formation of tooth enamel is defective in vitamin A deficiency, and dentine formation is abnormal, often resulting in striking deformities of the teeth. It is believed that deficiency of this vitamin does more damage during the formative period of teeth than other vitamin deficiencies.

One international unit (I.U.) is equivalent in activity to 0.0006 mg of β-carotene. Adults require about 5,000 units (3 mg) daily. During pregnancy and lactation the requirement rises to 6,000 units daily.

vitamins d

This vitamin is now well known as the "sunshine vitamin" and as the antirachitic vitamin. It was first believed that cholesterol was the substance in food that became antirachitic after ultraviolet irradiation. Later it was shown that ergosterol was the precursor of this vitamin. For some time it was believed that vitamin D was a single substance, but subsequent work has shown that at least 10 different substances possess vitamin D activity. They all have the same general chemical structure, and attention is called to their similarity to the sex hormones and to the adrenocortical hormones. Two that are outstanding result from the ultraviolet irradiation of ergosterol and 7-dehydrocholesterol. When ergosterol is irradiated, one of the successive products formed is *calciferol,* or viosterol (vitamin D_2). Irradiation of 7-dehydrocholesterol results in the formation of vitamin D_3.

Ergosterol → (Irradiation) → Calciferol (vitamin D$_2$)

7-Dehydrocholesterol → (Irradiation) → Vitamin D$_3$

In the discussion that follows vitamin D will be used as a collective term for vitamins D$_2$ and D$_3$.

The vitamin is fairly stable to heat and oxidation. The ultraviolet rays of sunshine convert 7-dehydrocholesterol in the skin into vitamin D. Today foods are irradiated to increase the content of vitamin D, or they may be fortified with the vitamin itself.

Vitamin D participates in the metabolism of calcium and phosphorus. It must be evident that to have the proper deposition in growing bones, there must be a normal concentration of these elements in the blood. Among other factors that determine the amounts of calcium and phosphorus in the blood are (a) the dietary intake, (b) the intestinal absorption, and (c) vitamin D. It is believed that the main function of vitamin D concerns the proper intestinal absorption of calcium and phosphorus. It has been shown that even though the diet contained calcium and phosphorus in the proper proportion for bone formation, rickets developed in the absence of vitamin D.

In rickets the bones are not sufficiently calcified, making them soft, weak, and easily deformed. Joints become knobby and the ribs become beaded (rachitic rosary), or the ribs become deformed, giving a pigeon-breasted appearance. Deficiency of vitamin D also affects the formation of teeth, a natural parallel, since teeth and bones have a similar composition. Rickets does not occur in adults after bone formation is complete, but under certain conditions (in women, for instance, after several pregnancies) deficiency results in demineralization (osteomalacia), particularly in the spine, pelvis, and lower extremities. As the bones soften, weight causes bowing of the legs, vertical shortening of the vertebrae, and flattening of the pelvic bones.

The international unit of vitamin D possesses the activity of 0.000025 mg crystalline calciferol. The daily requirement naturally varies with the amount of bone growth, and hence with age. But studies show that 400 units per day should be adequate for infants and growing children. Premature infants on an artificial diet need 600 to 800 units daily given in the form of cod-liver oil. Adult requirements have not been too well established, but 135 units daily may be beneficial to those persons not exposed to the sun. During pregnancy and lactation the requirement rises to 800 units per day.

vitamins e

This is a group of vitamins with varying degrees of potency, which are called *tocopherols* because they are antisterility factors in animals. There are three forms: alpha, beta, and gamma; but α-tocopherol is the most potent. Its formula has been established, and it has been synthesized.

The use of this vitamin in treating habitual abortions has been investigated, but the results are not conclusive.

Table 18-3 VITAMINS

Vitamin	Source	Positive Action	Result of Deficiency
Thiamin (B_1)	Yeast, cereal grains	Coenzyme for oxidation in the tissues	Beriberi; nerve paralysis; muscle atrophy
Riboflavin (B_2)	Yeast, liver, cereals, dairy products	Part of yellow enzyme, for metabolism of carbohydrates	Lesions on lips; visual difficulties
Nicotinic acid (niacin)	Liver, yeast, meat, peanuts	Part of coenzyme in carbohydrates metabolism	Pellagra; nervous, gastrointestinal, and mental disorders
Pyridoxine (B_6)	Yeast, grains, liver	Coenzyme	
Folic acid	Liver, kidney, yeast, green leaves	Necessary for blood formation	Some anemias
Cyanocobalamin (B_{12})	Liver	Necessary for hemoglobin formation	Pernicious anemia
Biotin	Liver, kidney, egg yolk		Egg-white injury
Ascorbic acid (C)	Citrus fruits, green vegetables	Formation of intercellular substance	Fragile capillaries; brittle bones and teeth; scurvy
Vitamin A (opthalamin)	Fish-liver oils, butter, egg yolk	Anti-infective, health of peripheral tissues	Night blindness; infections; xerophthalmia
Vitamins D	Cod-liver oil, egg yolk, butter	Deposition of calcium and phosphorus	Rickets
Vitamins K (menadione)	Alfalfa, cabbage, kale, tomatoes	Aids blood coagulation	Prolonged clotting time

vitamins k

These vitamins are derivatives of 1,4-naphthoquinone.

One of the naturally occurring ones, vitamin K_1, has a methyl group in the 2-position and the phytyl group ($-C_{20}H_{39}$) in the 3-position.

Vitamin K_1

2-Methyl-1,4-napthoquinone

The simpler compound, 2-methyl-1,4-naphthoquinone, has three times the potency of vitamin K_1 and has been proposed as the reference substance for defining the unit of vitamin K activity.

The outstanding result of vitamin K deficiency is prolongation of the clotting time of the blood. Vitamin K is necessary for the formation of prothrombin by the liver. Vitamin K, being fat-soluble, cannot be absorbed from the digestive tract in the absence of bile. In adults vitamin K deficiency is most common when the person is suffering from some condition in which bile does not reach the intestine in adequate amounts. Such conditions could be severe liver disease or obstructive jaundice. In these cases postoperative hemorrhage is frequent and dangerous. One way around this difficulty is to administer a water-soluble substance having vitamin K activity, which can be absorbed from the intestinal tract without bile salts. Another way is to administer bile salts along with the vitamin K.

This vitamin is synthesized by intestinal bacteria, and it is unlikely that any deficiency results from inadequate intake alone. Since the intestinal tract is sterile in the newborn infant, absence of this synthesis probably accounts for the low thrombin level and the resulting hemorrhages from injuries at birth. This condition may be counteracted by injecting water-soluble vitamin K intramuscularly upon delivery or by giving the mother prophylactic doses (1 to 2 mg daily) for a week prior to expected confinement. Low prothrombin level sometimes occurs during treatment with nonabsorbable sulfonamide drugs or oral streptomycin and is explained by a striking reduction in intestinal flora caused by these drugs (Table 18-3).

Study Exercises

1. Define enzyme, substrate, proenzyme, coenzyme, and hydrolysis.
2. How is the specificity of enzymes explained?
3. What are the chief components of saliva?
4. What substances are acted upon by the enzymes of saliva?
5. What is formed in each case from the substances in question 4?
6. What is the composition of gastric juice?
7. How is hydrochloric acid formed in the body?
8. What are three functions of hydrochloric acid in the stomach?
9. What is the function of pepsin? What are the products formed?
10. What is the function of gastrin?
11. What fat digestion occurs in the stomach? What are the products?
12. Analyze the word carboxypolypeptidase. What does it mean?
13. What is the composition of bile?
14. How does bile function in the digestive process?

15. What enzymes are found in intestinal juice?
16. What enzymes are found in pancreatic juice?
17. What are the end products of digestion of carbohydrates, fats, and proteins?
18. What deviations from the normal might one expect in the absence of bile?
19. Define digestion in chemical terms.
20. What hormones are specifically involved in the process of digestion?
21. Name the important hormone-secreting tissues.
22. What is the function of thyroxin? What abnormalities occur as the result of oversecretion?
23. What are the results of undersecretion of thyroxin?
24. What is the function of parathyroid hormone?
25. Name two hormones secreted by the pancreas. What are the functions of each?
26. What are some of the disadvantages of the present methods of administering insulin?
27. Why cannot insulin be administered by the oral route?
28. What recent attempts have been made to overcome the deficiencies of the present insulin therapy?
29. What hormone is secreted by the adrenal medulla? What are some of its effects?
30. Give three uses of adrenaline in the hospital.
31. Name three hormones secreted by the adrenal cortex. What are their effects in the body?
32. What hormones are secreted by the gonads? What are their effects?
33. Why is the pituitary gland called the "master gland?" Show how the pituitary influences other glands in the body.
34. Which vitamins are soluble in water and which in fat?
35. Give the function of two fat-soluble vitamins.
36. Name four water-soluble vitamins and give the function of two of them.
37. What are three deficiency diseases? State the deficient vitamin in each case.
38. In general, what foods are the best sources of vitamins?
39. How can the daily requirements of ascorbic acid be determined?
40. Why are vitamins D called the "sunshine vitamins?"
41. What is the effect of vitamin K deficiency?
42. How can the absorption of vitamin K be improved?
43. How is vitamin K used in the hospital?

chapter nineteen BLOOD

Ideas for preview or review

The blood is simultaneously an example of a solution, a colloidal mixture, an emulsion, and a suspension.

pH is an expression of the concentration of positively charged hydrogen ions.

Glucose, urea, and cholesterol are examples of nonelectrolytes.

A blood pH less than 7.35 is associated with acidosis.

A blood pH greater than 7.45 is associated with alkalosis.

The blood is not merely a circulatory liquid; rather, it might be called a circulating tissue. It is a tissue because it is made up of cells capable of metabolic changes. The blood might be likened to the transportation system of a community made up of many industries requiring specific raw materials and needing a system for disposing of waste products. For instance, food materials are transported from the intestines, oxygen is brought in from the lungs, hormones are added to the system at their points of secretion, and finally the waste products are carried away to be eliminated in various ways. Incidently, the community maintains a system of defense in the white corpuscles, which resist invading bacteria, and certain regulatory measures operate to keep things functioning smoothly, such as by regulating the water balance of the body, maintaining a constant pH in the system, and distributing heat to maintain a fairly constant body temperature. Thus it may be seen that the blood is a vital part of this complicated system functioning in the body, and its composition is remarkably constant in health. In disease the composition changes in one way or another, and detection of these changes provides an important clue in the diagnosis of a pathologic condition.

COMPOSITION

Blood is composed of a fluid, the *plasma*, and suspended structural components, chiefly *corpuscles*. The plasma makes up about two-thirds of the volume of blood, the corpuscles the other third. The structural components include the red corpuscles (erythrocytes), the white corpuscles (leukocytes), and the blood platelets. The corpuscles can be counted with use of the proper equipment. The average number of red corpuscles is about 5,000,000 per cu mm. The white corpuscles are less numerous, averaging 7,000 per cu mm, and the average number of platelets is 400,000 per cu mm.

Plasma (Extracellular Fluid)

Plasma contains about 90 per cent water. Of the substances present, proteins are the most abundant, 6 to 8 per cent, including fibrinogen, serum globin, and serum albumin. Fibrinogen (about 0.4 per cent) is the component of blood responsible for the production of fibrin, which is the immediate cause of clotting. This becomes evident in the failure of blood to clot after the fibrin has been removed. Removal of the fibrin is accomplished by whipping or stirring the blood immediately after it is drawn. The stirring rod becomes coated with a spongy mass of fibrin. Blood so treated is called *defibrinated blood*.

Glucose is the only carbohydrate normally in the plasma, and the concentration fluctuates between 0.08 and 0.12 per cent. Fats constitute about 0.2 per cent of the plasma, and the turbid appearance of blood is due to this emulsified fat. The amino acids average about 0.006 per cent. The concentrations of these nutrient substances in the blood seem surprisingly low, but it must be remembered that they are absorbed very gradually from the intestine and the tissues remove them from the blood at practically the same rate. Oxygen, both combined with hemoglobin and dissolved in the plasma, inorganic salts, and water complete the list of normal nutrients of the blood.

The waste materials resulting from the various metabolic processes going on in the body are continually transported by the blood to be eliminated. Considering the function of the blood as a transportation system, an exhaustive listing of all the substances present in plasma or which might be there under varying circumstances would include nearly all substances concerned in the biochemistry of man.

Red Corpuscles

In general, erythrocytes may be considered as a concentrated solution of hemoglobin in a fragile capsule. Hemoglobin is the most abundant constituent of blood, constituting about 14 per cent of the total blood and about 32 per cent of the red corpuscles. Hemoglobin is a conjugated protein, and during mild hydrolysis with dilute mineral acid, it splits into a protein, *globin*, and an iron-containing compound, *hematin*. This reaction, carried out in the presence of sodium chloride and glacial acetic acid at a slightly elevated temperature and producing the brown ferric hematin hydrochloride, is used in medicolegal work to determine if stains are blood.

Hemoglobin is the oxygen-carrying substance in the blood for it readily combines with oxygen to form oxyhemoglobin. Its capacity for absorbing oxygen from air is 60 times as great as that of plasma alone, in which the gas is merely dissolved. This reaction,

Oxygen + hemoglobin $\rightleftharpoons$ oxyhemoglobin

is easily reversed, and oxygen is given off upon the slightest decrease in the oxygen content of the blood. The reaction is driven from right to left by carbon dioxide, and thus it is that the faster carbon dioxide is produced by the tissues, the faster it releases oxygen for further oxidation reactions. Oxyhemoglobin is easily reduced by hydrogen sulfide, which accounts for the high toxicity (0.07 per cent) of the inhaled gas. Carbon monoxide and nitric oxide combine directly with hemoglobin to form quite stable compounds and therefore are highly toxic, with death resulting from tissue anoxia.

The color of hemoglobin is the dark-purplish or bluish color characteristic of venous blood, and 5 Gm per 100 ml produces cyanosis. Oxyhemoglobin has the bright red color of arterial blood. The color of blood compared with arbitrary standards indicates the condition of the blood; e.g., a *color index*[1] of 1 indicates that each red cell contains a

---

[1] The color index represents the relative quantity of hemoglobin in each red blood cell as compared with the standard of 100 per cent hemoglobin (16.6 Gm per 100 ml × 6) in a red blood cell count of 5 million per cu mm blood. The percentage of hemoglobin (relative to the standard) found in a sample divided by the percentage of red blood cells (relative to the standard count) gives the color index. By short-cut mathematics, the number of grams of hemoglobin per 100 ml blood multiplied by 6 and divided by twice the first two digits of the red-blood-cell count gives a quotient, which is the color index.

normal amount of hemoglobin. In pernicious anemia, the color index is greater than 1 because there is destruction of red blood cells and those which remain contain an increased amount of hemoglobin. In anemia caused by hemorrhage, it is less than 1 because there has been a great loss of red blood cells and the new cells contain an amount of hemoglobin below normal. Liver extract contains a substance necessary for the formation of new red blood cells, and its injection is often ordered in pernicious anemia. Carboxyhemoglobin is bright red in color, and the flushed appearance of the victim indicates the cause of the poisoning.

Any destruction of the red corpuscles is called *hemolysis*. This has been discussed in the case of hypotonic salt solutions, but anything that destroys the capsule can cause hemolysis; i.e., it can be dissolved by ether or bile salts, or it can be mechanically broken by the pressure changes in alternate freezing and thawing of blood. Snake venoms have a powerful hemolyzing effect. Bloods of different types may cause hemolysis when they are mixed, hence, the importance of testing and cross-matching the blood of a donor with that of a recipient before a transfusion.

The appearance of hemolyzed blood is distinctly different from that of normal blood. It becomes transparent and of a deep red color, and because of this transparency it is said to be "laked."

White Corpuscles

The leukocytes are the scavengers of the body. There are several types of white corpuscles, the most abundant being polymorphonuclear leukocytes, so called from the irregular-shaped or subdivided nucleus. They are the most actively ameboid of all the white corpuscles and make up about 70 per cent of the total white blood cells.

The white corpuscles have the power of ameboid movement, which enables them to pass out through the blood vessels into the lymph and tissues on occasion. The wall is not itself ruptured as these cells pass through. They apparently flow through interstices, which they find or temporarily produce between the cells that compose the thin capillary walls. The polymorphonuclear cells engulf foreign particles, especially bacteria, and destroy them through their bactericidal properties. The protection of the body from infecting bacteria is favored by the tendency of these leukocytes to collect in large numbers in any infected region. Dead leukocytes are very abundant in the pus discharged from such areas. The number rapidly increases to 10,000 to 16,000 per cu mm in appendicitis and to 50,000 or more per cu mm in pneumonia. In leukemia the white-blood-cell count may reach 500,000.

Platelets

The platelets seem to function in the clotting of blood. In the process of clotting there must be fibrinogen, prothrombin, thromboplastin, and calcium ions present to accomplish the process. The reaction of clotting may be represented as

Prothrombin + thromboplastin + calcium → thrombin
Thrombin + fibrinogen → fibrin

Fibrin precipitates as a mesh of fibers, which traps the blood cells and fluid to form a clot. Blood in contact with a wound clots more quickly than it does flowing from an artery without contact with injured tissues. Wounded cells and blood platelets are believed to be the source of something that contributes to the clotting. This substance is the thromboplastin named above. At least one compound, *cephalin,* a phospholipid, has marked thromboplastic properties.

It is obviously important that clotting should not occur in the circulation. Since the substances named above must coexist to accomplish clotting, the absence of one of them would prevent it. In other words, there must be certain inhibitors present to prevent the clotting. *Heparin,* produced in the liver, is such a substance, which inhibits prothrombin. *Hirudin,* prepared from the salivary glands of the leech, is used in experimentation or in surgery to check clotting; it inhibits thrombin. Attention is called to the delay in clotting caused by vitamin K deficiency. The cause of the slowness of clotting in hemophilia is not known, but it is interesting to note that the breakdown of platelets is slow in hemophilia.

The destruction of platelets can produce clotting inside the blood vessels and may result from a number of causes. For instance, a needle protruding into a blood vessel is quickly surrounded by a clot (thrombus). A clot may develop in a blood vessel when its inner lining becomes roughened by disease, by infection, by bruising during surgical opations, or by hardening of the arteries. If this clot is formed in a small artery or is moved by the circulation until it lodges in one, it will plug the artery and thus shut off the supply of blood to that locality. The seriousness of this obviously depends upon the locality affected; e.g., if the plug lodges in the heart or certain parts of the brain, it may be fatal. In order to reduce the hazard of clotting, heparin (sodium salt) or dicumarol is used separately or they may be used together as needed. Heparin is administered by injection, and its action is rapid but not lasting. Dicumarol, on the other hand, is given orally and is slow in action, but its effect is long-lasting. Since the clotting time is slowed by these drugs, their use should be cautiously controlled by laboratory data.

BLOOD ANALYSIS

As has been pointed out, the composition of the blood is remarkably constant. Hence, any deviation from the normal for each component is significant. The usual determina-

Table 19-1 NORMAL COMPOSITION OF THE BLOOD

Substance	Normal Range, mg per 100 ml.	Significance	
		Increased in	Decreased in
Nonprotein nitrogen	25–38	Nephritis	
Urea nitrogen	8–20	Nephritis	
Creatinine	1–2	Nephritis	
Uric acid	1.5–4	Gout	
Glucose	80–120	Diabetes mellitus	Hyperinsulism; insulin shock
Cholesterol	150–250	Nephritis	
Chlorides (as NaCl)	570–620	Nephritis	Gastrointestinal disturbances
Calcium	9.5–11.5	Hyperparathyroidism	Hypoparathyroidism; rickets
Phosphates	2.5–3.5 (adults) 3–5 (children)	Tetany	Rickets
Sodium	315–340	....	Addison's disease
Potassium	16–22	Addison's disease	
Carbon dioxide capacity	50–70*	Alkalosis	Acidosis

*Expressed as volumes per 100 ml.

tions, normal range, and significance are shown in Table 19-1.

The procedures for the analysis of blood are beyond the scope of this text. The usual tests require 5 ml each of oxalated blood or serum. The blood is usually collected from a vein in the arm before the patient is given breakfast. If plasma is to be used for the test, a small amount of potassium oxalate is added to the tube and thoroughly mixed with the blood to prevent clotting. If serum is to be tested, the blood is allowed to clot.

Study Exercises

1. What are the principal constituents of whole blood?
2. How does plasma differ from whole blood?
3. What is the difference between plasma and serum?
4. Why does the color of venous blood differ from that of arterial blood?
5. Why is carbon monoxide poisonous?
6. How does hydrogen sulfide act as a poisonous gas?
7. What is the function of (a) red corpuscles, (b) white corpuscles, (c) platelets, and (d) fibrinogen?
8. Why is hemoglobin a conjugated protein?
9. What use is made of the color index of blood?
10. Calculate the color index if a patient's blood contains 14 Gm hemoglobin per 100 ml blood and the red-blood-cell count is 4 million.
11. Under normal conditions why does blood not clot in the body?
12. What is hemophilia?
13. Why is blood analysis so important?
14. What is arterial thrombosis?
15. Why does blood clot?
16. Why should the presence of oxalate ions prevent the clotting of blood?

chapter twenty METABOLIC WASTES

Ideas for preview or review

The kidneys and the liver are the master chemists of the body. They compensate in many ways, and frequently over long periods of time, for the abnormal functioning of other areas in the body.

The primary consideration in the chemical control of the body is not "This substance is undesirable; excrete it," but rather "This substance is above normal concentration; excrete it" or "This substance is in short supply; retain it."

With very few and rare exceptions, the so-called waste products found in the urine are normal constituents of body fluids.

Consult an anatomy and physiology text for information on the structure and function of the kidneys and liver.

The excretion of waste products from the body is accomplished by a number of different paths: (*a*) the lungs provide for the elimination of gaseous or volatile wastes; (*b*) the kidneys, by secretion of urine, eliminate the bulk of the soluble waste; (*c*) the skin, which contains the sweat glands, serves in a minor way to excrete soluble waste; (*d*) the liver excretes cholesterol and bile pigments; (*e*) the intestinal epithelium excretes a number of inorganic substances foreign to the body, which might therefore be called waste; (*f*) such glands as the salivary, mammary, and tear glands may excrete foreign substances.

The purpose of these excretory functions is to preserve the constant composition and pH of the blood. The lungs and kidneys are particularly important in this respect.

THE LUNGS

The amount of carbon dioxide that can be absorbed by the blood from the cells is truly remarkable. Yet, in spite of this, the pH of the blood remains so constant that the variation is hardly detectable. For instance, arterial blood has a pH of 7.44, and that of venous blood may be 7.32. The same amount of carbon dioxide that was absorbed to produce the venous blood, if dissolved in water, would change the pH from 7 to 4.5. The cause of this constancy lies in the action of *buffers*. Carbon dioxide is continually being formed in the cells as an end product of metabolism, and as the concentration becomes greater than that in the lymph and blood, there is a steady diffusion of the gas in the direction of the blood. In the blood plasma these reactions may take place:

$$CO_2 + H_2O + NaCl \rightleftharpoons NaHCO_3 + HCl$$

$$CO_2 + H_2O + Na\text{ proteinate} \rightleftharpoons NaHCO_3 + H\text{ proteinate}$$

Within the blood cells these reactions take place as the HCl and CO_2 enter:

$$HCl + K_2HPO_4 \rightleftharpoons KH_2PO_4 + KCl$$

$$HCl + KHb \rightleftharpoons HHb^* + KCl$$

$$HCl + KHbO_2^* \rightleftharpoons HHbO_2^* + KCl$$

*The symbol Hb represents hemoglobin, HHb means hemoglobin as an acid, and KHb indicates the potassium salt of hemoglobin. Similarly, $HHbO_2$ and $KHbO_2$ refer to oxyhemoglobin in its two forms.

$$H_2CO_3 + K_2HPO_4 \rightleftharpoons KH_2PO_4 + KHCO_3$$

$$H_2CO_3 + KHb \rightleftharpoons HHb + KHCO_3$$

$$H_2CO_3 + KHbO_2 \rightleftharpoons HHbO_2 + KHCO_3$$

The entry of CO_2 into the corpuscles is aided by the reaction within the cells.

$$CO_2 + H_2O \rightleftharpoons H_2CO_3$$

which is catalyzed by the carbonic anhydrase existing there but not in the plasma.

In the lungs two things must happen: the blood must give up its carbon dioxide, and it must take on the oxygen required for further metabolism. Since the concentration of carbon dioxide is greater in the blood than in the alveolar air, the gas readily diffuses out of the plasma into the lungs. This reverses the reactions given above. At the same time the concentration of the oxygen in the inspired air is greater than that in the blood, which causes the oxygen to react.

$$O_2 + HHb \rightarrow HHbO_2$$

Since the oxyhemoglobin is a stronger acid than hemoglobin, it drives the CO_2 and HCl out of the corpuscles.

$$HHbO_2 + KHCO_3 \rightarrow KHbO_2 + H_2CO_3$$

$$H_2CO_3 \xrightarrow{\text{Carbonic anhydrase}} H_2O + CO_2 \uparrow$$

$$HHbO_2 + KCl \rightleftharpoons KHbO_2 + HCl$$

As the HCl arrives in the plasma, it reacts with $NaHCO_3$.

$$HCl + NaHCO_3 \rightarrow NaCl + H_2O + CO_2$$

Of the volatile substances removed from the body via the lungs, water is the most common. About 20 per cent of the water lost from the body leaves by this route. Upon occasion other substances might be exhaled, for instance, alcohol, oil of garlic, or acetone.

THE KIDNEYS

The function of the kidneys is to filter out of the blood the nonvolatile wastes and concentrate the urine. Incidentally, they maintain the constancy of the blood pH (respiration plays a part in this as shown above) and regulate the osmotic pressure of the blood.

The blood enters the kidney through the renal artery, which branches out into smaller and smaller capillaries. This mass of capillaries is called the *glomerulus*. The membrane that surrounds it is called *Bowman's capsule*, which opens into a long tubule. Several of these tubules are connected to larger collecting tubules, which deliver the urine to the bladder (Fig. 20-1).

There is a difference of opinion regarding the formation of urine, but the theory most commonly accepted states that *all the noncolloidal constituents of the blood plasma, together with a part of its water, filter through the glomerulus*. As this solution proceeds through the tubules, some of the water, salts, glucose, amino acids, and other constituents normally required by the body are reabsorbed through the walls of the tubules and restored to the blood. Waste products such as urea, uric acid, and creatinine are not reabsorbed and appear in the urine. The substances that are reabsorbed are never taken up completely, which explains why a trace of something that appears to be glucose is always present in normal urine. The useful substances are restored to the blood almost completely, unless they are present in concentrations above the capacity of the tubules to absorb them, which is called the *renal threshold*. This is a definite value for each substance. For instance, glucose will be excreted by the kidney if present in the blood in a concentration above 0.16 per cent, i.e., 160 mg per 100 ml blood. Because of the reabsorption of water, the urine is concentrated. It has been estimated that more than 60 liters plasma pass through the kidney to produce 1 liter urine. The difference is the amount reabsorbed.

volume of urine

A normal adult excretes on the average about 1,200 ml urine in 24 hr. Obviously, this amount will vary with individuals, since it

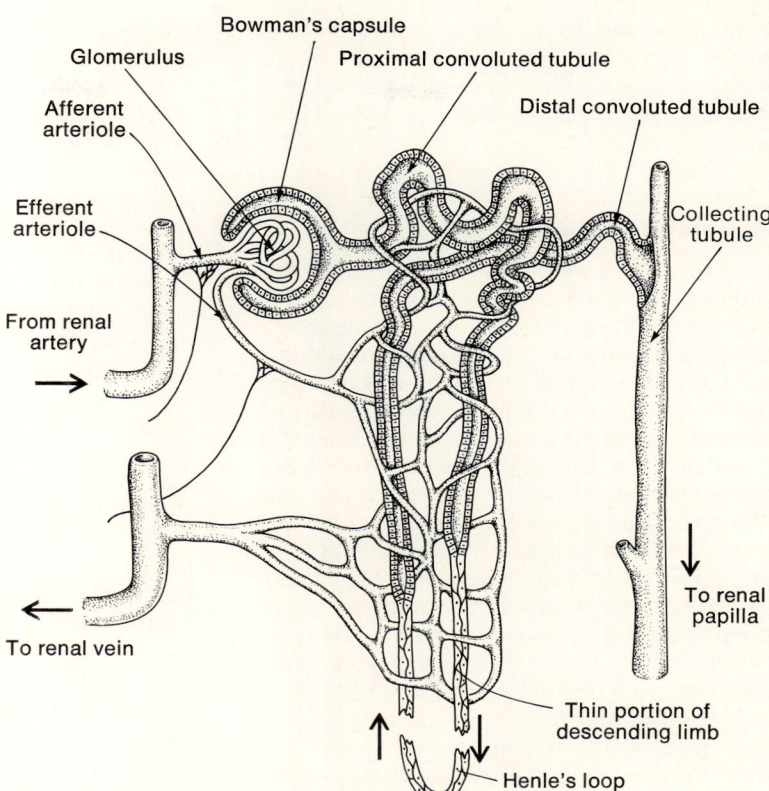

Fig. 20-1 The renal unit. (*From Russell M. DeCoursey, "The Human Organism," McGraw-Hill Book Company, New York, 1955.*)

is affected by a number of factors, for instance, the concentration of the blood caused by the passage of water into the tissues, a decrease of the blood flow through the kidneys, the amount of water ingested, and the amount of sweating. In general, any condition that reduces the amount of water available for excretion decreases the volume of urine produced (oliguria). On the other hand, an increase in the volume of urine may be caused by an excessive intake of water or by such a disease as diabetes. Any substance that causes an increase in the volume of urine (*polyuria*) is called a *diuretic*. In diabetes mellitus glucose may be considered a diuretic, since the volume of urine may be increased manyfold. Diabetes insipidus, a disorder of the pituitary gland, is characterized by an extreme polyuria associated with a tremendous thirst. This condition results from a diuretic activity of the anterior lobe of the pituitary gland unchecked by the antidiuretic hormone of the posterior lobe. In short, the kidney fails to reabsorb water from the glomerular filtrate in the normal way. The volume may increase to 30 liters per day, which means the excretion per hour of the amount of urine that normally would be excreted in a day. Urea has a marked diuretic action, and the volume of urine produced tends to be proportional to the urea excreted. Since urea is the chief end product of protein catabolism, a mild polyuria accompanies a high-protein diet.

Many inorganic salts have a marked diuretic effect, for instance, potassium or sodium acetate. Organic mercurial compounds and caffeine are used as diuretics in medical practice.

appearance of urine

Normal urine is yellow in color, but since the amount of coloring agent excreted each day is fairly constant, the color will vary inversely with the volume, i.e., from a light-straw color with a high volume to a deep-amber color with a low volume. Naturally, the presence of abnormal-colored constituents will alter the color of urine. Bile pigments excreted in the urine color it greenish-yellow or brown and also produce a yellow foam. Hemoglobin will color the urine reddish or brownish depending upon its source. Normal freshly voided urine is usually transparent, but after a heavy meal it may be turbid due to the precipitation of phosphates, since the urine may be basic then. As urine becomes basic upon standing due to the action of bacteria, phosphates may precipitate. Pus and material cast off from the tubules of the kidney (kidney casts) form a sediment in the urine during some infections involving the kidneys.

characteristics of urine

The specific gravity of a solution is a function of its concentration. If the volume remains constant, the specific gravity increases with an increase in the amount of solute present; if the amount of solute is held constant, the specific gravity increases with the reduction in volume of the solution. It must be evident, then, that the total solids in a 24-hr sample of urine may be calculated if we know the specific gravity and the volume for the period. According to Long's formula,

$$\text{Solids} = 2,600 \, (\text{sp. gr.} - 1) \times \frac{\text{volume}}{1,000}$$

For instance, the total solids in a sample of urine in which the specific gravity is 1.018 and the volume is 1,250 ml would be 58.5 Gm (2,600 × 0.018 × 1.250). The specific gravity of urine varies from 1.003 to 1.040, with the average about 1.018. The variability in the relative water and solid content results from the action of the kidneys.

The reaction of urine is also variable and depends upon the ingestion of potentially acidic or basic substances. Urine is commonly acidic, varying between a pH of 4.8 and 7, with about 6 as the average. The figure is higher for vegetarians, who may even produce a basic urine. For limited periods the figure may vary considerably from the average for the day. For instance, after a heavy meal the urine shows what is called the "alkaline tide" for 2 or 3 hr. This is caused by the withdrawal from the blood of the materials necessary for the production of the hydrochloric acid for gastric juice. To maintain the acid-base balance in the blood the urine is basic during this period.

The osmotic pressure of the urine varies widely as would be expected from the varying of the concentration of the urine. The variability of the osmotic pressure is of interest because it demonstrates nicely the effort of the kidneys to maintain a constant osmotic pressure of the blood.

composition of urine

Since the urine is the chief means of eliminating waste products of metabolism, it should be evident that the composition of the urine could be an important check on the normal functioning of the body and variations from the normal composition could be a powerful diagnostic tool. The physiologic normal values for urine are as follows:[1]

Measure	Value
Average 24-hr volume	1,200–1,500 ml
Reaction	Faintly acidic

[1] "The Merck Manual," 10th ed. *By permission from Merck & Co., Inc., Rahway, N.J., 1960.*

Metabolic Wastes

Specific gravity	1.005–1.022
Color	Amber
Urea	20–30 Gm
Uric acid	0.6–0.75 Gm
Total nitrogen	10–16 Gm
Ammonia	0.5–15 Gm
Chlorides	10–15 Gm
Phosphate	2–4 Gm
Creatinine	0.3–0.45 Gm
Total sulfur	1–3.5 Gm
Total solids	50–70 Gm

Urea is the chief nitrogen-containing constituent of urine, containing from 60 to 90 per cent of the total nitrogen. Since it is the principal end product of protein metabolism, the amount of urea in the urine will vary with the amount of protein in the diet. Urea is rapidly and completely hydrolyzed to ammonia and carbon dioxide in the presence of urease.

$$CO(NH_2)_2 + H_2O \xrightarrow{\text{Urease}} 2NH_3 + CO_2$$

The ammonia produced is titrated[2] against a standard acid solution, and the amount of urea present is calculated from the amount of ammonia produced.

Uric acid results from the metabolism of nucleoproteins and occurs in the urine in the form of urates. Nucleoproteins are present in cell nuclei in food as well as in the body, and it must follow, then, that a diet including large amounts of cell nuclei such as liver, kidneys, and sweetbreads must produce an increased amount of urates in the urine. Gout is a disease characterized by an increase in the uric acid content of the blood and a corresponding decrease in the urine, resulting in the deposition of uric acid or urates in the joints. The amount of uric acid in the urine is also decreased by an impairment of the kidneys as in nephritis. Since the urates are the least soluble substances in urine, they are the most difficult to eliminate and therefore a change in the distribution between the urine and the blood readily indicates a faulty kidney functioning. Also, because of their low solubility, the urates may crystallize out as kidney stones, especially when the urine is highly acidic. Obviously, the treatment in such cases would involve the production of less uric acid through diet control and the drinking of large amounts of water to dilute the urine. Finally, uric acid excretion may be increased when there is an excessive destruction of nuclear tissue as during convalescence from pneumonia and in leukemia, in which there is a proliferation of the white cells that pour out into the blood stream, increasing the leukocyte count.

Creatinine is the anhydride of creatine and shows less variation than the other substances found in urine. This relative constancy immediately suggests that its concentration is independent of the amount of protein eaten. The amount of creatinine is increased by the ingestion of creatinine, which occurs in meat and meat extracts, in which case the creatinine seems to have merely passed through the body unchanged. Excretion of creatinine is increased during fevers, diabetes, and starvation. It is interesting to note that, although the amount of creatinine excreted is proportional to the extent of the muscular development (size of muscle mass) of an individual, it does not vary with muscular activity nor with athletic training.

Ammonia occurs in urine in the form of such salts as ammonium urate and chloride. Any time that large amounts of acidic products must be excreted by the kidneys, ammonia is formed and produces salts. This is part of the body mechanism to maintain the proper acid-base balance and at the same time spare the loss of sodium and po-

---

[2] Titration is a process in volumetric analysis using a standard solution for the determination of the amount of some substance in another solution.

tassium, which are needed in the buffers of the blood. The excessive acidic products may result normally from the ingestion of acid-forming foods but more often from the inability to oxidize some of the acids formed in metabolism. An increase in the amount of ammonia excreted is considered an index of acidosis resulting from starvation or diabetes mellitus and other diseases.

Chlorides are second only to urea in the amount excreted in the urine daily, and this amount varies with the amount of sodium chloride in the diet. In nephritis the amount of sodium chloride excreted in the urine is reduced, which increases the amount in the blood and tissues. This increase in the osmotic pressure causes more water to be transferred to the tissues, resulting in edema. For this reason a salt-free diet is prescribed.

Phosphates in the urine result from the metabolism of such phosphorus-containing foods as casein, nucleoproteins, and phospholipids. The amount excreted depends upon the amount of those foods ingested and will vary further under abnormal conditions. For instance, in such diseases as rickets and osteomalacia, resulting from vitamin D deficiency, the quantity of phosphates excreted will be increased. The type of phosphate excreted depends upon the pH of the blood; that is, if the blood is too acidic, NaH_2PO_4 is excreted, but if too basic, Na_2HPO_4 is found in the urine. This emphasizes once more the ability of the kidneys to maintain the acid-base balance of the body.

Sulfur is excreted in the urine in three forms, which are generally reported in urinalysis as the "total sulfur as sulfate." About 80 per cent of the total sulfur is in the form of inorganic sulfates, and since the sulfur results from the metabolism of sulfur-containing amino acids, the amount varies with the diet. About 10 per cent of the total sulfur falls under the heading of *ethereal sulfates*. Indole and skatole, which result from putrefaction in the intestines, are detoxified in the liver to form nontoxic substances. For instance, indole forms indoxyl potassium sulfate, which is called indican.

An increase in the amount of indican excreted indicates an increase in the amount of putrefaction in the colon. The remaining 10 per cent or the total sulfur is called *neutral* or *unoxidized sulfur,* which means that sulfur atoms still have a valence of -2, and is found in such compounds as sulfides.

Abnormal Constituents of Urine

Under certain pathologic conditions various substances appear in the urine. Identification of these substances provides a valuable aid in diagnosis.

The term *diabetes mellitus* is derived from Greek and Latin words and signifies the passage of sugar in the urine. Persistent glycosuria is presumed to be evidence of this disease. It must be remembered, however, that any time the renal threshold for glucose is exceeded, glycosuria results. The ingestion of too much carbohydrate at one time produces alimentary glycosuria; high emotion may cause the secretion of adrenaline into the blood stream and result in emotional glycosuria; and in certain kidney disorders renal glycosuria results. The determination of the blood sugar gives more conclusive evidence for the diagnosis of diabetes. Benedict's solution is used for the qualitative test for glucose, whereas Fehling's solution serves better quantitatively. However, by using comparative color charts, it is possible to obtain a fairly adequate quantitative idea with Benedict's solution. Lactose is also a reducing agent, and since lactose may appear in the urine of a woman during lactation, a positive Benedict's test at that time may not indicate glycosuria. Fermentation of the urine with yeast, producing carbon dioxide bubbles, will indicate that the sugar present is glucose.

Albumin, being colloidal, is not normally found in the urine because of the inability of the kidneys to pass such large molecules. However, in nephritis the glomeruli are affected, and protein does escape into the urine. Such tests as Heller's test and the heat test serve to show the presence of albumin. In the latter test it must be remembered that phosphates may cause a turbidity at this higher temperature. However, the addition of acetic acid clears the turbidity when it is caused by phosphates, but albumin remains insoluble.

Acetone bodies appear in the urine during starvation and diabetes and result from the faulty oxidation of fatty acids. This name refers to the mixture of acetone, β-hydroxybutyric acid, and acetoacetic acid, which occur together. A positive test for any one of the three proves the presence of the acetone bodies. Of these three, acetone is the one most easily detected. If to a small amount of urine a few drops of sodium nitroprusside are added and the mixture is made basic, a ruby-red color denotes the presence of acetone or creatinine if the latter is present in sufficient quantity. However, if an excess of acetic acid is added, the color is intensified if acetone is present but changes to yellow if only creatinine is present. Bile in the urine can usually be detected by the color of the urine. When the bile duct is obstructed as in jaundice, bile appears in the blood and is removed by the kidneys. A chemical test for bile consists in putting a layer of urine over one of concentrated nitric acid. At the junction of the two layers various colored rings will develop (green, blue, violet, red, and brown) as the bile pigments become oxidized by the nitric acid.

Blood may appear in the urine under two conditions: it may result from a lesion in the kidneys or in the urinary tract (hematuria), or it may follow any circumstances in which there is a "laking" of the red blood cells and the blood pigment is excreted (hemoglobinuria). Conditions that may cause the latter situation are such diseases as scurvy, malaria, or purpura; extensive burns; injection of hypotonic solutions; or transfusions in which the donor is not of the proper blood group. While appreciable amounts of blood are detected by the color imparted to the urine, smaller quantities are detected by the benzidine test. To the urine is added a small amount of a saturated solution of benzidine in glacial acetic acid, followed by an equal amount of hydrogen peroxide solution. A greenish-blue color indicates the presence of hemoglobin.

Kidney Efficiency Tests

There are several methods used to determine the efficiency of the kidneys.

phenolsulfonphthalein test

Phenolsulfonphthalein (phenol red) is a nontoxic dye, which is injected intravenously. Urine is collected at intervals, and the amount of dye in the urine is determined colorimetrically after the red color is produced by adding sodium hydroxide. Normally from 60 to 85 per cent of the dye will be excreted in 2 hr.

urea clearance test

In this test, after the patient is given a light breakfast, two 1-hr specimens of urine are collected and a blood specimen is obtained at the end of the first hour. A comparison of the urea content in the three specimens permits a calculation of the ability of the kidneys to eliminate urea. An ability less than 65 per cent of that of normal kidneys is indicative of a diseased condition.

concentration test

This test and another one directly opposite (dilution test) are considered to be most valuable for showing a diseased condition, because they show what the kidneys can do rather than what they are doing. During the concentration test the patient is allowed no fluids for the 24 hr required, beginning with the previous evening. Urine specimens are collected every 3 hr beginning at 8:00 A.M., using a separate container for each specimen. Urine voided during the night to 8:00 A.M. is collected in one container. The volume and specific gravity for each specimen is determined. Normally the specific gravity of at least one sample will be from 1.025 to 1.030. A diseased kidney will produce urine of a lower specific gravity.

In the dilution test the patient drinks 1,500 ml water, and then urine specimens are collected every half hour. At least one of these specimens should normally have a value of 1.003. Because of the simplicity of these tests they are frequently used.

THE SKIN

When we recall that the skin contains about two million sweat glands and the sebaceous glands, surely we may class it among the excretory organs. Sweat is composed chiefly of water, about 99 per cent, and sodium chloride. Urea is contained in sweat to the extent of about 0.08 per cent. This amount may seem insignificant, but when we consider the amount of sweat excreted in a day, the actual amount of urea excreted does become significant. For instance, under ordinary circumstances the amount of perspiration that we unconsciously excrete (insensible perspiration) amounts to about 600 ml per day. In this volume the amount of urea excreted would be 1.5 per cent of the total urea produced in the body. During profuse sweating the volume may increase to 3,000 ml and under these conditions the urea excreted would be about 10 per cent of the total produced. The percentage of urea excreted is greatly increased during such conditions as uremia when the kidneys excrete little or no urine. This is another instance of compensation within the body; i.e., if waste is not eliminated via the kidneys, the skin attempts to correct the situation. The chief function of sweat, of course, is to provide a cooling effect on the body due to the heat of evaporation.

THE LIVER

While it might be argued that the liver acts as a secretory rather than an excretory organ, if we consider the elimination of various substances that would cause physical disorders if not eliminated, then the liver is truly an excretory organ. For instance, bile pigments result from the hydrolysis of hemoglobin and are excreted in the feces. An accumulation of these bile pigments produces the skin color characteristic of juandice. Further, the liver excretes cholesterol, and if it is not sufficiently eliminated, the cholesterol concentration in the blood may rise to the point where it is deposited in the larger arteries. It is believed by some that such a deposition is at least contributory to arteriosclerosis. Finally, inorganic salts are passed from the blood, through the liver, into the bile.

THE INTESTINAL EPITHELIUM

In the intestines various inorganic substances, which may be foreign to the body or which may be present in excess of the bodily needs and are therefore classed as wastes, are excreted. Selenium represents a foreign substance 20 per cent of which is excreted by the intestinal tract. Selenium-contaminated grains grown in certain areas of the country and its increased use in industry have made selenium toxicology a subject for U.S. Public Health Service study. Excess calcium and magnesium are excreted, at least in part, through the walls of the intestines. During saline cartharsis water is excreted through these structures.

GLANDULAR SECRETIONS

In addition to their normal constituents, tears, milk, and saliva may excrete foreign substances. The transmission of undesirable products of the digestion of certain foods to an infant by way of breast milk is well known to nursing mothers. Various drugs readily appear in the saliva. It is an interesting experiment to drink a dilute solution of potassium iodide, rinse the mouth, and then make a test for the iodide ion in the saliva at regular short time intervals. It is a well-known fact, of course, that tears are salty.

FECES

The residue of the digestive processes such as the undigested and indigestible parts of food (cellulose, seeds, etc.), the remains of intestinal secretions, and any insoluble inorganic salts are contained in feces. Of the 25 to 50 Gm dry fecal matter eliminated daily about one-fourth to one-half is bacteria, both dead and living. The normal brown color of feces is due to urobilin, formed by the reduction of bilirubin. Highly colored foods, such as beets and blueberries, may change the color of the feces. On the other hand, abnormal color and character of the feces are important because they indicate abnormalities in the digestive tract. For example, feces that are clay-colored and of the consistency of clay indicate faulty fat digestion or absorption; tarry feces are caused by old blood and indicate bleeding, from an ulcer, for instance. It is an interesting fact that, in general, the character of the feces is independent of the nature of the diet; that is, there is no significant difference in the feces whether the diet is exclusively carbohydrate or protein, all of which confirms the fact that feces are derived chiefly from materials inherent in the gastrointestinal tract.

Study Exercises

1. Name six routes for the excretion of wastes from the body.
2. What are buffers?
3. In general terms, how is carbon dioxide eliminated from the cells where it is produced?
4. What is the mechanism by which carbon dioxide is set free in the lungs?
5. What volatile substances are eliminated by the lungs?
6. What is the function of the kidneys?
7. What is the mechanism by which urine is formed?
8. Explain the difference between oliguria and polyuria.
9. List several diuretics and indicate whether they occur in the body or are administered.
10. What is the "alkaline tide?"
11. What is the immediate cause of gout?
12. What is the composition of kidney stones?
13. An increase in the amount of ammonium salts in the urine indicates what condition?
14. What is the consequence of a reduction in the amount of sodium chloride excreted in the urine?
15. What is indicated by an increase in the amount of indican excreted in the urine?
16. A positive test for a reducing sugar in the urine may not indicate diabetes mellitus. Why not?
17. What is Heller's test for albumin in the urine?
18. The presence of acetone bodies in the urine is caused by what condition?
19. Distinguish between hematuria and hemoglobinuria.
20. Briefly describe a kidney efficiency test.
21. Why should the liver be classed as an excretory organ?
22. Which glands excrete foreign substances?

Chapter twenty-one MECHANICS

Ideas for preview or review

Watching a champion golfer, swimmer, diver, or bowler, the sport appears to be very easy. Close observation of the champion shows us that his body takes maximum advantage of the laws of physics. The average person generally does not. However, effectively or ineffectively, we apply these principles in body movements. Most of the material in this chapter is familiar through experience, but it is desirable to learn the scientific terminology and to examine these principles with the aim of wider, more effective application.

Energy is the ability to do work.

Work is force applied through a distance.

Force is a push or a pull.

A machine is a device that transmits force so as to do work.

In mechanics we are concerned with the relationships between matter, forces, and motion. The movements of the body and body parts, the alignment and positioning of the body, physical stability and equilibrium, etc., fall into this category.

EQUILIBRIUM

We say that a chemical reaction has reached equilibrium when the speed of the forward and reverse reactions is equal. This system is not at rest. Much activity is going on, but the results of the two reactions cancel each other out and there is no measurable change. A physical system is in equilibrium when all the forces (pushes and pulls) acting upon it cancel each other out. A 200-lb man, sitting on a chair, exerts a downward force of 200 lb on the chair. The chair, on the other hand, must be pushing upward or exerting an upward force of 200 lb. If the upward force were less than 200 lb, the man would find himself on the floor.

Equilibrium may be stable or unstable. We know from experience that it is possible to stand a half dollar on its rim. We also know that in this position a very slight force will upset the coin, and it will fall over. Once the coin is lying on its side, it is in stable equilibrium.

CENTER OF GRAVITY

The mass or weight of an object is distributed throughout the object. If the object is symmetrical, one can easily find the point around which the weight is equally distributed. With a 12-in. ruler, we would expect that half the weight would lie between the front edge and the 6-in. mark and the other half between the 6-in. mark and the 12-in. mark. We can test this by suspending the ruler at this point. If we have found the point around which the weight is evenly distributed, the suspended ruler will remain stationary and will not rotate. If we suspend the ruler at the 5-in. mark, we discover that the ruler swings so that the 12-in. side drops down. We can conclude that the weight of the ruler appears evenly balanced at the 6-in. mark. This is the center of gravity for this object. If the object is asymmetrical, the center of gravity can be found by experiment i.e., by trying different points, one can discover the point from which the object can be suspended without tending to rotate.

TORQUE

A torque or moment of force produces or tends to produce rotation. It is the product of the force times the radius through which the force acts. To open a swinging door easily, one pushes at the edge of the door, away from the hinge. Pushing close to the hinge requires a much larger force to open the door. At the outer edge of the door, the effective force applied at right angles to the door is the force times the distance between the door hinge and the edge of the door.

STABILITY

A body is in stable equilibrium when all the forces acting on the body and the torques acting on the body cancel out or equal zero. Contributing to stability are a broad base and a low center of gravity. The lower the center of gravity, the greater must be the force applied to upset the object. Could we say that women are more stable than men? In the ideal female form, the center of gravity is lower than it is in the ideal masculine form. A broad base increases the stability of an object.

Consider the problem of learning to walk with crutches. The patient's center of gravity must be kept within the base made by the crutches. If the center of gravity moves outside the base, the patient will tend to topple over.

LEVERS

We usually think of machines as complicated pieces of equipment, such as an adding machine or a flying machine. Actually, some machines are very simple. They do work, which is the application of force through distance. Levers are examples of simple machines. A lever consists of a fulcrum, or wedge, and a bar. A force is applied at one point on the bar which transmits it to the resistance at some other point on the bar. Levers are divided into three classes depending upon the relative positions of the force, the resistance, and the fulcrum.

In a lever of the first class, the fulcrum is between the force and the resistance. This is the situation in a seesaw or teeter-totter, scissors, scales or balance, and a hammer when used to pull nails.

In a lever of the second class, the resistance is between the fulcrum and the effort. An example is the wheelbarrow, where the front end is the fulcrum, the effort is at the handles, and the resistance is the load in the barrow. In the swinging door, the hinge is the fulcrum, our hand on the outer edge is the effort, and the resistance of the door is in between.

In a lever of the third class, the fulcrum is again at the end. But this time, the resistance is at the opposite end, and the effort is applied between. Eyebrow tweezers, forceps, and nutcrackers are levers of the third class (Fig. 21–1).

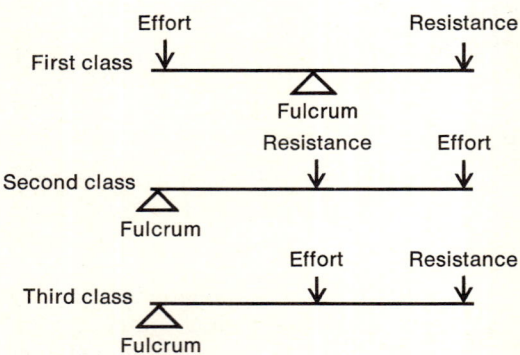

Fig. 21-1 Levers.

Mechanics

Children know that if a lightweight child wishes to play on the seesaw with a heavier child, the heavier child must sit closer to the center and the lighter child must sit at the end. In this way, the lighter child can balance the heavier. Unknowingly, these children are applying the fundamental law of the lever. A smaller force acting through a longer distance can counterbalance a larger force acting over a shorter distance. In the form of an equation, force 1 × distance 1 is equal to force 2 × distance 2.

$$F_1 \times D_1 = F_2 \times D_2$$

This machine, the lever, gives us a *mechanical advantage*. It enables us to use a 5-lb effort to lift a 100-lb load. If the 100-lb load is 1 ft from the fulcrum, our 5-lb effort would need to be applied at least 20 ft away. The long effort arm gives us an increase in force. We see this in pliers, bone shears, and pump handles. With a short effort arm and a longer resistance arm, the speed of the motion can be increased. This is the case with scissors or paper shears.

Levers operate in the body. Lifting a weight in the hand is an example. The weight is the resistance, the elbow is the fulcrum, and the muscles of the upper arm (brachii) supply the effort. Generally speaking, the body is designed for speed and variety of motion rather than increased force.

INCLINED PLANE

The inclined plane or wedge is another simple machine. Here again, a smaller effort is applied over a longer distance. Consider the effort involved in sliding a 50-lb object up a ramp 5 ft high and 10 ft long. This amounts to raising the 50-lb object 5 ft. The relationship is: effort × length of the plane is equal to resistance × height of the plane. The effort needed in this case would be 25 lb. If the plane were 25 ft long, the effort needed would be 10 lb. From these examples, we can see that a long plane gives the greatest advantage. Here the effort is expended over a longer distance. Ramps are good examples of inclined planes.

PULLEYS

Still another simple machine is the pulley. Pulleys consist of wheels with bands, belts, or ropes running around them. A single fixed pulley simply changes the direction of the force applied. Single fixed pulleys are used in applying traction to a fractured bone. The downward direction of the gravitational pull on a weight hung on a pulley is changed to a horizontal pull on the bone. One type of traction using this principle is Buck's extension.

Movable pulleys and combinations of more than one pulley give a mechanical advantage. This advantage depends upon the number of ropes in the pulley system.

VECTORS

Forces with directions are *vector quantities*. When we speak of a downward force or a horizontal pull, we are speaking of vectors. When two or more forces act on an object, they may reinforce each other, they may tend to cancel each other out, or their net effect may be some intermediary result. If two persons cooperate to lift an object by pulling on it vertically, each exerting a force of 10 lb, the result is an effective force of 20 lb. If an object is pulled to the east by a force of 10 lb and pulled to the west with a force of 10 lb, the result is zero; there is no resultant force, and the object remains stationary. If the object is pulled to the east with a force of 5 lb and simultaneously pulled to the north with a force of 5 lb, the object will move in a direction somewhere between these two forces; the force with which it moves will be less than the mathematical sum of the two forces. The direction in which the object will move can be determined by adding up the vectors. In other words, these two forces have a net effect of one force on the body. This is called the *resultant*. The force can be calculated, or it can be determined by diagram. To determine the resultant force by diagram, the forces are drawn to scale with their directions. The directions are indicated by arrowheads. The lines are drawn arrowhead to arrow tail in the appropriate directions. The resultant force can be determined by measuring the length and the direction of a line drawn from the beginning of the first vector to the arrowhead of the last. A modification of this procedure is called the *parallellogram method*. In this method the two vectors are laid out from the same point, and the parallelogram is constructed. The resultant is computed by measuring the length and angle of the diagonal (Fig. 21–2).

If the forces act at right angles, they form the two legs of a right-angle triangle. Their resultant will be equal to the hypotenuse and can be calculated from the equation

$$a^2 + b^2 = c^2$$

where $c =$ hypotenuse

For the object pulled east with a force of 5 lb and north with a force of 5 lb,

$$5^2 + 5^2 = c^2$$
$$25 + 25 = c^2$$
$$c^2 = 50$$
$$c = 7.07 \text{ lb} \quad \text{resultant}$$

The direction of the force is northeast.

TRACTION

The principal purpose in our discussion of pulleys, vectors, etc., is to give the nurse a background necessary for the intelligent care of the patient in traction. Regardless of how complicated the traction apparatus may appear to be, the principles upon which all these things operate are very simple. When a bone has been fractured, the contraction of the muscles may pull the sections of bone out of alignment. The process of realigning the bone pieces is called *reduction* of the fracture. If the muscles attached to the fractured bone are not too forceful, the physician, by exerting his own strength, can pull against the contracted muscles and realign the fractured bone manually.

The muscles associated with the femur are the most forceful muscles in the body. When the femur is fractured and the muscles contract, pulling the fractured femur out of alignment, it is sometimes advisable to use a steady pull for days to realign the fractured bone. This pulling is *traction*. We have established that a body pulled in one direction will remain stationary if it is simultaneously pulled by an equal force in the opposite direction. This opposite force is *countertraction*. The countertraction is usually the weight of the patient. Without sufficient countertraction, the patient's body would be pulled in the direction of the traction.

A moment's thought will enable us to see that the direction of the pull must be maintained unchanged if the desired results are to be maintained. One way to do this is to immobilize the patient. A pin can be inserted in the distal end of the femur (at the knee) and a pull applied directly at this point, or straps can be attached by adhesive tapes and a pull applied directly. In both cases, the mobility of the patient is seriously limited.

Like the object that was pulled northeast as a result of two forces, one north and one east, it is possible to exert an effective force on the distal end of the femur by the application of forces on other areas of the limb. If a force is exerted on the ankle in the direction of the foot of the bed and another force is applied upward under the knee, the resultant force will have a direction in between these two forces. This principle is applied in Russell's traction. In this apparatus, a frame is set above the bed. Four single fixed pulleys are used. These pulleys change the directions of the ropes and therefore the directions of the forces. One pulley is attached to the frame above the patient's knee. Another pulley is attached to the frame above the foot of the

Mechanics

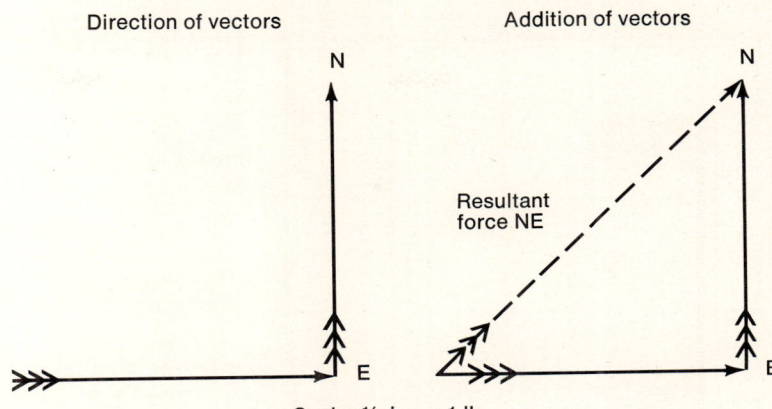

Fig. 21-2 Vectors.

bed. The two remaining pulleys are attached singly beyond the foot of the bed on a level with the patient's foot. A sling placed under the patient's knee is attached to a rope, which runs up vertically around the pulley above. The rope continues in a horizontal direction and runs over the upper pulley at the foot of the bed. Its direction is thus changed to downward. The rope runs down to the level of the patient's foot, around one pulley, through a loop or pin attached to the patient's foot, and out and over the last pulley. The rope ends in a downward direction. Attached to this end of the rope is a freely hanging weight (Fig. 21-3).

The amount of the weight suspended on the rope is determined by the physician. If the weight hanging on the rope is 10 lb, all the parts of the rope exert a force of 10 lb. If this were not the case, the rope would not be stationary. The forces on the patient's limb are exerted at the knee and the foot. The force on the knee is 10 lb. This force is in an upward direction. There are two parts of the rope attached to the foot. Each part exerts a force of 10 lb. The force on the foot is 20 lb in a horizontal direction. By diagram or by mathematics, we can determine that the resultant force is approximately 22 lb and its direction starts from the knee.

The nurse is careful to observe that the weight is always free swinging—not on the floor, not tangled in anything. The weight must hang freely, or the forces will be changed. The nurse checks to see that the countertraction is sufficient. If the patient slips down in the bed so that his foot rests against the foot of the bed, there is no pull on his foot but rather a push against the foot by the bed. It may be necessary to raise the level of the foot of the bed by putting the legs of the bed on blocks, to prevent this slipping.

This type of traction permits a certain amount of motility to the patient. Usually, an overhead trapeze is supplied to facilitate motion and permit the patient a certain amount of exercise. This is particularly important from the point of view of the rehabilitation and comfort of the patient.

The patient may move, linen may be changed, and pillows may be changed as long as the alignment of the femur remains in the proper plane.

ELASTICITY

An important property of many materials including tissues is elasticity, which may be defined as the *ability of a material to return*

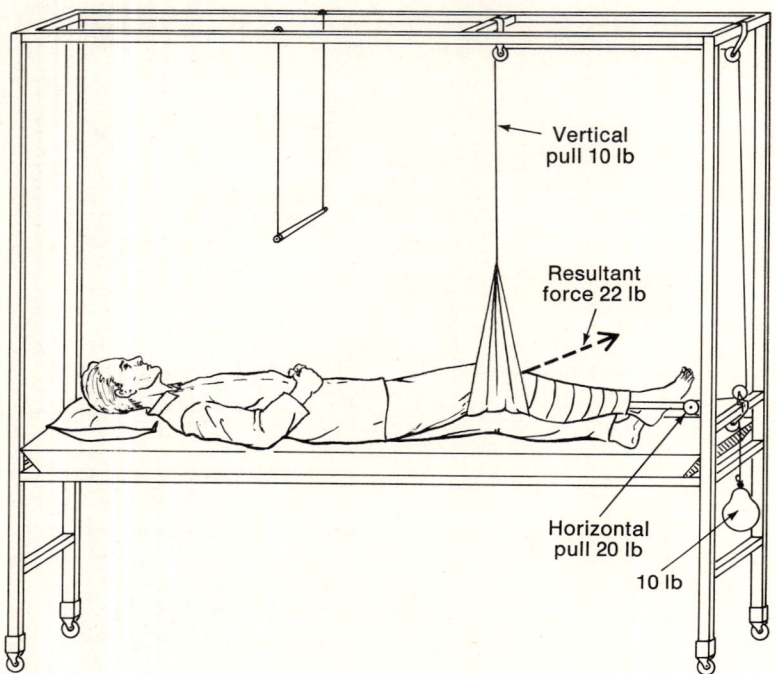

Fig. 21-3 Russell's traction.

more or less to its original shape after a deforming force has been removed. A deforming force can be so great that the material cannot return to its original shape, in which case the material has exceeded its *elastic limit*. Many tissues contain elastic fibers which enable the tissues to return to their original form after they have been under strain. An important concept in understanding respiration is that when the lungs expand during inspiration, they are under strain and tend to return to the undistended state. This is described as the "elastic recoil of the lungs." In some pathological conditions, such as emphysema, the elastic recoil of the lungs is diminished and, as you might expect, the patients have difficulty exhaling. The elasticity of bones in young children compared to the loss of this property upon aging explains the difference in the number of fractures in these two age groups.

MECHANICS AND THE BODY

In good body mechanics, the best way to accomplish work or motion is the way that requires the least amount of effort.

Work with a broad base.

Use large muscles in preference to small ones. The working muscles are the muscles of the arms and legs, not the muscles of the back.

Keep the body parts properly aligned, with the centers of gravity over the base.

Work with weights and carry objects close to the body (the center of gravity).

When positioning pillows to support a body part, place them as nearly under the center of gravity of the part as possible.

Use all your body weight in pushing or pulling, not just a few muscles.

Push in preference to pulling.

Squat in preference to bending.

Slide in preference to lifting.

For the recumbent patient, keep the body parts in alignment to prevent contractures.

Supply support when and where needed. Remember that forces applied to tissues can alter their shape and growth.

Study Exercises

1. In physical terms, define equilibrium, center of gravity, and stability.
2. What would be the effect on the center of gravity in each of the following situations: plaster of paris cast on the right leg, a cast on the left arm, carrying a full dinner tray, and carrying a valise in the right hand?
3. Which would be less fatiguing, carrying a bucket of water balanced on top of the head or carrying the bucket in the hand?
4. Identify the three classes of levers. Give two examples of each.
5. A student bends forward from the hips to pick up a heavy object. Considering the body as a lever, in this position, what part of the body acts as the fulcrum, where is the effort applied, and what muscles do the lifting?
6. What will be the direction and magnitude of the resultant force in each of the following situations? a pull north of 5 lb and a pull south of 4 lb; a pull north of 5 lb and a pull west of 4 lb; a pull north of 5 lb and a pull east of 3 lb.
7. A patient with a fractured femur in Russell's traction asks the nurse if she can remove the pillows beneath her head. What can the nurse tell the patient?

chapter twenty-two WAVES

Ideas for preview or review
 Sound and light are forms of energy.
 Energy rays and energy waves are the same. It is simply a question of usage: we speak of X rays and gamma rays; we would be equally correct to say X waves and gamma waves.
 If more detail is desired about the ear and the eye, consult an anatomy text.

Waves are a kind of motion called *periodic motion*. They are a way of transmitting energy. If we are fishing in a boat in the middle of a lake and a speed launch passes on the north, the wash of the launch when it reaches our boat causes it to ride up and down on the waves, although the swells formed by the launch are traveling to the south shore of the lake. This type of surface wave is a kind of *transverse wave,* in which the particles that make up the wave move perpendicular to the direction of the wave. Energy from the sun is transmitted in transverse waves. When the particles of which the wave is composed move in the same direction as the wave motion, we have longitudinal or *compression waves*. Sound waves are of this type.

Waves can be described by three characteristics: length, frequency, and amplitude. Wavelength is the distance from a point on one wave to the corresponding position on the next wave. Frequency is the number of waves that pass a given point in 1 sec. One can see the relationship between these two properties. If the wave is a long wave, not many of them can pass a given point in 1 sec; if they are short waves, many can pass. Consequently, the shorter the wavelength, the higher the frequency. The frequency of a wave determines the pitch. The higher the frequency, the higher the pitch.

The height of the peak of a wave is the *amplitude.* Amplitude determines the loudness or intensity of a sound. When the amplitude is large, the sound (if in the audible range) is loud; if small, the sound is soft. Very high and very low notes are not heard at low amplitudes.

SOUND WAVES

Sound waves arise from vibrations and travel through some kind of medium. They cannot pass through a vacuum. As the sound wave passes through a medium (solid, liquid, or gas), the molecules on the forward part of the wave are pushed closer together, causing higher pressure, and those behind are left farther apart, causing lower pressure. We have then, alternating areas of higher pressure, lower pressure, higher pressure, etc., giving rise to a compression or pressure wave.

hearing

Sound may also be considered as the sensation transmitted by the hearing apparatus of the ear and interpreted by the brain. If there was an explosion in the middle of a desert and no one around to hear it, would we have sound or just energy of vibration (compression waves)? If a patient has had a head injury that has destroyed the hearing centers in the brain, his ears and auditory nerves may be perfectly normal but the patient is deaf. Conversely, if the auditory nerve and brain centers are normal but the eardrum has lost its ability to vibrate, sound can be conducted through the head bones. Recall how the dentist's drill sounds when the teeth are being drilled. In this case, the vibrations of the drill are transmitted through the teeth and jaw to the ear. Some types of hearing aids make use of this principle of bone conduction.

The unit for describing the loudness of sound is the decibel. The human ear has a range up to approximately 130 decibels. The audible range is usually considered to fall between 20 and 17,000 cycles per second (cps). Frequencies below 20 cps are called *infratones*, and those above 17,000 cps are called *ultratones, supertones,* or *hypertones*. The range of ultratones extends to 2 hundred million cps. A thousand cycles is a kilocycle; a million cycles is a megacycle. The ultratones then can reach 200 megacycles per sec. Most human speech falls in the range of 300 to 3,500 cps.

Sound waves, like other waves, can be reflected, refracted, transmitted, and absorbed. They travel in straight lines and can bounce off a surface (be reflected), giving rise to an echo. Echoes of echoes are called reverberations and, along with echoes, are a great disadvantage in theaters and halls. These can be minimized by the use of sound-absorbing materials. Drapes, rugs, furnishings, and special-composition building materials can greatly reduce this nuisance.

If the waves pass from one medium to another in which the velocity is different, the path of the wave is deflected from a straight line. This is called *refraction*. This property is important when dealing with the application of ultrasound waves.

Impaired Hearing. The instrument used to test hearing is the audiometer. This instrument is set at selected frequencies, and the sounds are intensified until the patient can detect them. Some hearing losses can be compensated by louder sounds. Other losses are spotty. Certain sounds may be audible and others inaudible. This latter condition has been described as "Swiss cheese hearing." Many older patients have some kind of hearing impairment, and many are sensitive about this defect. Some persons will pretend they understand when they have heard incompletely, and some will interpret for themselves. Remember, hearing really occurs in the brain. Speak directly to the patient with imparied hearing. Don't turn your head down or away. Sound travels in straight lines. Also, many patients are quite adept at lip reading. If the patient does not understand you, repeat your idea in different words. The tendency is to repeat the same word in a louder voice. This increased loudness will not help the lip reader or the patient with spotty hearing, whereas using different words may enable both these persons to understand you. Be alert also to the presence of extraneous sounds that may make difficulties for the wearer of a hearing aid. The person with normal hearing can pick the desired sounds from a background of noise, but the hearing aid intensifies all the sounds within its range.

Ultrasound

Waves with frequencies much above 20,000 cps are short waves and have some properties similar to other short radiations, such as X rays. Like other sound waves, they do not pass through a vacuum and only with difficulty through air. Their velocity is four times greater in solids than in liquids.

Ultrasonic waves were used during the First World War to locate submarines. The principle on which this procedure worked was the echo principle, like that now used in radar. When the velocity of the ultrasound waves in water and the length of time it took for the echo to be recorded were known, the position of a submarine could be calculated. At this time it was noted that sonic and ultrasonic frequencies of high intensities in water were destructive to living cells and organisms.

It had been discovered that crystals without a center of symmetry developed positive and negative charge when they were compressed. This was called the *piezoelectric effect* (*piezo* means "pressure"). Later the reverse condition was found to be true; that is, when the crystal is subjected to an electric field, it expands and contracts very rapidly, thus setting up vibrations of very high frequency well beyond the range of human hearing.

These ultrasound waves have found a limited use in medicine. The apparatus consists of a suitably cut quartz crystal covered by electrodes. When high-frequency oscillation voltage is used, powerful vibrations are set up in the quartz disk. The vibrations thus generated can radiate into the adjoining medium. The quartz crystal is mounted in a hand applicator called a sound head, or treatment head, and the head is brought in close contact with the tissue to be treated. In this country, the most popular frequency used is 1,000,000 cps, or 1 megacycle.

Several theories have been proposed to explain the effects produced when tissues are exposed to ultrasonic energy. One is called a mechanical effect. Individual cells are vibrated, giving a kind of massage at the cell level. This may cause stimulation of blood and lymph vessels. Cell membranes are affected, with an increase in intercellular osmosis, giving results similar to those obtained with treatments of procaine hydrochloride.

Another explanation is based on a thermal effect. Temperature increases of from 2 to 4°C with a depth of 2 in. are not unusual. Overdoses of ultrasonic energy are easily observed by the reddening of the tissues due to this thermal effect.

In addition, there is a physiochemical effect from ultrasonic energy that suggests a change in the colloidal nature of protoplasm with shifts from the gel to the sol condition.

Still other investigators claim neurotropic effects, indicating action on neurons and neural pathways. There is also the possibility that these waves may alter the mitotic processes in cells and thus have long-range or delayed effects.

Although several or all of the above explanations may be correct depending on the conditions, it is generally accepted that in the hands of an expert and with proper dosage the use of this energy can produce markedly desirable effects with no observable accompanying damage.

Ultrasonic energy is finding an application in the treatment of degenerative and dystrophic diseases with painful conditions of the joints (with or without arthritic deformities), in the lossening of adhesions, and in the relief of painful symptoms. It has also been useful in cases of neuralgia and in cases involving inflammatory conditions in soft tissues. In this last condition it is as effective as X rays. Other conditions treated with ultrasonic energy are follow-up treatment of fractures, prolapsed intervertebral disk, peripheral circulatory disorders, asthma, septic skin conditions, and boils.

Ultrasonic energy can be used as a diagnostic tool. One possibility is the application of the radar or echo technique. Calculi and other structures, which can bounce back the ultrasonic waves, can be detected in soft tissue and their size and position determined. Another application is the development of the hyperphonogram, or sonograph. This is a sonic shadow image due to the difference in absorption of sonic waves by different tissues. Ultrasonic waves of very low intensity are passed through the skull. The receiver registers the transmitted waves as shadows on a photographic film. By moving the source of the waves and the photographic film, the size and shape of the brain and ventricle, tumors, scars, and metallic pieces can be detected.

Until the full explanation of the effects of ultrasonic waves have been determined, physicians urge caution in their use particularly in the treatment of infants, pregnant women, and persons with heart disease or malignant conditions.

LIGHT WAVES

Light waves comprise a very small band of what are called *electromagnetic radiations*. These are energy waves ranging from very long wavelengths, such as we find in radio waves, to those of extremely short lengths, the cosmic rays (Table 22-1). The length of radio waves can be 10,000 meters. Some cosmic rays are a trillionth of a millimeter, and others are still shorter. Television waves are shorter than radio waves, about 10 meters in length. Shorter still are radar waves; these are about 1 cm in length. Heat waves vary from 1 to 1/1,000 mm. The heat waves are also called *infrared*. All these waves are invisible to the human eye, and all but heat waves are completely undetectable by the human senses. The nerve endings in the retina of the eye react to waves between 7 and 4 ten thousandths of a millimeter. Waves falling in this range are *visible light*. The longer waves are red, and the shorter ones are violet. In between, there are the other colors of the rainbow or spectrum. In decreasing wavelength, they are red, orange, yellow, green, blue, and violet.

Shorter than visible violet is ultraviolet, shorter still are X rays, gamma rays, and cosmic rays. Ultraviolet, X, gamma, and infrared rays are used in therapy. Ultraviolet lamps are used in the treatment of certain skin disorders; X rays are used for diagnostic purposes as well as for the destruction of tissue; and gamma rays are also used to destroy tissue. Infrared lamps are used because the infrared rays, when absorbed by tissue, cause heating in the tissue. Painful joint conditions, such as arthritis, are relieved by this method.

Light waves are transverse waves. Like sound waves, they can be reflected, refracted, transmitted, and absorbed. In addition, they can travel through a vacuum. Light waves, like the other electromagnetic radiations, travel at the rate of 186,000 miles per sec. Smooth, polished surfaces *reflect* light very effectively. The light rays bounce off the surface, and the angle from which they leave the surface (angle of reflection) is the same as the angle at which they impinge upon the surface (angle of incidence). Flat, smooth, polished surfaces make good reflectors and are used as mirrors. An uneven surface scatters the light, causing a diffuse reflection (Fig. 22-1).

The ophthalmoscope uses this principle of reflection. Light striking a mirror is reflected into the patient's eye and is then reflected back to the doctor's eye through a hole in the mirror. This device makes the interior of the eye, the retina, and the blood vessels visible. Specially shaped mirrors are widely used in dentistry. Mirrors, properly placed, permit a recumbent patient to read.

Light rays can be *absorbed* by different materials. If all the light is absorbed, the object appears black. If no light is absorbed, all the rays are reflected and the object appears white. Dyes, paints, and pigments have the ability to absorb certain wavelengths and to reflect others. If all the wavelengths are absorbed except red and red is reflected from the object, the object appears red.

Table 22-1 ELECTROMAGNETIC RADIATIONS

Type of wave	Wavelength
Long-wave radio	10,000 meters
Short-wave radio	10 meters
Television	1 meter
Radar	1 cm
Infrared (heat)	1-0.001 mm
Visible light	0.0007 (red)-0.0004 (violet) mm
Ultraviolet	0.00001 mm
X rays	0.00000001 mm
Gamma rays	0.0000000001 mm
Cosmic rays	0.000000000001 mm

Waves

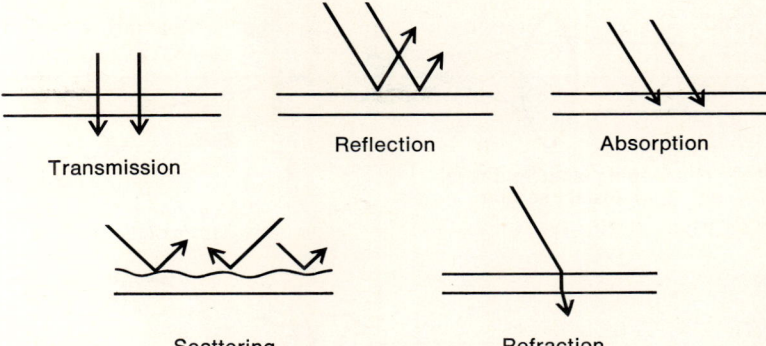

Fig. 22-1 Paths of lightwaves.

The speed of light varies as it travels through different media. If the light passes from one medium to another, for example, from air through water, in a perpendicular direction, the light will pass directly through, but if the rays strike the surface at an angle, they will be bent away from the perpendicular. This is called *refraction.* In passing through a prism, the colors of which white light is composed are refracted by different amounts. As the light emerges from the prism, it is separated into its constituent colors. Sometimes after a storm raindrops act as tiny prisms, forming a beautiful rainbow.

lenses

Lenses are pieces of transparent material, usually glass or plastic, whose surfaces have been curved to bend (refract) rays of light in a desired manner. If the material is curved so that the center is thicker than the edges, the lens is *convex.* If the edges are thicker than the center, the lens is *concave* (Fig. 22-2).

Parallel rays of light passing through a convex lens are refracted and cross at a point. This is the *principal focus.* The distance from the principal focus to the center of the lens is the *focal length.* A lens that causes light rays to meet is a *converging,* or *positive, lens.* If parallel rays of light passing through a concave lens are spread apart, it is a *diverging,* or *negative, lens* (Fig. 22-3).

The camera consists essentially of a lens, an aperture or diaphragm, a film or photographic plate, and a device for focusing the image. Light rays from objects at a distance are considered parallel, and the camera can be set in advance; i.e., the distance between the film and the lens can be predetermined.

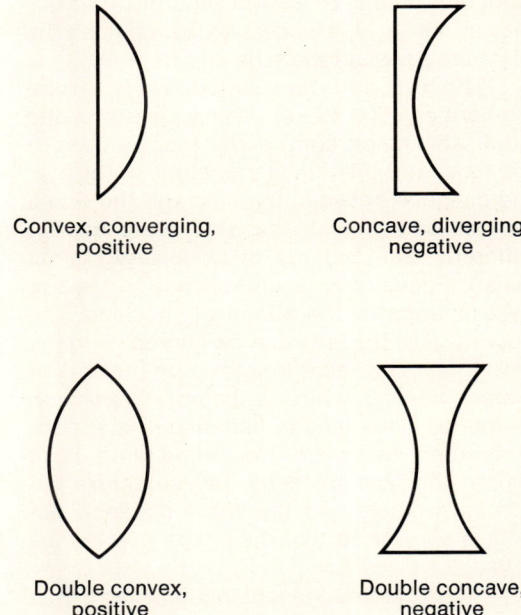

Fig. 22-2 Concave and convex lenses.

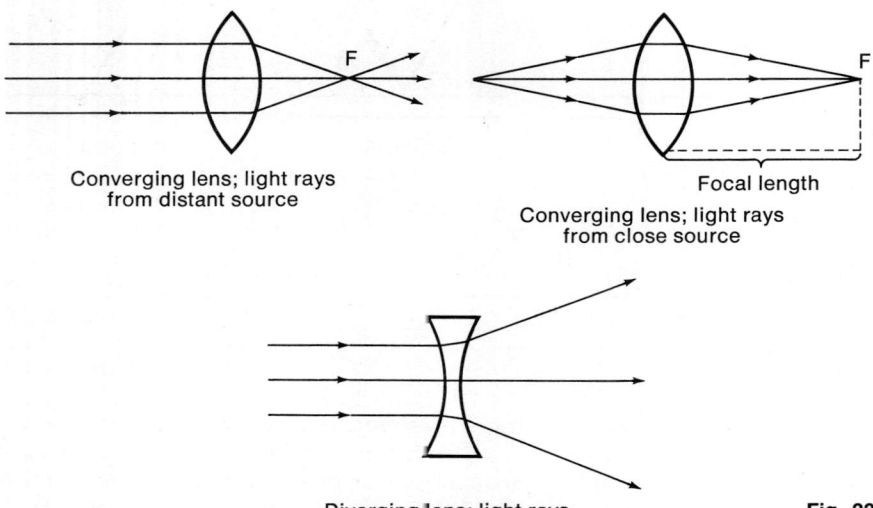

Fig. 22-3 Effect of lenses on light waves.

Rays of light striking a lens from a source close to the lens are not parallel, and their point of convergence will be beyond the principal focus. To take a picture of an object up close, it is necessary to increase the distance between the lens and the film.

The eye and the camera have several similarities (Fig. 22-4). The aperture is the pupil, the diaphragm is the iris, the lens is the lens, and the film is the retina. In the eye, the distance between the lens and the retina cannot be varied to accommodate for the different requirements of close and distant vision. Focusing is accomplished in the eye by a change in the shape of the lens. The curvature of the lens can be altered, with the result that the focal length is changed. For distant viewing, which is the normal position of the lens, the lens is flattened. Distant objects then focus on the retina. For close vision, the lens thickens, the curvature becomes greater, and the focal distance becomes shorter so that the image falls on the retina.

In the camera the lens remains the same, and the distance between the lens and the film can be changed. In the eye the distance between the lens and the retina remains the same, and the shape of the lens is altered. If the lens does not *accommodate* properly so that the image is focused on the retina, impaired vision results.

If the eye lens tends to retain the thickened or more rounded shape, the individual can see objects at close range but distant vision is blurred. In this case, the lens is too converging or convex. This condition is nearsightedness, or *myopia*. The correction would be to place an appropriately diverging lens in front of the overconverging lens. This correcting lens would be concave.

When the eye lens fails to thicken and become more convex for focusing on objects up close, the person can see distant objects clearly but images of nearby objects appear indistinct. This is farsightedness, or *hyperopia*. Since the eye lens does not converge enough, the remedy would be to place a converging or convex lens in front of the eye to make the correction (Fig. 22-5).

We have confined our description of refracting power to the lens. Actually, the cornea, the aqueous humor, and the lens are involved in this process. *Astigmatism* is usually due to a defect in the shape of the cornea. As a result, vertical and horizontal rays of light reflected from an object are not focused at the same point on the retina, and the image is indistinct. Correction can be made by a lens ground to compensate for the defective shape.

Clouding of the lens or cornea is known as *cataract*. The protein material of which these structures are composed alters from transparent to opaque. This process has been compared to cooking egg albumin, a change from clear to white. The opaque material can be removed, and the refractive power lost can be replaced by appropriate lenses.

binocular vision

Each eye, being in a different position, sees a different image; the right eye sees one image and the left eye sees another. Sight, like hearing, occurs in the brain, and these two images are fused into one image with depth, or three dimensions. It is extremely difficult to judge distances and depth using only one eye. This can be easily demonstrated by holding two pencils, one in each hand, at arm's length. Keeping the arms stiff and one eye closed, it is difficult to bring the tips of the pencils quickly together. Inability to estimate depth and distance properly is *aniseikonia*. In this condition, the size or the shape of the images in each eye differ, and the brain errs in interpreting the actual position of the object. This could be a serious handicap for automobile drivers.

If the images in each eye do not fall in the same position on each retina, double vision *(diplopia)* results. This is easily seen by looking at an object and pressing one eyeball slightly sideways. If the condition is not too severe, the individual may correct for the defect by squinting *(strabismus)*. Diplopia may lead the brain to ignore one image. Crossed eyes or divergent eyes or inability to focus both eyes simultaneously can cause double vision.

the retina

The retina is the light-sensitive area in the eye. It consists of nerve endings in the form of rods and cones and contains a light-sensitive substance: rhodopsin, or visual purple. Light striking rhodopsin causes a chemical change, and the purple color bleaches out. Light causes a photochemical or a photoelectric change in the retina, which triggers the nerve impulse. The rhodopsin is resynthesized in dim light. Experience has taught us that in passing from a brilliantly lighted area to a dimly illuminated one, we are momentarily blinded. Presently the eyes adapt to the difference in luminosity. Inability to make this adaptation is night blindness *(nyctalopia)*. Persons with this defect experience difficulty in distinguishing ob-

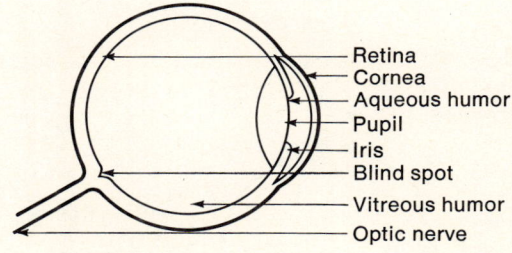

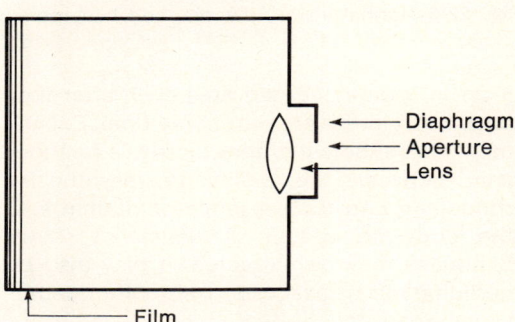

Fig. 22-4 Comparison of the eye and the camera lens.

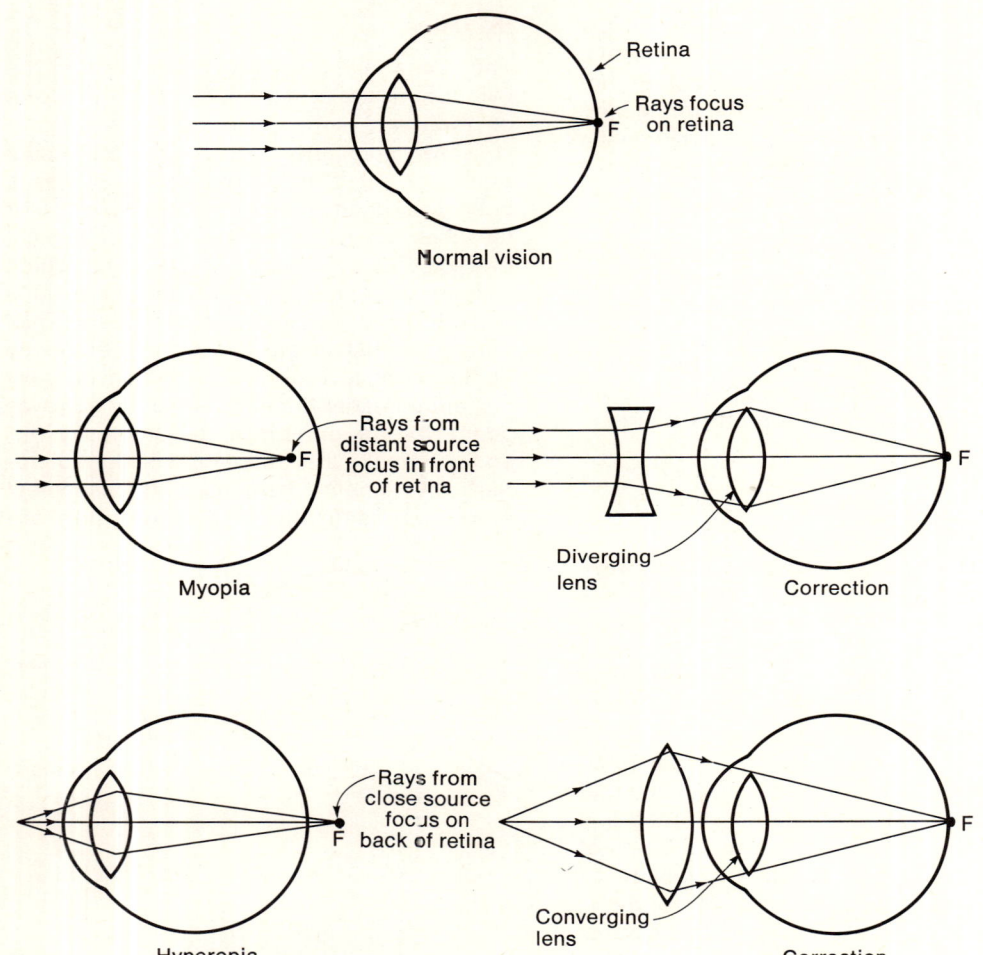

Fig. 22-5 Normal vision, myopia, and hyperopia.

jects in a dimly lighted area even after they have been in the area for some time. Experiments have indicated that there is a relationship between the ability to resynthesize rhodopsin and the presence of vitamin A in the body. In vitamin A deficiency, night blindness may be expected. It may also be hereditary or be associated with other pathologic conditions, such as glaucoma (increase in pressure in the eye).

At the point at which the optic nerve enters the eye, there are no nerve endings. Light striking this part of the retina is not perceived. This is the blind spot. Here again, the brain interprets for us to give us a complete picture.

Study Exercises

1. Describe three characteristics of waves.
2. What are sound waves?
3. What is the difference between sound and hearing?
4. What are reverberations?
5. What are two uses of ultrasound in medicine?
6. Distinguish between reflection and refraction.
7. What is a concave lens? What is its effect on light?
8. What is a convex lens? What is its effect on light?
9. What is the visual defect in nearsightedness? How may it be corrected?
10. What is the visual defect in hyperopia? How may it be corrected?
11. What is the relationship between vitamin A deficiency and night blindness?

chapter twenty-three IONIZING RADIATIONS

Ideas for preview or review

In science, we find many areas that can be considered from several different points of view, each correct but each with a different focus. To different persons, ionizing radiations and radioactive materials mean different things. The medical specialist may be concerned primarily with their effects on normal and abnormal tissues; the chemist and physicist may want to know how they can be used to help explain the mysteries of matter and energy; the geologist may want from them information about the past history of the earth; the military man may consider them as weapons; the astronaut may think of them as fuel for interplanetary travel; and so on.

These rays and radioactive materials can be harnessed. They can impart energy to systems, and energy implies work. We find in the literature references describing the energy of these rays in terms of work units. Such a unit is the erg. Work includes both force and distance, and an erg is equal to a force of 1 dyne acting through a distance of 1 cm. A dyne in turn is the unit of force that would cause a mass of 1 Gm to be accelerated 1 cm per sec per sec. Another unit that is used is the electrostatic unit. This is a measure of the energy involved in the repulsion of like charges.

In the material that follows, we have focused on the biologic aspects of these rays and have attempted, in nontechnical language, to answer the questions: What are ionizing radiations? How are they formed? What are their effects on tissues? How are they used in the hospital? What precautions should be observed when using them?

Ionizing radiations are of two kinds: energy waves and streams of particles. The energy waves are electromagnetic radiations of short wavelength and high frequency. Heat rays, visible light, and ultraviolet rays are also electromagnetic radiations but of longer wavelength and lower frequencies. The ionizing rays are X rays and gamma rays. All the electromagnetic radiations travel with the speed of light (186,000 miles per sec), and the amount of energy a wave possesses depends upon its wavelength, amplitude, and frequency. X rays and gamma rays are high-energy waves and, as a result, are very penetrating. Streams of particles can transmit energy from one place to another. Their penetrating power depends upon their speed and their size. Think of the force of impact of the bullets from a machine gun. Protons, electrons, neutrons, mesons, positrons, etc.—a host of particles—can be formed from disintegrating nuclei. But we are interested particularly in two kinds of particle streams, namely, alpha particles, which are positively charged, and beta particles, which are negatively charged. Ions, we recall, are charged particles. Ionizing radiations produce charges on particles, some of which lose electrons and become positively charged; when the liberated electrons attach themselves to other particles, these in turn become negatively charged.

X RAYS

Current electricity is the flow of electrons through a conductor. The electrons move from the area in which there are the greater number of electrons (negative) to the area in which there are fewer electrons (positive). This difference in charge is the *potential,* and the pressure under which the electrons move is the *voltage.* At high voltages, the electrons can be forced to move through a vacuum from the negative pole, the *cathode,* to the positive pole, the *anode.* These streams of electrons are called *cathode rays.* Wilhelm Röntgen discovered that when cathode rays strike matter, penetrating rays are given off. By placing a metal target in the path of the cathode rays, X rays are formed. Depending upon the speed of the cathode rays and the material of which the target is made, X rays of different wavelengths can be made. Those of very short wavelength are extremely penetrating and are referred to as *deep* or *hard* X rays. X rays of longer wavelength are less penetrating and are called *soft* X rays. X rays, like other kinds of radiation can be reflected and absorbed. Different kinds of materials absorb X rays to different degrees. When Röntgen was studying the properties of these rays, he discovered that they would pass through a shield of black paper and that they could affect a photographic plate. He is reported to have taken an X-ray picture of his wife's hand in which the bones and her ring were clearly visible but the fleshy part was not. Soft tissue does not absorb X rays to the extent that bones and other radiopaque substances do. Barium sulfate and Urokon, an iodine compound, are examples of substances through which X rays cannot penetrate. These radiopaque materials are used to outline soft tissues and make them visible in X-ray photography. Barium sulfate by mouth and barium sulfate enemas are used to outline the gastrointestinal tract prior to an X-ray study of that system for diagnostic purposes.

The X-ray table is designed so that photographic plates can be placed under the body part to be photographed while the X-ray machine is placed in position directly above. If, instead of photographic plates, a fluorescent screen is used, we have a *fluoroscope.* The screen is coated with material that glows when subjected to X rays. The patient is placed between the screen and the X-ray tube. The physician can then examine the internal organs by the shadows cast by the X rays.

Let us bear in mind that in these diagnostic procedures the ionizing effect of X rays is an undesirable concomitant effect and should be kept to a minimum by well-shielded equipment and careful calculation of exposure time.

RADIOACTIVITY

Shortly after Röntgen's discovery of X rays, Henri Becquerel detected penetrating rays given off by uranium. These rays had the same kind of effect on the photographic plate as did X rays, they could pass through a shielding of black paper, and they appeared to be spontaneously emitted indefinitely. Pierre and Marie Curie undertook to discover the cause of this phenomenon. In the process, Marie Curie discovered a much more potent source of these rays, namely, radium.

We know today that radioactivity is the result of the disintegration of the nuclei of atoms. Let us review very briefly the theory of atomic structure.

The nucleus or core of the atom contains the heavy particles, the protons and the neutrons. Each of these particles has a weight of 1, but the protons carry a positive charge whereas the neutrons have no charge. The number of protons in the nucleus of an atom is the atomic number of the element. The total number of protons plus neutrons is the atomic weight of the atom. All the atoms of an element have the same atomic number, but they may have different weights. Atoms with the same atomic number but different weights are *isotopes.*

Uranium has atomic number 92. There are several isotopes; one has atomic weight 238 and another has atomic weight 235. This means that each of these forms of uranium has 92 protons in the nucleus and each has the same chemical properties, but one has 146 neutrons and the other has 143. The question arises: How can 92 like charges exist in such close proximity? Why does the nucleus not fly apart? In the case of the heaviest elements, the nuclei are unstable and shoot off particles. The atomic number and atomic weight are altered, and we have other elements formed. The elements formed by this spontaneous disintegration are called *daughters,* and the process is called *natural radioactivity,* or *radioactive decay.*

Uranium, element 92, changes to element 90, thorium. Thorium changes to radium, element 88. Radium changes to radon, element 86, which in turn changes to polonium, element 84. Finally a stable element is formed as a result of these transformations. This stable element is lead. These daughters and various related granddaughters are formed by the loss of nuclear material.

Subatomic Particles

alpha particles

Alpha particles resemble the nuclei of helium atoms. They have a weight of 4 and a charge of +2. This means that they are composed of 2 neutrons and 2 protons. The symbol used to designate this particle is $_2He^4$. The preceding subscript gives the charge, and the superscript gives the weight. Uranium, weight 238, atomic number 92, on losing an alpha particle, changes to atomic number 90, atomic weight 234. This element is thorium.

beta particles

Beta particles or rays are electrons. They have negative charges and are so light in comparison with the weight of protons, neutrons, and alpha particles that we usually consider them to be weightless. However, it has been calculated that 1,839 electrons are equal in mass to 1 proton. The loss of a beta particle from the nucleus causes practically no change in atomic weight but produces an increase in atomic number of 1. Iodine isotope, weight 131, atomic number 53 ($_{53}I^{131}$), loses an electron or beta particle and changes to xenon, one of the inert gases, weight 131, but atomic number 54 ($_{54}Xe^{131}$). Why does the atomic number increase with the loss of a negative charge? Remember that the electron is lost from the nucleus. The atomic number is the number of protons or the positive charge on the nucleus. The loss of an electron from the disintegration of a neutron frees 1 more positive charge, and the atomic number is therefore increased by 1.

gamma rays

Gamma rays, as we have previously indicated, are energy waves that are similar to X rays. The early investigators of X rays were fortunate because their machines could deliver only rays of relatively long wavelength, and they suffered little bodily damage due to ignorance. Gamma rays, on the contrary, are of very short wavelengths and are very penetrating. Present day X-ray machines can produce X rays whose wavelength compares favorably with that of gamma rays.

Half-life

How long does the radioactivity of these substances last? Becquerel noticed that his uranium continued to emit rays for a long time. To understand the way these compounds decay, let us recall a geometric progression[1]. If we decide to walk to a door in such a fashion that we would first walk half the distance, then half the remaining distance, then half the remaining distance, etc.,

---

[1] In a geometric progression or series, the elements in the series differ by a factor. In this case, the factor is one-half.

we realize that we would never reach the door. Why? Because no matter how close we got to the door, we would always have half the remaining distance to go.

The rate at which these radioactive substances decay follows a similar pattern. Eventually, the amount of radioactivity remaining would be infinitesimally small. We therefore speak of the length of time it takes for half the material to decay. This is called the half-life of the element. Radium has a half-life of about 1,600 years. If we had 1 Gm radium in 1900, by the year 3500, $1/2$ Gm would have changed into daughter elements and $1/2$ Gm radium would remain. In another 1,600 years, one-quarter of the original gram would disintegrate and $1/4$ Gm radium would remain. The rest would have changed into other elements. By the year 5700, one-eighth of the original gram of radium would still be emitting rays. For all practical purposes, the patient with a radium implant is receiving the same amount of irradiation on the last day of implant as on the first, and the nurse must exercise the same precautions throughout the treatment. Radon, on the other hand, has a half-life of 3.85 days. This means that 3.85 days after a radon implant, half the original material has decayed. In 7.7 days, one-fourth remains; In 11.5 days, one-eighth remains. In 2 weeks approximately fifteen-sixteenths of the element has distintegrated. If we compare the rates of decay of these two naturally radioactive elements, we realize that the patient with a radon implant, unlike the patient with a radium insert, offers no hazard a few days after implant.

Artificial Radioactivity

In the Middle Ages, the alchemists searched for a way to change base metals into gold. The chemist today can do exactly that. By bombarding atoms with alpha particles, protons, neutrons, electrons, etc., both stable and unstable isotopes can be formed. The unstable ones are radioactive and decay into stable elements by the emission of sub-atomic particles. These isotopes are made by bombarding the nuclei of atoms with high-energy particles. This can be accomplished in the *cyclotron.* The cyclotron consists essentially of a circular or spiral track passing between powerful electromagnets. The bullet particles are shot into the track, and every time they pass the magnets, they are given an extra kick so that they move faster and faster and their energy builds up to millions of volts. These high-energy particles strike the target atoms. Some of the target atoms may be split, some may lose protons or neutrons, and some may capture the bullets. In this way a variety of isotopes can be formed.

atomic reactors

Radioactive isotopes are also obtained as by-products of atomic piles or atomic reactors. The atomic pile is a device to control the chain reactions of atomic fission. Becquerel's sample of ore was not a rich source of uranium. When the atoms of fissionable material are separated in space or mixed with inert material, the chances of the emitted particles from one nucleus striking another fissionable nucleus are remote. When, however, the material is more concentrated, a particle from one nucleus can strike another nucleus, causing it to disintegrate. Particles from this nucleus cause other atoms to break up, and the reaction proceeds faster and faster. As the atoms disintegrate, a vast amount of energy is released. The amount of material needed to sustain a violent chain reaction is called the *critical mass.* In an atomic bomb, the fissionable material is kept in separate compartments until it is time to set it off. At that point, a mechanical device drives the material together, the critical mass is reached, and the reaction runs away. We have an explosion. The atomic pile can control the rate of atomic reaction in such a way that the reaction can be sustained but at the same time prevented from running away. This is accomplished by movable rods, which can be raised or lowered into the reaction chamber. The reaction chamber contains the fissionable material. To accelerate the reaction, the rods are pulled out. The particles from atomic disintegration are able to strike other fissionable atoms. When the rods are lowered into the chamber, some of the particles are trapped in the rods and are not free to strike other fissionable atoms, and the reaction slows down. These reactors are finding a use in industry as a source of power. The heat generated by atomic disintegration is extracted in a heat exchanger and used in the conventional ways to drive turbines, power submarines, etc. Many of the by-products are radioactive substances with long half-lives.

Detection and Measurement

There is no known way to alter the rate of decay of radioactive material. No one knows how to stop the radioactivity. No one knows how to slow it down or to speed it up. Freezing it, burning it—nothing that we know of has any effect. This means that we must live with atomic wastes and protect ourselves against them until their radioactivity has naturally dissipated.

measuring radiation

Radioactivity can be detected by various devices. The best known is the Geiger counter. This instrument can detect the presence of a single ionizing radiation. It consists of a tube with a small amount of an inert gas, such as helium or argon, under very low pressure. Running through the tube are two electrodes, a central wire and a metal tube. The wire passes through the length of the tube. When the apparatus is in operation, a relatively high voltage is applied to these electrodes. An ionizing radiation, entering the tube, hits the molecules of inert gas, causing them to form ions. The positively charged ions are attracted to the cathode and the negatively charged ions to the anode. In this way a current is set up, which can be amplified and used to activate a sound-producing device or to move a needle on a scale. The sounds formed are clicks, and the intensity of the radiation is recorded as counts per minute. As the intensity increases, the instrument clicks more and more rapidly. With a rich source of radiation, it chatters like a machine gun. In addition to the audible measurement, dials or devices for making permanent records of the amount of radiation can be attached to the counter. The Geiger counter is used by prospectors in the search for radioactive ores, by persons measuring the amount of radiation in an area (monitoring), and, in the hospital, to discover the location and amount of radioactive material in a patient receiving radioactive isotopes.

dosimeters

A very simple device for measuring radiation is the *film badge.* This is made of a metal holder about the size of a packet of matches, into which is inserted a strip of film bearing a number or code for the purpose of identification. The film badge is worn by a person who might be exposed to radiation. It is worn on the body part closest to the suspected source of radiation, usually on the chest or attached to a band at the wrist. Recall Becquerel's discovery of the rays from uranium. Ionizing rays affect the photographic film, and when it is processed, the amount and kind of radiation that has reached the film can be read. The person wearing the badge can learn the amount of radiation to which he has been exposed in a given period of time. In an operating room in which uterine radium implants were a common procedure, a nurse was observed wearing her film badge on a belt on the back of her uniform. She was using it as a safety pin. What she apparently failed to appreciate was the fact that some of the ionizing radiations are absorbed by tissue and her badge would record only the radiation that succeeded in passing through her body and not the total radiation to which she had been exposed.

Other devices for measuring radiation are the ionization chamber and the scintillation counter. These are more accurate than the film badge and can be read immediately. They can be made in small sizes. A very popular one is about the size of a fountain pen and can be clipped to the breast pocket and carried just like a pen.

The technique of measuring the amount of radiation to which an individual has been exposed is called *personnel monitoring,* and the instruments mentioned above for the detection of the quantity and quality of radiation exposure are referred to as *dosimeters.* They measure the dose of radiation.

units of measure

Several different units are used to measure amounts of radiation. One is the *curie,* named for Marie Curie, the discoverer of radium. A curie is the amount of radiation formed by 1 Gm radium. This is a relatively large unit. One thousandth of a gram of radium emits a millicurie (mc) of radiation, and one millionth of a gram emits a microcurie (μc). X rays, passing through air, cause the molecules in the air to break up into charged particles. The quantity of X rays that forms one electrostatic unit of electricity in 0.001293 Gm air is called the *roentgen* (r) after Wilhelm Röntgen, the man who discovered X rays. The roentgen is the unit used to designate dosages of X rays and gamma rays. One thousandth of a roentgen is a milliroentgen (mr).

Both these units refer to radiation or ionization in air. Persons working with biologic materials are interested in the amounts of radiation absorbed by different tissues or the relative destructive effects of different radiations on various tissues. Some of the factors that would affect the amount of radiation actually received by a tissue are the distance between the tissue and the source of the radiation, the energy of the rays, the kind of ray (alpha, beta, gamma, X), the type of material between a radiating source and the tissue (water, air, rubber, bone, lead, etc.), and the type of tissue itself. As Röntgen discovered with his original X rays, soft tissue does not absorb X radiation to the extent that bone does. We have here a question of the difference between exposure and absorption.

Three other units of measure are in current use. RBE (relative biologic effectiveness or radiation biologic effect) is a measure of the reaction of a tissue to the absorption of equal amounts of different kinds of ionizing radiations. The *rad* (radiation absorbed dosage) measures the amount of energy imparted to a particular tissue as a result of exposure to ionizing radiation. This is an energy measurement rather than an ionization measurement and is equal to the absorption of 100 ergs of energy per Gm.

Another unit is *rem* (roentgen equivalent man). It is equal to the rad multiplied by the RBE of the particular tissue. In other words, this unit includes the amount of energy absorbed by a specific tissue and the reaction of the tissue to the radiation. In this way the biologic damage to the tissue can be assessed. At the present time, the roentgen is used to describe dosages of radiation in this country, and the rad is used extensively in other parts of the world.

External and Internal Radiation

An important distinction exists between external and internal radiation. As the name implies, internal radiation is caused by radioactive material in body tissues. Elements such as radium and strontium are bone seekers and, when ingested, deposit in the bones, replacing the normal calcium. From this vantage point, the surrounding tissue is irradiated. Rays that would be relatively harmless if they reached the body from the external environment can be extremely powerful when released within body tissues.

When the body is exposed to radiation from a source outside the tissues, it is subjected to external radiation. Exposure might result in a heavy dose of radiation during a short period of time to the whole body, as might occur in an atomic explosion. All during his lifetime, an individual is exposed to radiation from naturally radioactive rocks and soil and to cosmic rays (background radiation). Exposure results from the inhalation or ingestion of radioactive debris from atomic fallout and from medical and dental procedures involving X rays, fluoroscopy, and radioactive substances.

Of the ionizing radiations, gamma rays are the most damaging and the most penetrating. Hard X rays are comparable in their effects. These rays can be stopped by concrete blocks and by lead of appropriate thickness. Most gamma rays are completely absorbed in passing through 9 in. concrete, and this is the thickness required if concrete is used as a shielding material. The heavy alpha particles can travel about 2 in. through air. They lose their energy by collision with air molecules. They can be stopped by a few sheets of paper and by the outer layer of the skin. The light negatively charged beta particles can travel several feet through air. Aluminum sheeting, $1/16$ in. thick, completely absorbs them. If the sorce of the beta radiation is very close, these high-speed electrons can penetrate a few millimeters through skin. We see from this that external irradiation with alpha and/or beta particles presents no hazard, except possibly to the skin, and only then if the radiating source is practically in contact with the skin. If the alpha or beta emitter is incorporated into the tissues, as occurs following the ingestion of strontium[90], the tissue immediately adjacent to the radiating source will be subjected to ionization (internal radiation).

Biologic Effects

Rapidly growing tissues are more susceptible to radiation damage than are older, slower-growing tissues. One characteristic of tumors is their rapid growth pattern, and tumors are more sensitive to ionizing radiations than neighboring normal cells. In an adult, the ova or spermatozoa are formed relatively rapidly and are particularly susceptible to the deleterious effects of ionizing radiations. Bone marrow and organs requiring a rich blood supply are also highly susceptible. The embryo, the fetus, the infant, and the child, because their growth rate is rapid, are more vulnerable than the adult.

The ionizing rays, entering a cell, may knock electrons out of atoms. These electrons, in turn, become attached to other atoms or molecules. In this way, pairs of charged particles are formed. The ionizing ray may cause a molecule to break up, and the fragmented particles may rearrange themselves in a combination different from the original molecule. Different arrangements of atoms in a molecule impart different properties to the compound. The fragmented molecules may unite with other radicals to form entirely different compounds. Water molecules, H—O—H, may be changed to hydrogen peroxide, H—O—O—H; saturated compounds may become unsaturated; unsaturated compounds may double up with others; and relatively small molecules may unite to form larger combinations (polymerization). A moment's thought about the delicately balanced chemical composition of the normal cell will enable us to understand that the presence of these unnatural substances in the cell will either poison the cell to some degree, change its rate of growth, change the character of its daughter cells, or kill it. Enzymes in the cell may be altered by ionizing radiation or by the chemical action of the strange compounds formed from other cell substances. The modified enzymes interfere with the growth patterns of the cells. Some cells may fail to divide and, instead, grow to enormous sizes before they eventually die. In other cells, the rate of growth may be accelerated. If the rate of cell division is markedly increased, the result is abnormally rapid growth of the tissue.

genes

Every somatic cell contains a set of chemical directions enabling it to reproduce other cells like itself. The directions for the development of a new individual are contained in the genes located in the chromosomes of the egg and sperm. When the gonads are irradiated, changes may occur in the genes. When these directions are changed, the daughter cells are defective or deformed. Genetic changes produce either dominant and apparent or recessive and hidden changes in the offspring. The body modification controlled by the recessive gene will not become apparent until an egg and a sperm, each containing the same defective gene, unite. Some investigators contend that the presence of one defective recessive gene modifies, to some extent, the effect on the developing individual of the accompanying normal gene.

Mutations in genes are classified as lethal, sublethal, and beneficial. A *lethal* gene kills the organism outright. *Sublethal* genes impair cell functions, weaken the individual, and reduce his or her ability to reproduce. Sublethal genes are transmitted from parent to offspring for several generations until they are either bred out or united with another sublethal gene of the same type, eventually causing death. Theoretically, some mutations could produce beneficial or desirable changes in the offspring. On this premise, research on the effects of irradiation on plants and seeds is being carried out to discover whether new strains can be developed that are larger, more resistant to disease, faster growing, etc. Unfortunately, the limited evidence so far available on the results of ionizing radiations on the genetic changes in human beings indicates that in human beings these changes are all undesirable.

genetic effects

Irradiated individuals may be unable either to produce sex cells or to produce normal sex cells. In this way an individual may become sterile or less fertile. The number of miscarriages and stillbirths is higher among irradiated women than in the general population. Some defective genes, from sources other than irradiation, exist in the general population, and these may be responsible for many of the birth defects. It has been estimated that 4.5 per cent of infants are defective in some way. Irradiation of the parents or the unborn child increases the likelihood of either mental or physical defects in the child. In addition, there is reduced vigor, increased susceptibility to disease, and diminished reproductivity due to sublethal genes.

somatic effects

In some way, as yet unexplained, ionization of the tissues accelerates the aging process, leading to a shortened life span. Exposure of the skin to X or gamma radiation may lead to the development of a reddened area on the exposed part. This is correctly described as *erythema*. Nurses are cautioned against referring to this development as a "radiation burn," because patients sometimes conclude that the "burn" is the result of negligence. When, for therapeutic reasons, the dose of radiation must be high, erythema is a normal although undesirable side effect. Irradiation may lead to greater susceptibility to disease and injury to growing tissue. Cancers or leukemia may develop. Acute radiation sickness, as in exposure to atomic explosion, causes prostration, vomiting, diarrhea, and loss of appetite. These symptoms are followed by reduced white blood cell count, loss of hair, anemia, hemorrhages, and generalized infection.

If an adult is exposed to whole body radiation of 700 r, the result is almost certain death. A dose of 500 r of whole body radiation is called lethal dose 50 (LD_{50}) and indicates that 50 per cent of the exposed individuals are not expected to survive. Symptoms of nausea and vomiting follow whole body exposure to 200 r, and doses as low as 50 r administered in a short period of time to the whole body cause detectable changes in the blood picture. Let us emphasize that these figures refer to the exposure of the whole body to radiation during a short period of time. Small areas of the body can receive doses as high as 10,000 r with minimum damage to the rest of the body. The ionizing beam is accurately focused on the tissue to be irradiated, and other parts of the body are shielded from the rays by materials that prevent their passage. Rubber sheeting impregnated with lead is one such shielding device. Ingenious methods have been developed to reduce to a minimum the amount of radiation received by the normal body wall when it is necessary to irradiate a deep tissue with an external source of radiation. If, for example, a massive dose of X rays is to be administered to a malignant tumor in the abdominal cavity, the patient can be positioned in a rotating chair. While, by careful focusing, the X rays continually impinge on the malignant tissue, the chair is rotated so that different portions of the abdominal walls are exposed for a short period. If the treatment begins with the patient facing the X-ray source, the rays would pass through the ventral body wall. Then as the patient is rotated, the right lateral wall would be irradiated, then the dorsal wall, and then the left lateral wall, and again the ventral aspect would be exposed. All this while the rays are concentrated on the malignant tissue.

An individual may be exposed to small doses of ionizing radiation over long periods of time with minimum damage. It must be borne in mind, however, that radiation effects are cumulative. At the present state of knowledge the recommended maximum permissible level of whole body radiation exposure is set at 0.3 r per week. As more knowledge is accumulated, this figure will probably be revised. One very practical suggestion in this area considers the age of the individual. It is recommended that the maximum permissible exposure to total body radiation per year be set at 5 rem multiplied by the individual's age less 18 years [5 rem $\times$ (age $-$ 18)]. A 50-year-old person would be permitted 160 rem per year, and a 20-year-old would be allowed only 10.

With the exception of the casualties from atomic bombings, the Japanese fishermen too close to atomic testing in the Pacific, and a very, very few nuclear accidents, there are not many instances of healthy persons who have experienced radiation damage. The histories of the workers who painted numerals on watch dials using a luminous paint that contained radium have provided scientists with much information in this area. These workers pointed their brushes by touching them to the tongue and lips and, in so doing, ingested the radium. As a result, their tissues were subjected to internal radiation, and many of them developed the symptoms metnioned above. A few early workers with X rays experienced radiation damage. In these cases, *ignorance* of the properties of radium and X rays was the real cause of the injury. The hundreds of workers in atomic installations as well as roentgenologists and X-ray technicians give evidence that the observance of recommended precautions in the handling and use of substances and apparatus producing ionizing radiation can enable an individual to work safely in this field.

The nurse who cares for a patient with a diagnosed, communicable disease uses isolation technique and observes all the precautions necessary to prevent the spread of the illness. The *danger* to personnel from communicable disease exists when the condition is undiagnosed. In much the same way, knowledge of the nature of radiation and the observance of the recommended precautions enables the nurse to work safely with patients receiving radiation therapy.

precautions

Our discussion of the properties of ionizing rays should enable us to deduce the appropriate precautions to be observed.

a. Spend as little time close to the source of radiation as possible. Work at a distance. Radiation is dissipated according to the *law of inverse squares.* Have you ever observed, in focusing a home movie projector, for example, that if you set the screen near the projector, your picture is small, but as you move the screen farther away from the projector, the picture can be focused larger and larger but the brightness diminishes? It is easy to understand that the same amount of light that illuminates the small screen is spread out more to cover the large screen and will therefore be less brilliant. When the distance is doubled, the picture expands to four times its original size. This means that the same amount of light covers four times the original area and is diluted by a factor of 4. In other words, if our original picture was 1 sq ft in area and our second picture was 4 sq ft in area, each square foot of the second picture receives only one-fourth the amount of light received by the original picture.

The same thing happens with ionizing radiations. A nurse working at the bedside of a patient receiving therapy in the form of radioactive implant or isotope may be exposed to a certain amount of radiation. If the nurse moves back 2 ft, the exposure is reduced to one-fourth; 3 ft, to one-ninth; 4 ft, to one-sixteenth, etc. The skilled nurse will plan nursing care so that the amount of time spent in very close proximity to the patient is reduced to a minimum. The patient will be encouraged to do as much for himself as possible. (This is a good idea not only from the point of view of reducing the exposure of the nurse but also for the rehabilitation of the patient.) Generally speaking, patients receiving these treatments may share a room with other patients, because the beds are positioned sufficiently far apart.

b. Use the appropriate tools for handling radioactive materials. These instruments are designed with long handles so that the materials can be kept at a distance from the worker. The reason for this is explained above.

c. Wear protective clothing and use shielding devices. Rubber aprons impregnated with lead, goggles if the individual does not wear glasses, rubber gloves, and lead or lead-lined containers are some of the items in common use.

d. Wear the dosimeter assigned. Don't borrow another. It is very important that individuals working with radioactive materials have a running measure of the amount of radiation to which they have been exposed and the time in which this exposure has been accumulated. With this information, persons can be safeguarded from overexposure. Long before a nurse would approach the permitted limit, the hospital administration would change the assignment.

Decontamination

In the event of an atomic accident or a nuclear explosion, the problem of removing radioactive material from an area would become acute. Since there is presently no known way to alter the rate of radioactive decay, two alternatives are possible: either to remove the radioactive material and store it until the radioactivity has dissipated or to dilute the material until the degree of radioactivity falls within permissable limits. The choice of procedure will depend upon the circumstances and the use to which the area is to be put. If people were to use the area for very short periods of time, for example, as a passageway, decontamination would not need to be as thorough as it would if the area were to be used by the same individuals for hours.

Material to be stored on land is first sealed in containers and then either buried in prepared excavations or abandoned in caves or unused mines. For burial at sea specially designed steel boxes lined with at least 4 in. of concrete are used to contain the radioactive material for long periods of time. The ocean floors have been mapped to show areas covered with mud or slime. The concrete-lined steel boxes are cast into the sea over these areas. It is to be hoped that the boxes will sink into the ocean floor and that by the time they disintegrate, radioactive decay will have progressed to the point that danger to the flora and fauna of the sea will be minimal.

The choice of method for the removal of radioactive material from surfaces depends upon the type of surface and the location. The surface, if noncombustible, may be flamed and the smoke and ash collected, it may be scoured, or it may be vacuum cleaned. If the level of radiation is high, the collected debris should be sealed in a lead container and stored for aging. If the debris can be diluted, as in a large body of water, or if it can be safely spread over a large area, hosing methods may be used which include water or steam. Other substances may be mixed with the water or steam to facilitate the removal of radioactive material. These additives may be detergents, foaming agents, or complexing agents. The latter, such as carbonates, oxalates, and citrates, react chemically with the surface material to form compounds that are more easily removed. In these removal methods, constant monitoring is necessary to make certain that the radiation levels in the debris and diluted materials are kept within safe limits.

COMMONLY USED RADIOACTIVE MATERIAL

Radium

The element radium (Ra) is a metal with atomic number 88. The naturally occuring radioactive form of the element has an atomic weight of 226. Any element formed with 88 protons in the nucleus, either in the process of radioactive decay or as a product of nuclear reaction, is radium.

All together there are about 13 isotopes, whose half-lives vary from that of Ra^{221}, which is 30 sec, to that of Ra^{226}, which is 1,620 years. Radium occurs in group 2 of the periodic table. In descending order, group 2 contains beryllium, atomic number 4; magnesium, 12; calcium, 20; strontium, 38; barium, 56; and radium, 88. The chemical and physical properties of radium resemble those of the metals above it in the table. Like them, it has a valence of 2 and reacts with acids to form salts. Radium used in the hospital is not the pure metal but one of its salts. The salts usually used are radium sulfate and radium chloride. Recall that "barium" enemas and "barium" meals used to outline the digestive tract for X-ray pictures consist, not of the element barium, but of a white salt: barium sulfate. Since calcium, strontium, and radium are closely related chemically, we can understand why strontium and radium are bone seekers. Their properties are similar to those of calcium and they can replace calcium in bones.

Radium emits gamma rays, high-energy beta particles (referred to as hard betas), and alpha particles. As radium disintegrates, its first decay product is the inert gas radon. Radium salts used in the hospital are enclosed in platinum cases, which prevent the escape of radon gas and also stop the alpha and beta particles. Radium is used for its gamma radiation, and these high-energy waves can pass through the platinum cases. Therefore radium is stored until the moment it is to be used in lead cases.

When in use, the radium is placed in an *applicator*. This is a container made for the part of the body that is to be treated. In the radiology department, the desired amount of radium is put into the appropriate applicator; this in turn is placed in a properly labeled lead chest. The chest is transported to the operating room on a long-handled, wheeled platform. When it is time to insert the applicator, the chest is opened and the applicator removed with long forceps held at arm's length.

Radium finds its greatest use in the treatment of cancer in those parts of the body which are readily accessible, for example, the skin and the uterus. In many cases of cancer of the cervix, radium is used in conjunction with surgery both pre- and postoperatively. The preoperative use of radium helps to reduce both the malignant tissue and the amount of bleeding.

Radium applicators are usually removed by the physician at the bedside, with the nurse assisting. The physician removes the applicator, cleans it, and puts it back into the lead chest (which is in readiness), and the chest on the platform is returned to the radiology department.

Occasionally one reads a news item in the paper about a search for lost radium in the city dump or the hospital refuse. The search is, of course, made with a Geiger counter. But how did the radium get to the city dump? Obviously, someone was careless. Very probably, at the time the applicator was removed, dressings and bandages were taken from the patient and the applicator was discarded with the dressings. In the presence of alert, informed, careful nurses, mistakes like this do not occur.

Radon

Radon, atomic number 86, atomic weight 222, is an inert gas found in group 0 of the periodic table along with the other inert gases: helium, neon, argon, krypton, and xenon. When a radium atom loses an alpha particle from its nucleus, an atom of radon is formed. Since the half-life of radon is 3.85 days, it in its turn decays rather rapidly.

Radon is used in the hospital enclosed in tiny gold cases called radon *seeds*. These are prepared by pumping the gas from a radium source into gold tubes of very small diameter. The tubes are pinched into minute pieces in such a way that the ends are sealed. The name "seeds" refers to their size. They look like golden grains of wheat.

Radon seeds are inserted directly into malignant tissue and are frequently left in place. For example, in cancer of the tongue, the seeds can be inserted into the tissue. Because of their small size they do not interfere with the motility of the organ, and because of their short half-life the radiation falls off rapidly.

Radioactive Isotopes

These substances, formed in nuclear reactors, are finding a wide use in research and are used in the hospital for both diagnosis and therapy (Table 23-1). Recall that the isotope has the same chemical properties as its nonradioactive twin element but that it is easy to detect and trace by its ionizing radiations. Different body tissues select from the blood stream those substances needed by them for repair, growth, and function.

iodine

Thyroid tissue picks up iodine to use in the manufacture of the hormone thyroxin. When the radioactive iostope of iodine (I^{131}) is administered by mouth, it follows the same path as ordinary iodine and is absorbed by thyroid tissue. I^{131} emits beta and gamma rays, and the amount of uptake can be measured very easily by passing the probe of a sensitive scintillation counter over the body. Sometimes cells from a malignant growth

Table 23–1 RADIOACTIVE ELEMENTS USED IN MEDICINE

Element	Symbol	Half-life
Cobalt	Co^{60}	5.3 yr
Gold	Au^{198}	2.7 days
Iodine	I^{131}	8 days
Phosphorus	P^{32}	14.3 days
Radium	Ra^{226}	1,600 yr
Radon	Rn^{222}	3.85 days
Sodium	Na^{24}	14.8 hr

migrate from one part of the body to another. If this happens, in the case of malignant thyroid tissue, the location of the migrant tissue (metastasis) can be determined with the counter. The diagnostic dose is very small and is easily excreted from the body. Radioactive iodine in larger doses can be used to destroy malignant thyroid tissue. Iodine^{131} has a half-life of 8 days and is rapidly absorbed from the gastrointestinal tract and eliminated from the body, principally in the urine. Amounts used for diagnostic purposes are so small that the patient need not be hospitalized. The urine can be flushed into the sewage system using a relatively larger amount of water than usual. Therapeutic doses of radioiodine are larger than those used for diagnosis but in most cases present very little hazard. When we remember that half the radiation is gone in 8 days and that the isotope is excreted from the body, we realize that radiation from the patient decreases rapidly. If, for some reason, a patient receiving iodine therapy is nonambulatory, the nurse would handle urine and urine-contaminated materials with rubber gloves. In addition to use of I^{131} in the treatment of malignant conditions of the thyroid, it is sometimes administered in cases of Graves' disease (exophthalmic goiter).

The hormone thyroxin stimulates metabolism, particularly the rate of oxidation in the body. In some cases of chronic congestive heart disease, a normally active thyroid keeps the body's rate of oxidation higher than is desirable for the impaired heart. By administering I^{131}, part of the thyroid gland can be destroyed and the amount of thyroxin reduced. In this way the rate of oxidation in the body can be slowed down.

Some promising research has been done to expand the use of I^{131} as a diagnostic tool. It has been incorporated into the protein molecule serum albumin. In this way, the protein is tagged and can be followed. This new compound is used in studying and measuring movements of blood flow and to locate brain tumors.

phosphorus

Phosphorus exists in tissues in the form of both organic and inorganic compounds. We think of calcium phosphate, part of the solid structure of bones and teeth, and of the sodium phosphate buffer system in the control of blood pH. Two organic phosphorus compounds important in enzyme function are adenosine diphosphate (ADP) and adenosine triphosphate (ATP). Phospholipids, proteins containing phosphorus, such as casein, and the various nucleic acids are widely distributed in protoplasm. Erythrocytes (red blood cells) in man contain 3 mg phosphorus per 100 ml. One of several radioactive isotopes of this element, P^{32}, is a beta emitter with a half-life of 14.3 days. It gives off no gamma radiation and affects both white and red blood cells. It has been used in the treatment of chronic leukemia, a disease in which there is a tremendous increase in the number of white corpuscles. A large excess of red blood cells and an increase of blood volume is characteristic of a condition called polycythemia vera. Radioactive sodium phosphate has been used, with some success, in these cases also.

The absorption of P^{32} by normal brain tissue is very much less than the absorption by brain tumors. The tumor absorbs from 5 to 100 times more P^{32} than the neighboring normal tissue. This affords a method for correctly determining the location and size of the tumor. The P^{32} is injected into an artery going to the brain and is absorbed by the tumor. A specially constructed probe counter of very small diameter (about 3 mm) can be used during surgery. The probe can be inserted into the tissue and, from the differences in intensity of radiation, the limits of the tumor can be discovered. We realize, of course, that in brain surgery the object is to remove the least possible amount of tissue and, at the same time, remove the complete tumor.

P^{32}, being a beta emitter, would not be used as a source of external radiation because the electrons are stopped or deflected so easily. The exception to this is in some cases of skin cancer. The isotope is absorbed in blotting paper, and the paper is attached to the affected skin area. In this way, only the immediately adjacent tissue is irradiated. With a half-life of 14.3 days, the isotope lends itself to research studies. Red blood cells have been induced to pick it up as part of their phosphorus content and in this way have been tagged. They can then be traced in the circulatory system, their number estimated, and blood volume computed. In other studies, much information has been accumulated about the way the body synthesizes and metabolizes such compounds as phospholipids and the inorganic phosphates of bones and teeth. When taken internally, phosphorus, like iodine, is excreted from the body. Although it emits only beta rays, the same precautions in handling excreta are observed.

sodium

Radioactive sodium with atomic weight 24 has a half-life of 14.8 hours. It affords a quick and painless method of determining the location of areas of restricted circulation. Sodium chloride, containing the radioactive Na^{24} in place of the stable sodium, is injected into the artery. The radiosodium is carried by the blood stream and can be traced by a counter. Where the count is high, the blood circulation is good. In areas in which the blood vessel is constricted, the circulation will be poor and the count low. In this way, the exact point of constriction can be determined. This method can be adapted to measure circulation time and patterns of blood flow.

The volume of extracellular fluid and the amount of sodium in the fluid can be determined by injecting a known amount of radioactive sodium into the body. After sufficient time for the mixture to reach equilibrium has passed, a sample of fluid is withdrawn and analyzed. From the degree of dilution of the Na^{24}, the volume of fluid can be calculated, and from a comparison of the concentrations of the radiolabeled sodium with the natural unlabeled sodium, the amount of unlabeled sodium can be computed. The determination of the amount of sodium in the fluid is important in the case of a patient on a salt-free diet. In some of these patients, the sodium decreases to the point where the kidneys are affected. It is then necessary to add salt to the diet. This procedure answers the question: When must salt be included in the diet, and how much?

gold

Radioactive gold, Au^{198}, is a beta emitter with a half-life of 2.7 days. Unlike the isotopes discussed above, which become involved in the metabolism of the body and, if administered in larger than tracer amounts, require care in the disposal of body wastes, radiogold lost from the body would be in drainage fluid. In some cases of lung cancer fluid escapes into the pleural cavity (effusion), and sometimes after a patient has received massive doses of X rays, edema (abnormal amount of fluid) may develop. Au^{198} is injected into the affected part. The result is a diminution in the amount of fluid and decrease of pain. If any of the gold escapes through the site of injection, a deep pinkish-purple color will show on the dressings. To handle these, the nurse wears gloves, uses forceps, and deposits them in a specially marked container for disposal. As in the care of any patient receiving radioisotope therapy, the nurse wears a dosimeter. Aside from the possibility of fluid loss at the point of injection, excreta pose no problem.

Au^{198} is inserted into malignant tumors in the form of very thin wires. The wires are inserted into nylon sutures and are stitched into the tumor. Au^{198} eases pain, reduces excess fluid, and may also arrest the development of the tumor.

strontium

One reason many persons are concerned about the above-ground testing of nuclear weapons is that the blast scatters nuclear debris and radioactive iostopes into the atmosphere. Some of the particles are carried by the winds to points distant from the original blast area and eventually return to earth in the form of rain or snow *(fallout)*. Conceivably, these substances could become incorporated into plant or animal tissue and be eaten by human beings. Strontium^{90} is one of the radioactive iostopes formed in atomic detonations. Many of the others do not offer the same hazard as Sr^{90}, either because they have short half-lives or because they are, if ingested, rapidly excreted from the body. Sr^{90} has a half-life of 28 years and a biologic life of 18 years. The biologic life of an isotope in the length of time the element remains in the body. In addition, strontium, like radium, deposits in the bones when ingested. Since the biologic life of Sr^{90} is so long and its half-life is longer, the reason for concern is obvious. Governments have established monitoring systems around the world to make certain that the amounts of Sr^{90} in fallout does not become a hazard.

Sr^{90} is a beta emitter, but when the radiation is internal, beta rays can destroy or alter nearby tissues. In research, the isotope is used to study the healing process in bone and as a possible treatment of bone cancer. In medicine, it has been used as a source of external radiation for the reduction of birthmarks if they are less than 2 mm in depth. In addition, malignant growths of the eye are treated with Sr^{90} built into the end of a probe.

cobalt

Cobalt^{60} has been described as the "poor man's radium." It is manufactured in atomic reactors and can be obtained at a fraction of the cost of radium. Its half-life is 5.3 years. It emits gamma radiation of homogeneous energy and soft beta rays. The beta rays from Co^{60} can be filtered out by steel, aluminum, and even plastic, whereas those from radium require platinum as a filtering material. Co^{60} is used in the so-called cobalt bomb. This is an external source of gamma rays that can be focused on cancerous areas deep in the body. The isotope is also fashioned into fine threads or wires and small, wafer-thin disks for use at the site of the malignant tissue. One of the advantages of Co^{60} over radium is that it can be used near bone tissue. Even though a radioactive material such as Co^{60} is inserted into the body, the radiation is external. The isotope does not become a part of the metabolism of the body, and there is no problem with body wastes.

Isotope Studies

A brief word about some of the exciting research work being done with isotopes.

Zinc^{65} is being applied to the study of malignant leukemia. This malady has been described as cancer of the white blood cells. One of the peculiarities of this condition is that the diseased white blood cells contain less zinc than normal cells.

Iron^{59} is incorporated into the hemoglobin of red blood cells to discover the secret of their formation: exactly where they are formed in the body, how they are made, and how they are destroyed.

Sulfur^{35}, which is radioactive, and nitrogen^{15}, which is not, are utilized to tag amino acids and to learn how they are involved in metabolism.

One need not be overly optimistic to hope that the results of these and many other studies will lead to a wider and deeper understanding of the chemistry of life.

Study Exercises

1. Define cathode rays, soft X rays, hard X rays, alpha particles, beta particles, and gamma rays.
2. What is meant by ionizing radiations? What is meant by natural radioactivity? How could an element be made artificially radioactive?
3. What is a fluoroscope? Why are barium enemas and barium meals given before fluoroscopy?
4. What is meant by the half-life of a radioactive element? A nurse knows that the half-life of radium is about 1,600 years and that of radon is 3.85 days. How would this knowledge affect the planning of nursing care for a patient with a radium implant and for another with a radon insert?
5. What is meant by personnel monitoring? What are dosimeters? Describe two types of dosimeters.
6. Distinguish between the following units: curie, roentgen, RBE, rad, and rem.
7. What is the difference between internal and external radiation?
8. Differentiate between lethal and sublethal mutations.
9. What are some of the somatic effects of overexposure to ionizing radiations?
10. List four types of precautions to be observed when working with ionizing radiations.
11. How and for what purposes are radium, radon, and radioactive cobalt used in the hospital?
12. How can I^{131} be used to locate metastasized thyroid tumors?
13. What is meant by radioisotope therapy?
14. What extra precautions must be observed for a patient receiving isotope therapy as compared with a patient who has a radium implant?

SUGGESTED REFERENCES

Arnow, L. E.: "Introduction to Physiological and Pathological Chemistry," 7th ed., Mosby, St. Louis, 1966.

Bachman, C. H.: "Physics," Wiley, New York, 1955.

Best, C. H., and N. B. Taylor: "The Human Body," 3d ed., Henry Holt, New York, 1965.

Best, C. H., and N. B. Taylor: "Physiological Basis of Medical Practice," 8th ed., Williams & Wilkins, Baltimore, 1966.

Bogert, L. Jean: "Fundamentals of Chemistry," 9th ed., Saunders, Philadelphia, 1963.

Brooks, Stewart M.: "Basic Facts of Body Water and Ions," Springer, New York, 1960.

Cantarow, A., and M. Trumper: "Clinical Biochemistry," 5th ed., Saunders, Philadelphia, 1955.

Cantarow, A., and Schepartz: "Biochemistry," 4th ed., Saunders, Philadelphia, 1967.

Flitter, H. H.: "An Introduction to Physics in Nursing," 5th ed., Mosby, St. Louis, 1967.

Gamble, James L.: "Clinical Anatomy, Physiology, and Pathology of Extracellular fluid," 6th ed., Harvard, Cambridge, 1954.

Garb, Solomon: "Laboratory Tests," 3d ed., Springer, New York, 1963.

Oser, B. L. (ed.): "Hawk's Physiological Chemistry," 14th ed., McGraw-Hill, New York, 1965.

Jensen, J. T.: "Introduction to Medical Physics," Lippincott, Philadelphia, 1960.

Kimball, A.: "College Text Book of Physics," 6th ed., Henry Holt, New York, 1954.

Kleiner, I., and J. Orten: "Human Biochemistry," 7th ed., Mosby, St. Louis, 1966.

McCue, J. J. G.: "The World of Atoms," Ronald, New York, 1956.

Metheny & Snively: "Nurses Handbook of Fluid Balance," Lippincott, Philadelphia, 1967.

Sacks, Jacob: "The Atom at Work," 2d ed., Ronald, New York, 1956.

Sackheim, George: "Practical Physics for Nurses," Saunders, Philadelphia, 1957.

Stearnes, Howard: "Fundamentals of Physics and Applications," 2d ed., Macmillan, New York, 1956.

Walker, Asimov and Nicholoas: "Chemistry and Human Health," McGraw-Hill, New York, 1956.

West, Todd, Mason, Burggess: "Textbook of Biochemistry," 4th ed., Macmillan, New York, 1966.

GLOSSARY

absorption, the passage of a substance through material, as the passage of fluid through the intestinal mucosa into the blood or lymph. The taking up of material or energy by a solid or liquid, as the ability of a black body to pick up light and heat rays and of bones to pick up X rays.

acetone bodies, substances formed by abnormal fat catabolism: acetone, acetoacetic acid, and β-hydroxybutyric acid.

acid, a proton donor.

acid-base balance, regulation of the alkalinity of the blood. The normal range of the pH of the blood is 7.35 to 7.45.

acidosis, more than the normal concentration of hydrogen ions in the blood. A blood pH of less than 7.35.

activator, metal necessary for enzyme function.

adhesion, the attraction between unlike molecules or materials.

adsorption, the adherence of molecules to a surface.

aeration, the mixing of air with a substance, as the spraying of water into the air.

albumin, a simple protein, such as ovalbumin, lactalbumin, and serum albumin.

alcohol, an organic compound with the characteristic group $-\overset{|}{\underset{|}{C}}-O-H$.

aldehyde, an organic compound with the characteristic group $-\overset{H}{\underset{}{C}}=O$.

aldose, a sugar containing an aldehyde group, e.g., glucose.

alkaline tide, the rise in the pH of urine after a meal.

alkaloids, organic nitrogenous heterocyclic compounds small amounts of which cause marked physiologic reactions. The alkaloids have slightly basic properties and are usually administered in the form of their salts.

alkalosis, higher than normal pH of the blood. A pH reading higher than 7.45.

alkyl group, open-chain group in organic compounds, as methyl, ethyl, propyl.

alpha particles, the nuclei of helium atoms. They have a weight of 4 and a charge of +2.

amide, organic compounds with the characteristic group $-\overset{O}{\underset{}{\overset{\|}{C}}}-NH_2$.

amine, organic nitrogen compounds. They are classified as primary, secondary, tertiary, and quarternary depending upon the number of groups substituted for hydrogen in the primary amine with the general formula CNH_2.

amino acid, organic compounds containing both the carboxyl group of an acid (COOH) and the amino group (NH_2). They have properties of both acids and ammonia.

amino group, $-NH_2$.

ammeter, an instrument used to measure the amount of electric current.

ampere, the unit of intensity of electric current. It is the amount of electric current produced by 1 volt acting through a resistance of 1 ohm.

amphoteric compound, a substance having both acidic and basic properties.

amplitude, the property of a wave that describes the height of the wave peak.

anabolism, metabolic processes which result in the building or repair of tissue or the synthesis of body substances.

analysis, chemical change causing the separation of compound substances into their constituents.

anhydride, a substance from which water has been removed.

Glossary

anion, a negatively charged particle.
anode, the positive pole in a cell or battery.
astigmatism, imperfect vision usually due to a defect in the cornea.
atmospheric pressure, the weight of the air at sea level. It is recorded as 15 lb per sq in., or 760 mm mercury.
atom, the smallest particle of an element involved in chemical change.
atomic number, the number of protons in the nucleus of an atom.
atomic weight, the sum of the protons and neutrons in the nucleus of an atom.
autoclave, an apparatus for sterilizing using steam under pressure.
"barium," a radiopaque salt, barium sulfate, used to outline the digestive tract. It is administered as a barium meal or a barium enema.
barometer, an instrument for measuring atmospheric pressure.
basal metabolism or basal metabolic rate (BMR), the rate of oxidation in the body when the stomach is empty and the mind and body are at rest.
base, a proton acceptor.
beta particles, electrons or streams of electrons.
bile, fluid secreted by the liver containing salts for the emulsification of fats.
boiling point, the temperature at which the vapor pressure of a liquid equals the atmospheric pressure.
buffers, substances which when present in a solution resist change in pH when acids or bases are added to the solution.
buoyancy, the upward force exerted upon an object submerged in a fluid.
caisson disease ("the bends"), distress and necrosis of tissue caused by too rapid change from higher to lower pressure. Nitrogen gas dissolved in the blood at the higher pressure forms bubbles at the lower pressure which become lodged in the capillaries, thus cutting off the circulation to the area served by the capillaries involved.
calorie, a unit of heat. The amount of heat needed to raise 1 Gm water 1°C.
capillary action, the ability of a liquid to rise in a tube when the adhesive forces are greater than the cohesive forces.
capillary tube, a tube of narrow bore used to measure capillary attraction.
casein, a conjugated protein which yields phosphoric acid as one of the products of digestion. It is found in milk.

catabolism, metabolic processes resulting in the breakdown or destruction of tissue.
catalyst, a substance which alters the rate of a chemical reaction.
cataract, a clouding of the lens of the eye due to an alteration in the protein of which it is composed.
cathartic, a substance which causes a fecal evacuation.
cathode, the negative pole in a cell or battery.
cathode rays, streams of electrons passing from a cathode through a vacuum.
cations, positively charged particles.
cell, a receptacle with electrodes and an electrolyte which generates an electric current.
charging, the restoration of the latent electricity to a cell or battery by passing an electric current through it.
coagulation, the process of becoming solid or jellified.
coenzyme, a substance which must be present for an enzyme to function.
cohesion, the attraction of like molecules to each other.
colloids, particles in a mixture larger than average molecules but smaller than visible suspended particles.
combination, chemical change resulting in the formation of compounds from elements.
combustion, a rapid chemical reaction resulting in the evolution of light and heat.
compounds, two or more elements chemically combined in definite proportion by weight.
compression waves (longitudinal waves), waves in which the particles of which the wave is composed move in the same direction as the wave.
concave lens, a lens with a depressed or hollow surface.
condensation, the change of state from gas to liquid.
conductance, the ability of a substance to transmit an electric current.
conductor, a substance which transmits energy easily, as a conductor of heat or electricity.
convection, transmission of heat by currents, such as air or water currents.
converging lens (positive), a lens which causes light rays to meet.
convex lens, a lens whose surface is arched.
countertraction, a back pull or resistance to traction.
covalence, combining of elements through sharing electrons.
crenation, the shrinking of cells.

critical mass, the concentration of radioactive material sufficient to sustain a chain reaction.

crystallization, the formation of definite characteristic shapes in solids.

curie, the amount of radiation formed by 1 Gm radium.

current electricity, flowing electricity; electricity in motion.

daughter elements, new elements formed in radioactive decay.

decay, the disintegration of radioactive elements.

decomposition, chemical change resulting in the formation of simpler materials from more complex ones.

decontamination, the freeing or removal of undesirable material.

dehydrolysis, a chemical reaction in which water is removed from the chemical composition of the substance.

deliquescence, the ability of a substance to absorb water and form a solution with itself.

denaturation, the alteration of properties of a protein, as by heat or acid.

density, the ratio of the weight to the volume.

detergent, a cleaning agent.

dialysis, the passage of dissolved material through a permeable membrane.

diffusion, the movement of molecules.

digestion, the chemical and physical process of changing ingested food so that it can be absorbed through the intestinal walls into the blood and lymph.

diplopia, double vision.

discharge, the equalization of different charges between two points.

distillation, changing a liquid to a vapor and collecting the condensed vapor.

diuretic, a substance which increases the flow of urine.

diverging lens, a lens which spreads apart rays of light.

dosimeter, a device for measuring small amounts of radiation.

double decomposition, chemical change in which two compounds react to form two new compounds.

dynamo, an apparatus for the conversion of mechanical energy into electrical energy.

edema, the accumulation of fluid in the tissues.

efforescence, loss of water of crystallization.

electricity, a form of energy consisting of a flow (current) or an accumulation (static) of electrons.

electrocardiograph, an instrument for recording changes in the electric potential of the heart muscle during the heart beat.

electrocautery, a method for searing tissue in which a wire is heated by an electric current.

electroencephalograph, an apparatus for recording electrical variations of the brain.

electrolysis, passage of an electric current through a compound resulting in the decomposition of the compound.

electrolytes, substances which, either dissolved or melted, will conduct an electric current.

electromagnetic radiation, energy waves.

electromagnetic force, potential difference.

electron, a practically weightless particle with a negative charge.

electrovalence, the combining of atoms through a loss and gain of electrons.

emulsion, a liquid mixture in which droplets of one liquid are distributed through another.

endothermic reaction, a chemical change in which energy is absorbed.

energy, the power to do work.

enzyme, an organic catalyst synthesized by living cells.

equilibrium, a condition in which opposing forces or rates are equal.

esters, organic compounds with the characteristic group

$$-\underset{\underset{}{}}{\overset{\overset{O}{\|}}{C}}-O-C$$

ether, organic compounds with the characteristic group C—O—C.

evaporation, change of state from liquid to vapor.

exothermic reaction, chemical change in which energy is liberated.

extract, a solution in which ethyl alcohol is the solvent.

fallout, the return to earth of radioactive debris resulting from atomic explosions in the atmosphere.

fermentation, decomposition of carbohydrates through the action of molds or bacteria.

fluidity, the tendency to flow.

force, the cause of the acceleration of moving bodies. A push or a pull.

frequency, the number of waves passing a given point in a given period of time.

fulcrum, a wedge on which or about which a lever turns.

gel, a semisolid colloid.

glycosuria, the presence of glucose in the urine.

ground, to connect a conductor to the earth as part of an electric circuit.

half-life, the length of time it takes for half of a sample of radioactive material to decay.

halogens, the family of elements consisting of fluorine, chlorine, bromine, and iodine.

heat, a form of energy due to the random motion of molecules.

hemolysis, the rupture of red blood cells.

heterocycles, organic ring compounds in which atoms other than carbon are involved in the ring structure.

hormones, metabolic regulators secreted by the endocrine glands.

hydrates, crystals containing water of crystallization.

hydrolysis, the chemical reaction of a substance with water.

hydronium ion, the hydrated hydrogen proton, H_3O^+.

hydrostatic pressure, pressure resulting from the weight of fluids at rest.

hyperglycemia, more than the normal concentration of glucose in the blood.

hyperopia, farsightedness.

hypertones, sound frequencies above 17,000 cps.

hypertonic solution, a solution having a higher osmotic pressure than another.

hypothermia, treatment involving the use of cold.

hypotonic solution, a solution with a lower osmotic pressure than another solution.

indicators, substances that change or develop color with changes in pH.

infrared rays, heat waves, longer rays than those of visible light.

infratone, sound frequencies below 20 cps.

inhibitor, a negative catalyst. It decreases the rate of a chemical reaction.

insulator, a nonconductor of heat or electricity.

interstitial fluid, liquid between the cells and the capillaries.

intracellular fluid, liquid inside the cell membranes.

ion, a charged particle.

ionization, the formation of charged particles.

ionization chamber, a device for measuring the degree of ionizing radiations.

isoelectric point, the pH at which a protein can ionize at both the carboxyl and the amino group.

isomers, organic compounds with the same molecular composition but different arrangements of atoms in the molecules and different properties.

isotonic solutions, two solutions with the same osmotic pressure.

isotopes, elements with the same atomic numbers but different atomic weights.

ketones, organic compounds with the characteristic group $C-\overset{\overset{\displaystyle O}{\|}}{C}-C$.

ketone bodies, substances formed in the abnormal catabolism of fats. Acetone, β-hydroxybutyric acid, and acetoacetic acid are the substances formed.

ketose, a sugar containing the ketone group.

ketosis, the accumulation of ketone bodies in the blood.

kindling temperature, the temperature at which a substance ignites and burns as a self-sustaining reaction.

kinetic energy, energy in motion.

laking, the liberation of hemoglobin from the red cells.

latent heat, the heat absorbed when a liquid evaporates or a solid melts.

lens, a transparent structure designed to change the direction of light rays.

lever, a bar used for transmitting and modifying force and motion.

liquid, matter with definite volume but no shape of its own. It takes the shape of the container.

macromolecules, molecules of great size.

magnet, a body able to attract iron.

mechanical advantage, the ratio of the work done by a machine to the force applied to the machine.

metabolism, all the chemical changes and energy changes occurring in the body.

molar solution, a solution in which 1 gram-molecular weight of solute is dissolved in 1 liter of solution.

mole, a gram-molecular weight of a compound.

myopia, nearsightedness.

neutron, a subatomic particle with a weight of 1 and no charge. It is found in the nucleus.

normal solution, a solution which contains in 1 liter 1 gram-equivalent weight of solute.

nyctalopia, night blindness.

oliguria, reduced amount and frequency of urine.

ophthalmoscope, an instrument for examining the interior of the eye.

osmosis, the passage of solvent molecules through a membrane.

osmotic pressure, the pressure that must be

applied to a solution to prevent the flow of water into it through a semipermeable membrane.

oxidant, an oxidizing agent. A substance which readily liberates oxygen.

oxidation, a chemical change involving the addition of oxygen or the loss of hydrogen or the loss of electrons or the numerical increase in the valence of an element.

pentose, a sugar with 5 carbon atoms in the molecule.

permeable membrane, a membrane through which both solute and solvent can diffuse.

pH, a system for describing the degree of acidity or alkalinity of a solution.

photosynthesis, the synthesis of compounds by plants in the presence of sunlight through the action of chlorophyll.

piezoelectric effect, the development of positive and negative charges on compressed crystals.

plasma, the liquid part of the blood.

polar compounds, with unequal distribution of electrons in the molecule.

polymers, compounds with the same percentage composition but different molecular weights.

polyuria, an abnormally large secretion of urine.

potential difference, the difference in charge between two places or objects.

potential energy, energy at rest, hidden energy, as the potential energy in food.

proenzyme, the inactive form of an enzyme. A substance from which an enzyme can be formed.

protective colloid, a substance which prevents the precipitation or coagulation of a colloid.

proton, a subatomic particle with a weight of 1 and a positive charge. It is found in the nucleus of the atom.

radioactivity, the spontaneous disintegration of the nuclei of atoms.

radioisotope, a form of an element which emits radiations (alpha, beta, or gamma rays).

redox, oxidation-reduction.

reduction, a chemical change involving the loss of oxygen or the addition of hydrogen or the gain of electrons or a numerical decrease in valence.

renal threshold, the limit of the capacity of the kidney tubules to reabsorb useful substances.

replacement, a chemical reaction in which an element and a compound form another element and another compound.

resistance, the opposition offered by material to the passage of an electric current.

respirator, an apparatus for artificial breathing.

respiratory quotient, the ratio of the volume of carbon dioxide formed to the volume of oxygen consumed when food is oxidized.

salts, compounds formed when the hydrogen of an acid is replaced with a metallic ion.

saponification, the process of making soap.

scintillation counter, a device for measuring radioactivity.

semipermeable membrane, a membrane which permits passage of solvent molecules but retains solute molecules.

siphon, a bent tube with arms of uneven length for transferring liquids.

specific gravity, the density of a substance referred to the density of water, which is considered to be 1.000. If the substance is a gas, its density is compared with that of air.

specific heat, the amount of heat needed to raise 1 Gm of a substance 1°C.

supertones, sound waves with frequencies above 17,000 cps.

surface tension, the close packing of molecules at a surface of interphase.

syneresis, the shrinkage of a colloid as it ages.

synthesis, the process of making more complex substances from simpler ones.

temperature, the intensity of heat.

tincture, a solution in which ethyl alcohol is the solvent.

triose, a sugar with 3 carbon atoms in the molecule.

torque, a force which produces or tends to produce rotation.

ultrasound, sound waves with frequencies above 20,000 cps. They are beyond the range of human hearing.

ultraviolet rays, light rays shorter in wavelength than violet light. They are invisible to the human eye.

urinometer, a device for determining the specific gravity of urine.

valence, the combining ability of an element.

vectors, forces with direction.

vesicant, a substance that causes blisters.

viscosity, the resistance of a liquid to flow.

volt, the unit of electromotive force.

voltmeter, an instrument to measure electricity in volts.

waves (rays), periodic or vibrating motion.

X rays, electromagnetic radiation of extremely short wavelength. They are very penetrating high-energy waves.

INDEX

Absolute temperature, 50
Absolute zero, 42
Absorption, 100
Acetaldehyde, 152–153
Acetamide, 165
Acetanilid, 36
Acetoacetic acid, 160, 192–193
Acetone, 154, 192–193
Acetone bodies in urine, 246–247
Acetylene, 143
Acid(s), 107–108, 154–156
    acetic, 107, 108, 111, 155, 192
        glacial, 155
    acetoacetic, 160, 192–193
    acetylsalicylic, 159
    amino, 165–167
    arachidonic, 187
    ascorbic, 228
    barbituric, 169
    benzoic, 155
    β-hydroxybutyric, 159, 192–193
    β-ketobutyric, 160, 192–193
    boric, 111
    butyric, 155, 186, 192
    caproic, 192
    carbolic, 111
    carbonic, 107, 108, 120, 127–128, 242
    citric, 111, 155, 158
    definition of, 106, 107
    diacetic, 160
    diprotic, 108
    folic, 227
    formic, 155
    hydrochloric, 111, 212

Acid(s):
    hydroxy, 158–160
    hypochlorus, 82
    ionization of, 108
    keto, 160
    lactic, 111, 158
    linoleic, 186, 187
    linolenic, 187
    malonic, 155
    nicotinic, 226–227
    nitric, 111
    nucleic, 168
    oleic, 186, 187
    organic, 155–156
    oxalic, 155
    p-aminobenzoic, 166
    palmitic, 186, 187
    pantothenic, 226–227
    phosphoric, 108, 111
    properties of, 107–108
    pyruvic, 160, 191
    salicylic, 159
    stearic, 186, 187
    sulfuric, 108, 111
    tannic, 201
    triprotic, 108
    uric, 245
    uses of, 111
Acid-base balance, 126–129
Acidosis, 126–130
    metabolic, 128–130
    respiratory, 126
Acrolein, 188
Acromegaly, 222

ACTH, 223
Addison's disease, 220
Addition, 146
Adenine, 168
Adenosinediphosphate (ADP), 215
Adenosinetriphosphate (ATP), 215
ADH, 223
Adhesion, 55
ADP, 215
Adrenaline, 219
Adrenals, 219
Adrenocorticotropic hormone, 223
Adsorption, 100
Aeration of water, 78
Agents, 35–36
  oxidizing, 35–36
  reducing, 35
Alanine, 166, 197, 204
Albumin, 202
  in urine, 247
Alcohol(s), 148–152
  benzyl, 150
  denatured, 149
  dihydric, 148
  ethyl, 148–149
  fermentation of, 148
  isopropyl, 150
  methyl, 148
  monohydric, 148
  polyhydric, 151
  specific heat of, 84
  tribromomethanol, 149
  trihydric, 148
Aldehyde, 152–153
  oxidation of, 152
  properties of, 152–153
Aldose, 175
Aldosterone, 223
Alimentary glycosuria, 182
Aliphatic compounds, 139
Alkalies, 109
Alkaline tide, 244
Alkaloids, 169–171
  atropine, 170
  caffeine, 170
  cocaine, 170
  codeine, 170
  coniine, 171
  curarine, 171

Alkaloids:
  ergotoxine, 171
  morphine, 170
  nicotine, 170
  papaverine, 170
  physostigmine, 170
  pilocarpine, 170
  quinine, 171
  reserpine, 169
  strychnine, 170
  theobromine, 170
Alkalosis, 128–130
  metabolic, 129–131
  respiratory, 128
Alkane(s), 139–141
Alkyl groups, 141, 142
Alkyl halides, 147–148
Alkyl radicals, 141
Alkyl sulfides, 172
Allergies, 220
Alpha particles, 271
Amide, 165
Amine, 163
  primary, 163
  secondary, 163
  tertiary, 163
Amino acid(s), 165–167
Amino group, 163
Aminopolypeptidase, 215
Ammeter, 45
Ammonia, 245–246
Ammonium salts, 133
  quaternary, 164
  in urine, 245–246
Ampere, 45
Amphoteric compounds, 196
Amplitude, 259
Amylase, 210, 213
  pancreatic, 213
  salivary, 210
Amyl nitrate, 157
Amylopsin, 215
Amytal, 169
Anabolism, 37, 191–192
  of CHO, 181–182
  of fat, 191–192
  of protein, 204
Analysis, 33
Androsterone, 221, 224

Index

Angstrom unit (Å), 18
Anhydride, 87
Aniline, 164
Anion, 17, 105
Aniseikonia, 265
Anode, 105
Anthracene, 145
Antidiuretic hormone, 223
Antidotes, 108
Archimedes' principle, 52
Aromatic compounds, 139
Arrhenius theory, 105–106
Artificial pneumothorax, 53
Aryl groups, 141
Aryl radicals, 141, 142
Aschheim-Zondek test, 223
Ascorbic acid, 228
Aspirin, 159
Astigmatism, 264–265
Atabrine, 171
Atmospheric pressure, 26
Atom, 2–3, 13–16
Atomic bomb, 272
Atomic by-products, 272–273
Atomic number, 13
Atomic pile, 272–273
Atomic weight, 10, 13
ATP, 215
Atropine, 170
Autoclave, 27
Avertin, 149

B-complex vitamins, 225–228
Barbiturates, 168–169
   Amytal, 169
   Luminal, 169
   Nembutal, 169
   Pentothal, 169
   Seconal, 169
   Veronal, 169
Barbituric acid, 169
Barium sulfate, 269
Barometer, 26
Basal metabolism, 207
Bases, 109
   properties of, 109
   uses of, 111
Becquerel, Henri, 270

Bends, 52
Benedict test, 153, 180
Benzaldehyde, 152–153
Benzene, 143
Benzidine test, 247
Benzoic acid, 155
Benzoin, 155
Benzyl alcohol, 150
Benzyl benzoate, 157
Beriberi, 226, 231
Bernoulli, Daniel, 64
Bernoulli principle, 60
Berzelius, Jöns, 3
β-carotene, 229
β-hydroxybutyric acid, 159
β-ketobutyric acid, 160
β-oxidation theory, 192–193
Beta particles, 271
Bicarbonate ion, 21, 107, 108, 120, 127–128
Bile, 213
   pigments, 213
   salts, 213
   in urine, 244, 247
Bilirubin, 213
Biliveridin, 213
Binocular vision, 265
Biological effects of radiation, 275–276
Biotin, 227
Biuret reaction, 202
Blood, 235–239
   analysis of, 238
   composition of, 235, 238
   functions of, 235
BMR, 207
Body, as a source of electric current, 47
Body electrolytes, 124–126, 133–134
   functions of, 130
Body fluids, 123–124
   disturbances, 131–134
   exchange of, 124
Body mechanics, 256
Boiling, 26–27, 85
Boiling point, 27
Bomb, atomic, 272
Bonds:
   double, 141–142
   hydrogen, 19–20
   single, 139–140
   triple, 143

Boric acid, 111
Bowman's capsule, 243
Boyle, Robert, 49
Boyle's law, 49–146
Bromine, 145–146
Brönsted and Lowry theory, 106
Bubble drainage, 65–69
Buffers, 119–121, 128, 241
Buoyancy, 52
Burns, 132, 247
Butane, 140
Butyric acid, 155, 186, 192

Caffeine, 170
Caisson's disease, 52
Calciferol, 229–230
Calcium, 218
Calorie, 40
Camera, 263–265
Capillary action, 55–56
Caproic acid, 192
Carbohydrases, 215
Carbohydrates, 175–183
   acid hydrolysis of, 179
   anabolism of, 181
   catabolism of, 182
   chemical properties of, 179
   classification of, 176–177
   composition of, 175
   enzymatic hydrolysis of, 179
   fermentation of, 180
   formation of, 178
   metabolism of, 181
   oxidation of, 179, 182–183
   physical properties of, 178
   reactions with iodine, 180–182
   respiratory quotient of, 183
   structure of, 181
Carbolic acid, 150
Carbon, 3
Carbon dioxide, 81, 241
   combining power, 129–130
Carbon monoxide, 79–81
Carbon tetrachloride, 148
Carbonic acid, 107, 108, 120, 127–128, 242
Carbonic anhydrase, 212
Carbonyl-hemoglobin, 80

Carboxyl group, 213
Carboxyhemoglobin, 80
Carboxypolypeptidase, 213, 215
Carrel-Dakin solution, 82
Casein, 196
Cast, plaster, 87–88
Catabolism, 37
   of CHO, 182
   of fat, 192–193
   of protein, 204
Catalyst, 36–37, 208
Cataract, 265
Catharsis, saline, 99–100
Cathode, 105
Cathode rays, 270
Cation, 17, 105
Caustics, 109
Cell, 46
   dry, 46
   storage, 46
   voltaic, 46
Cellulose, 178
Celsius, 40–41
Center of gravity, 251
Centigram, 8
Centimeter, 9
Cephalin, 190, 237
Cerebroside, 190
Change, 32
   chemical, 32
   physical, 32
   of state, 24–26
Charging, 46
Charles's law, 50
Chemical change, 33–36
   conditions causing, 36
      catalysts, 36
      energy, 36
   types of: analysis, 33
      combination, 33
      decomposition, 33
      double replacement, 33
      oxidation-reduction, 35–36
      redox, 35–36
      simple replacement, 33
      synthesis, 33
Chemical equations, 33–35
Chemical formulas, 3–4, 20–22

Index

Chemical "hot water bottle," 97
Chemical symbol, 3
Chloral, 153
Chloride ion, 16–17
Chloride retention, 133
Chloride shift, 130
Chlorides in urine, 245–246
Chlorination, 85
Chlorine, 82
Chloroform, 147
Chlorophyll, 178
Cholamine, 190
Cholecystokinin, 213
Cholesterinase, 208
Cholesterole, 150, 214
Choline, 165, 190
Chromoprotein, 196
Chyme, 213
    enzyme action on, 214
Chymotrypsin, 213, 215
Chymotrypsinogen, 213
Citric acid, 155, 158
Citric acid cycle, 216, 217
Clinitest, 180
Clorox, 82
Coagulation:
    of colloids, 101
    of proteins, 200–201
Cobalt, radioactive, 283
Cocaine, 170
Codeine, 170
Coenzyme, 209
Cohesion, 55
Cold, 42
    in therapy, 43
Collagen, 203
Colloid, 91, 92, 100–102
    absorption, 100
    coagulation of, 101
    osmotic pressure of, 101
    properties of, 100
Colloidal dispersions, 92
Colloidal solutions, 92
Colloids, protective, 101
Color index, 236
Combustion, 76–78
    spontaneous, 78–79

Common elements, 21
    valence numbers of, 21
Compounds, 4–5
    unsaturated, 18
Compression waves, 259
Concave lens, 263
Concentration of solutions, 92–97
Concentration gradient, 99
Concentration test, 247–248
Condensation, 25–26
Conductance, 103–104
Conduction of heat, 41
Conductivity, 103–104
Conductors:
    of electricity, 103
    of heat, 43
Coniine, 171
Conjugated proteins, 196, 201
Convection, 41–42
Converging lens, 263–264
Convex lens, 263–264
Coordinate covalence, 18
Corpus luteum, 221–222
Corticosterone, 219
Cortin, 220
Cortisol, 220
Cortisone, 220
Countertraction, 254
Covalence, 17
Creatine, 245
Creatinine, 245
Crenation, 98
Cresol, 150
Cretinism, 218
Critical mass, 272
Crystallization:
    heat of, 97
    water of, 87
Crystalloids, 100–101
Cubic centimeter, 9
Cuprous oxide, 153, 180
Curare, 170
Curarine, 170
Curie, Marie, 270
Curie, Pierre, 270
Curie (unit), 274
Current electricity, 45, 103, 105
Cyanocobalamine (B_{12}), 165, 227

Cyclic compounds, 139
   hydrocarbons, 143
Cyclopropane, 143
Cyclotron, 273
Cystine, 166
Cystinuria, 166
Cytosine, 168

Dakin's solution, 82
Dalton, John, 2, 3
Dalton's law of partial pressures, 50
Daughter elements, 271
Deamination, oxidative, 204
Debye-Hückel theory, 106
Decay, 78
   radioactive, 271
Decimeters, 7
Decomposition, 33
   of compounds of oxygen, 73
Decontamination of radioactive material, 278
Defibrinated blood, 236
Dehydration, 124
Dehydrocorticosterone, 219
Dehydrolysis, 37
Deliquesence, 88
Demerol, 171
Denaturation, 200
Denatured alcohol, 149
Density, 32
Deoxycorticosterone, 219
Derived protein, 196
Destructive distillation, 74
Detergent action, 55
Dextrin, 178
Dextrose, 176, 177
Diabetes insipidus, 223, 243
Diabetes mellitus, 192, 246
Diacetic acid, 160
Dialysis, 98–100
Diazines, 168
Dicumerol, 238
Diet, ketogenic, 160
Diethyl ether, 157–158
Diffusion of gases, 51, 241–242
Digestion, 211–215
   gastric, 211–213, 215

Digestion:
   intestinal, 213–215
   salivary, 210–211, 215
Dilution test, 247
Dipeptidase, 215
Dipeptide, 198
Diplopia, 265
Dipoles, 18
Diprotic acid, 108
Disaccharide, 177
Discharge, 44
Discharging, 46
Distillation, 84
   of liquid air, 72
Disubstitution, 145
Diuretic, 243
Diverging lens, 263–264
Divinyl ether, 158
DNA, 168, 205
Dosimeter, 273–274
   film badge, 273
   geiger counter, 273
   ionization chamber, 274
   scintillation counter, 274
Double decomposition, 33
Double displacement, 33
Double replacement, 33
Drainage bubble, 65–69
   closed chest, 53–54
   fistular, 131–132
   simple underwater, 53–54
Dreypak, 180
Dry ice, 81
Dual valence, 19
Dwarfism, 222
Dynamo, 46

Edema, 130
EEG, 47
Efflorescence, 88
Einstein's equation, 39–40
EKG, 47
Elastic limit, 256
Elasticity, 255
Electric current, 45
   body as source of, 47
Electric field, 44

Electricity, 43–46
　body as source of, 47
　cells, 46
　current, 45, 103, 105
　dynamo, 46–47
　generator, 46–47
　sources of, 46–47
　static, 43–47
Electrocardiogram, 47
Electrocardiograph, 47
Electroencephalogram, 47
Electrolysis of water, 72
Electrolyte balance, 130
Electrolyte imbalance, 136
Electrolyte losses, 131
Electrolyte retention, 133
Electrolytes:
　body, 123–134
　strong, 105–106
　weak, 105, 107
Electromagnetic radiations, 262
Electromotive force (EMF), 47–48
Electron, 13
　planetary, 15
　valence, 16–18
Electron microscope, 47–48
Electronic devices, 48
Electrovalence, 16–17
Elementary particles, 3, 13
Elements, 2
Elixir, 89
EMF, 47–48
Empirical formula, 20
Emulsifying agent, 102
Emulsion, 102
　breaking of, 102
　permanent, 92, 118
　temporary, 92
Emulsoid, 101
Endothermic reaction, 36
Energy, 39–48
　kinds of, 39
　　chemical, 39
　　electrical, 39
　　heat, 39
　　kinetic, 39
　　light, 39
　　magnetic, 39

Energy:
　kinds of: mechanical, 39
　　potential, 39
　　sound, 39
Enterokinase, 213
Enzymes, 208–209, 215
　action of, 209
　aminopolypeptidase, 215
　amylase: pancreatic, 213
　　salivary, 210
　amylopsin, 215
　carbohydrases, 215
　carbonic anhydrase, 212
　carboxypolypeptidase, 215
　cholinesterase, 208
　chymotrypsin, 215
　coenzymes, 209
　dipeptidase, 215
　enterokinase, 213
　glycogenase, 208
　hydrolase, 209
　hydrolysis of, 210
　isomerase, 210
　lactase, 213–215
　lecithinase, 208
　ligases, 210
　lipase, 215
　　gastric, 212, 213
　　pancreatic, 213
　lysases, 210
　maltase, 208, 211, 213–215
　oxidoreductases, 209
　pepsin, 211, 212, 215
　pepsinogen, 212
　peptidases, 214
　proenzymes, 208
　　pepsinogen, 208
　protease, 215
　　gastric, 211
　ptyalin, 212, 215
　steapsin, 215
　sucrase, 208, 214, 215
　synthetases, 210
　transferases, 210
　trypsin, 215
　urease, 208
Eosin, 172
Ephedrine, 220

Epinephrine, 165, 219
Equations, writing of, 33–35
Equilibrium, 56, 251
   dynamic, 26
Equivalent weight, 94
   milliequivalent weight, 96
Erepsin effect, 214
Ergosterol, 230
Ergotoxine, 171
Erythemia, 276
Erythrocytes, 289–291
Erythrotin, 227
Esterification, 156
Esters, 156–157
   medicinal uses of, 157
   properties of, 156–157
Estradiol, 221
Estriol, 221–222
Estrone, 221
Estrus, 221
Ethanal, 152
Ethane, 139
Ethanol, 148
Ether, 90, 157, 158
   diethyl, 90, 157, 158
   divinyl, 158
   mixed, 157–158
   simple, 157
Ethyl alcohol, 89, 148, 201
   action with protein, 201
Ethyl chloride, 147
Ethyl mercaptan, 171–172
Ethyl nitrate, 157
Ethyl nitrite, 157
Ethylene, 141
Evaporation, 24–26
Exchange of body fluids, 124
Exopthalmic goiter, 217
Exothermic reaction, 36
Explosion, 77
External radiation, 275
Extracellular fluid, 123–126
Extract, 89
Eye, 264–265

Fahrenheit, 40–41
Fallout, 283

Fat-soluble vitamins, 228–232
Fats, 185–189
   anabolism of, 191–192
   catabolism of, 192–193
   classification of, 185–186
   formation of, 186
   hydrogenation of, 189
   oxidation of, 187–188
   properties of, 187–189
   respiratory quotient of, 188
   saponification of, 189
   saturated, 186
   unsaturated, 186–187
Feces, 249
Fehling's test, 153, 180
Fermentation, 180
   of urine, 247
Fibrin, 237
Fibrinogen, 203, 237
Film badge, 273
Filtration, 84
Fistula drainage, 132–133
Fluid exchanges, 124
   input and output, 134
Fluidity, 56
Fluids, 23
   (See also Body fluids)
Fluorescein, 172
Fluorides, 110
Fluoroscope, 270
Focal length, 263
Folic acid complex, 227
Folin's reagent, 180
Follicle-stimulating hormone (FSH), 222
Force, 28, 252–253, 255–257
Formaldehyde, 152, 153
Formic acid, 155
Formula, 3–4, 20–22
   empirical, 20
   graphic, 20
   molecular, 20
   structural, 20
Fractional distillation of liquid air, 72
Fracture, reduction of, 254
Freezing, 24
Freezing point, 24
   depression of, 105
Frequency, 259

Fructose, 177
FSH, 222
Fulcrum, 252
Fuses, 46

Galactose, 177–179, 181
Gamma rays, 271, 274
Gas, 23, 24
   inert, 15, 16
Gas laws, 49–52
Gastric digestion, 211–213, 215
Gastric enzymes, 212, 213
Gastric juice, 211
Gastric lipase, 212, 213
Gastrin, 211
Geiger counter, 273
Gels, 101
Generator, 46
Genes, 276
Giantism, 222
Globin, 236
Globulins, 203
Glomerulus, 242–243
Glucagon, 219
Glucocorticoids, 223
Glucose, 153, 177
Glycene, 166
Glyceric aldehyde, 191
Glycerides, 186, 187
Glycerol, 151
Glyceryl butyrate, 186
Glyceryl nitrate, 151
Glycogen, 166
Glycogenase, 208, 214
Glycogenolysis, 220
Glycol, 151
Glycolipids, 190
Glycoprotein, 196
Glycosuria, alimentary, 182
Goiter, 217
Gold, radioactive, 282
Gonads, 221
Graham's law, 51
Gram, 9, 18
Gram atomic weight, 10
Gram equivalent weight, 94–95
Gram molecular weight, 10

Graves' disease, 217
Gravity:
   center of, 251
   law of, 1
   specific, 32
Ground, 44
Ground potential, 44
Group, 19
   alkyl, 141, 142
   aryl, 141
   characteristic, 138–139
Growth hormone, 222
Guanine, 168
Gypsum, 87–88

Half-life, 85–86
Halogens, organic, 147–148
Hard water, 85–86
   boiling of, 85–86
   permanent, 86
   softening of, 85–86
   temporary, 85–86
Hearing, 259–260
   impaired, 260
   "swiss cheese," 260
Heat, 40
   of condensation, 24
   of crystallization, 97
   of fusion, 24
   of solution, 97
   specific, 84
   in therapy, 43
   transfer of, 41–42
      conduction, 41
      convection, 41–42
      radiation, 42
   of vaporization, 24
Heat waves, 262
Helium, 81
Heller's test, 202
Hematin, 236
Hematuria, 247
Hemodialysis, 98
Hemoglobin, 203, 236, 241–242
   in urine, 247
Hemolysis, 98, 237
Hemorrhage, 132

Henry's law, 51–52
Heparin, 238
Heterocycles, 167–169
Hexamethylenetetramine, 164
Hexose, 175, 177
Hirudin, 238
Hofmann apparatus, 72
Homologous series, 140
Hopkins-Cole test, 202
Hormones, 208–224
    adrenaline, 219
    androsterone, 221, 224
    antidiuretic, 223, 224
    biologic action of, 224
    cholesystokenin, 213
    cortin, 219, 224
    epinephrine, 219, 224
    estradiol, 221, 224
    estriol, 221, 224
    estrone, 221, 224
    gastrin, 211
    glucagon, 219, 224
    insulin, 218–219, 224
    parathyroid hormone, 218, 224
    pregnanediol, 222, 224
    progesterone, 222, 224
    relaxin, 223
    secretin, 213, 224
    testosterone, 221, 224
    thyroxine, 216–218, 224
Human body, approximate composition of, 3
Hydrates, 87–88
Hydrocarbons, 139–147
    addition, 146
    formulas, 140–147
    hydrogenation, 147
    nomenclature, 141
    properties, 145–147
    saturation, 141
    substitution, 141–146
    unsaturation, 141, 143
Hydrochloric, acid, 111, 112
Hydrogen, 86, 107–108
Hydrogen bonds, 19–20
Hydrogen peroxide, 89
Hydrogenation, 147
    of oils, 189
Hydrolase, 209, 210

Hydrolysis, 37, 84
    of esters, 156, 188
    of salts, 118–119
Hydronium ion, 106
Hydrostatic pressure, 56–57
Hydroxides, 111
Hydroxy acids, 158–160
Hyperacidity, 212
Hyperbaric technique, 52
Hyperglycemia, 182
Hyperopia, 264, 266
Hyperthyroidism, 217
Hypertones, 260
Hypertonic solutions, 98
Hypochlorus acid, 82
Hypothermia, 42–48
Hypothyroidism, 217–218
Hypotonic solutions, 98

Imidazole, 167
Inclined plane, 253
Indican, 246
Indicators, 107, 116–117
    litmus, 107
    methyl orange, 107
    phenolphthalein, 107
Indole, 246
Infrared rays, 262
Infratones, 260
Inhibitor, 36
Insensible perspiration, 248
Insensible water loss, 134
Insulator, 43
Insulin, 218
Internal radiation, 274
Interstitial fluid, 123
Intestinal epithelium, 248
Intestinal juice, 214
    pH of, 214
Intracellular fluid, 123
Intravascular fluid, 123n.
Inversion, 179
Invert sugar, 179
Iodine, 217
    radioactive, 280–281
    reaction with carbohydrates, 180
Iodoform, 147

Iodostarch, 180
Ion exchange, 85
Ionic valence, 17
Ionization, 103–121
   of acids, 108
   Arrhenius' theory of, 105
   Brönsted theory of, 106
   Debye-Hückel theory of, 106
   degree of, 106
   ion-production for water, 114
   of water, 110
Ionization chamber, 274
Ionizing radiations, 269–283
   precautions with, 278
Ions, 105
   hydronium, 106, 107
Iron:
   radioactive, 283
   specific heat of, 84
Iron lung, 54
Isobutane, 140
Isoelectric point, 166
Isomerases, 210
Isomerism, 175–176
Isometric compounds, 138
Isopropyl alcohols, 89, 150
Isotomic solutions, 98
Isotopes, 15–16
   radioactive, 270, 273, 280–283

Jaundice, 248
Javelle water, 82
Juice:
   gastric, 211
   intestinal, 214
   pancreatic, 213
   salivary, 210

Kekulé, Friedrich, 143–144
Keratin, 203
Keto acids, 160
Ketones, 154
   oxidation of, 154
Ketose, 175
Ketosis, 160

Kidneys, 242
   efficiency tests, 247–248
Kilogram, 8, 9
Kiloliter, 8
Kilometer, 8
Kindling temperature, 76
Kinetic energy, 39
Kinetic molecular theory, 23–24
Knoop's theory of β-oxidation, 192–193
Krebs cycle, 216

Labarraque's solution, 82
Lactase, 213–215
Lactic acid, 158
Lactose, 177
Laking, 237
Latent heat, 24
Laughing gas, 79
Law(s):
   Boyle's, 49–50
   Charles's, 50
   of conservation of matter, 33, 40
   of conservation of matter and energy, 40
   Dalton's, 50–51
   of definite composition, 11–12
   Graham's, 51
   Henry's, 51–52
   of inverse squares, 278
   of the levers, 253
   Ohm's, 45–46
   of universal gas, 50
Lecithin, 190
Lecithinase, 208
Lens, 263–265
   accommodation, 264
   concave, 263
   converging, 263–264
   convex, 263
   diverging, 263–264
   negative, 263
   positive, 263
Leukocytes, 237
Lever, 252–253
   law of, 253
Levulose, 176
Lewis, G. N., 17

LH (luteinizing hormone), 222–224
Ligases, 210
Light, 262–263
    absorption, 262
    paths of, 263
    properties of, 262
    reflection of, 262, 263
    refraction of, 263
    visible, 262
    waves, 262
Lipase, 212, 213, 215
    gastric, 212, 213
    pancreatic, 213
Lipids, 185–193
    classification of, 185–186
    enzymatic hydrolysis of, 188
    formation of, 186
    hydrogenation of, 189
    hydrolysis of, 188
    metabolism of, 191–193
    oxidation of, 187–188
    properties of, 189
Liquid, 23
    supercooled, 23
Liquid air, 72
    distillation of, 72
Liquid mixtures, 91–102
    properties of, 92
Lister, 150
Liter, 8–10
Liver, 248
Long's formula, 244
Luminal, 169
Lungs, 241
Luteinizing hormone (LH), 222–224
Lyases, 210
Lysol, 150
Lysolecithin, 190

Macromolecules, 195
Magnet, 46
Malaria, 247
Malonic acid, 155
Maltase, 208, 211, 213–215
Maltose, 177
Mass, 1
Matter, 1

Mechanical advantage, 253
Mechanical energy, 39
Mechanics, body, 256, 257
Melting, 24
Melting point, 24
Membrane:
    permeable, 71, 98
    semipermeable, 97
Meperidine, 171
Mercaptan, 171
Mercurochrome, 172
Metabolic acidosis, 128
Metabolic alkalosis, 129–131
Metabolic regulators, 207–232
Metabolic wastes, 241–249
Metabolism, 37
    basal, 207
    of carbohydrates, 181–182
    of lipids, 191
    of proteins, 204
Metals, 2
    action with water, 86–87
    oxides, 87
Meter, 7, 9
Methane, 139
Methanol, 148, 152
Methyl alcohol, 148, 152
Methyl amine, 163–164
Methyl chloride, 147
Methyl mercaptan, 171
Methyl salicylate, 159
Metric and English equivalents, 9–10
Metric system, 7–10
Microcurie, 274
Microgram, 8
Micron, 8
Milk, 249
Millicurie, 274
Milliequivalent weight (meq), 96
Milligram, 8
Milligrams percent, 94
Milliliter, 8
Millimeter, 8
Millimicron, 8
Milliroentgen, 274
Millon's reaction, 202
Mineralocorticoids, 223
Minerals, 109–110

Index

Mixtures, 4–5
MKS (meter, kilogram, second) system, 7
Molar solution, 94
Mole, 10
Molecular weight, 10
Molecule, 3
Monitoring, personnel, 274
Monosaccharide, 177
Morphine, 170
Motor, electric, 47
Mucin, 210
Mustard gas, 172
Mutations:
   lethal, 276
   sublethal, 276
Myopia, 264
   figure, 266
Myosin, 203
Myxedemas, 218

Naphthalene, 144
Negative catalyst, 36
Negative lens, 263
Nembutal, 169
Neutralization, 117
Neutron, 3, 13
Niacin, 226–227
Niacinamide, 226
Nicotine, 170
Nicotinic acid, 226–227
Night blindness, 228, 265–266
Nitrazine, 117
Nitric acid, 111
Nitrogen, 79
Nitrogen heterocycles, 167–169
Nitrogen mustard, 172
Nitroglycerine, 151
Nitrous oxide, 79
Nonmetals, 2
Nonpolar, 18
Norepinephrine, 165
Normal solution, 94–95
Novocaine, 171
Nucleic acid, 168
Nucleoprotein, 196
Nucleus, 13
Nyctalopia, 265

Ohm, 45
Ohm's law, 45–46
Oleic acid, 155, 186–187
Oliguria, 243
Oncotic pressure, 126
Ophthalmin, 228
Ophthalmoscope, 262
Optical isomers, 176
Organic acids, 154–156
   properties of, 155–156
Organic compounds, 138–173
Orinase, 219
Osmosis, 97–98
Osmotic pressure, 101
   of colloids, 126
Osteomalacia, 230–231
Ovary, 221
Oxalic acid, 155
Oxidant, 35–36
Oxidation, 35–36
   of carbohydrates, 182
   of fats, 187–188
   of proteins, 201–202
   slow, 78
Oxidation-reduction, 35–36
Oxides:
   metal, 74, 87
   nonmetal, 74
Oxidizing agent, 35–36
Oxidoreductase, 209
Oxychloroquinone, 171
Oxygen, 71–75
   discovery of, 72
   preparation of, 72
   properties of, 73–74
   therapeutic uses of, 74–75
Oxygen test, 75
Oxyhemoglobin, 236, 241
Oxytocin, 223

p-aminobenzoic acid, 166
PABA, 166
Palmitic acid, 155, 186–187
Pancreas, 218
Pancreatic juice, 213
Panthothenic acid, 226–227
Papaverine, 170

Parathyroids, 218
Paregoric, 170
Partial pressure, 50–51
Pascal's principle, 52–53
Pellagra, 226
"Pelvic relaxation," 223
Pentaquinone, 171
Pentobarbitol, 169
Pentose, 175, 177
Pentothal, 169
Pepsin, 209, 211, 212, 215
   action of, 212
Pepsinogen, 208, 212
Peptidase, 214
Peptide link, 198
Peptone, 196, 198
Percentage solution, 93
Periodic motion, 259
Peritoneal dialysis, 99
Permanent hard water, 86
Permeable membrane, 91–98
Permutit process, 86
Personnel monitoring, 274
Perspiration, insensible, 248
pH, 115, 116
   in acid-base balance, 126–129
   of body fluids, 126–129
   of gastric juice, 212
   of intestinal juice, 214
   of urine, 244
Phenanthrene, 145
Phenobarbital, 169
Phenol, 150
Phenol red, 172
Phenolsulfonphthalein (PSP), 176
Phenolsulfonphthalein test, 247
Phenyl salicylate, 159
Phosphate in urine, 246
Phospholipid, 189
Phosphoprotein, 196
Phosphoric acid, 108, 111
Phosphorus, radioactive, 281–282
Photosynthesis, 178
Physical change, 32
Physical properties, 21–23
Phystostigmine, 170
Piezoelectric effect, 261
Pilocarpine, 170

Pituitary gland, 222–223
Plane-polarized, 176
Plasma, 123, 235
Plasmolysis, 98, 237
Plaster of paris, 87–88
Plaster cast, 87–88
Platelets, 237–238
Pneumothorax, 53
Polar bonds, 18
Polar compounds, 18
Polyhydric alcohols, 151
Polymer, 138
Polypeptide, 156, 198
Polysaccharide, 177
Polyuria, 243
Positive lens, 263
Potassium permanganate, 36
Potassium retention, 133
Potential, 270
   electrical, 43–44
   ground, 44
Potential difference, 43–44
Potential energy, 39
Pregnanediol, 222
Pressure, 49
   atmospheric, 26
   gas, 49
   hydrostatic, 56–57
      figure, 57–58
   vapor, 26–27
Priestley, Joseph, 72
Procaine, 171
Proenzyme, 208
Progesterone, 222
Prolactin, 222
Propane, 140
Properties:
   chemical, 32
   physical, 31–32
Protamine, 196, 218
Protease, 215
   gastric, 211
Protective colloid, 101
Protein, 195–206
   action with acid, 200
   action with alcohol, 201
   action with metallic ions, 199
   action with salts, 200

Index

Protein:
  amphoteric nature of, 197, 199
  anabolism of, 204
  catabolism of, 204
  chemical properties of, 199–202
  classification of, 196–197
  color tests, 202
  composition of, 195–196
  conjugated, 196, 201
  denaturation, 200
  derived, 196
  digestion of, 201
  formation of, 197–198
  heredity and, 204–205
  hydrolysis of, 201
  metabolism of, 204
  molecular weight of, 195
  oxidation of, 201–202
  properties of, 198–202
  simple, 196, 201
Protein sparers, 204
Proteose, 196, 198
Prothrombin, 237
Proton, 3, 13
PSP, 172
Ptomaines, 163–164
Ptyalin, 212, 215
Pulleys, 253
Pump, 63–64
Pupura, 247
Purine, 168
Pyrazine, 168
Pyridazine, 168
Pyridine, 167
Pyridoxine, 227
Pyrimidine, 168
Pyrrole, 167
Pyruvic acid, 160

Quaternary ammonium salts, 164
Quinacrine, 171
Quinine, 171

RAD (radiation absorbed dosage), 274
Radiation, 42, 274
  biological effects, 275–278

Radiation:
  electromagnetic, 262
  external, 275
  genetic effects, 276
  internal, 274
  ionizing, 269–284
  measuring, 273
  precautions, 278
  somatic effects, 276
Radioactive decay, 271
Radioactive decontamination, 278
Radioactive isotopes, 272, 280–283
Radioactivity, 271
  artificial, 272
  half-life, 271
  natural, 271
  subatomic particles, 271
Radioisotopes, 272, 280–283
Radium, 279–280
Radon, 279, 280
  seeds, 280
Rate of solution, 97
Rays:
  cathode, 270
  infrared, 42
RBE (radiation biologic effect), 274
Red corpuscles, 236–237
Redox, 35–36
Reducing agent, 35, 179
Reductant, 35, 179
Reduction, 35–36
  of fracture, 254
Refraction, 260
Regulator bottles, 65–68
Relaxin, 223
REM (roentgen equivalent man), 274
Renal threshold, 242
Renal unit, 243
Replacement, 33
Reserpine, 169–170
Resistance, 45
Respiration, 50
Respirator, 54
Respiratory alkalosis, 128
Respiratory quotient (RQ):
  of carbohydrates, 183
  of fats, 188
  of protein, 202

Resultant, 254
Retention:
   of chloride, 133
   of potassium, 133
   of sodium, 133
   of water, 133
Retina, 265
Rhodopsin, 228
Riboflavin, 226
Ribose, 177
Rickets, 230
Ringer's solution, 98
RNA, 168, 205
   messenger, 206
Roentgen, Wilhelm, 274
Roentgen (unit), 274

Salicylic acid, 159
Saline, catharsis, 99–100
Saliva, 210
   action of, 210–211
   composition of, 210
Salivary digestion, 210–211
Salol, 159
Salt, 109, 112–113
   depletion of, 132
   hydrolysis of, 118–119
Saponification, 157, 189
Saturated solution, 96
Saturation, 141
Scintillation counter, 274
Scurvy, 228
Seconal, 169
Secretin, 213
Semipermeable membrane, 97
Serum albumin, 125, 196
Serum globulin, 125, 196
Sex glands, 221
Simple underwater drainage (closed chest drainage), 53–54
Siphon, 57–59
Siphon suction apparatus, 65
Skatole, 246
Skin, 248
Sodium, radioactive, 282
Sodium benzoate, 36
Sodium ion, 16
Sodium retention, 133

Sodium salicylate, 159
Softening of water, 85–86
Solid, 23
Sols, 101
Solubility, 97
Solute, 91
Solution, 91–92
   concentration of, 92–97
   heat of, 97
   milligram percent, 94
   molar, 94
   normal, 94–95
   percent, 93
   properties, 91
   rate of, 97
   saturated, 96
   supersaturated, 96
   unsaturated, 96
Solvent, 91
Sorbitol, 176
Sound waves, 259–262
Specific gravity, 32
Specific heat, 84
   of alcohol, 84
   of iron, 84
   of water, 84
Spontaneous combustion, 78–79
Stability, 252
"Stanley steamer," 28
Starch, 177
States of matter, 23–28
Steapsin, 215
Stearic acid, 155, 186–187
Steroid nucleus, 101
Sterol, 150, 190–191
Strabismus, 265
Strontium, radioactive, 283
Strychnine, 170
Subatomic particles, 3, 13
Substance, 1–2
Substitution, 145
Substrate, 208
Sucrase, 208, 214, 215
Sucrose, 177
Suction, 59–69
Sulfadiazine, 117
Sulfanilamide, 165
Sulfhydryl groups, 171
Sulfur, radioactive, 283

Index

Sulfur compounds, 171
Sulfuric acid, 108, 111
Supersaturation, 96
Supertones, 260
Surface tension, 55
Suspension, 91–92, 102
Suspensoid, 101
"Swiss cheese hearing," 260
Symbol, 3
Syneresis, 101
Synthesis, 33
Synthetase, 210

Tannic acid, 201
Tears, 249
Temperature, 40, 42
   body, 42
Temperature scales, 40–41
   Celsius, 40–41
   Fahrenheit, 41
Temporary hard water, 85–86
Test(s):
   Aschheim-Zondik, 223
   basal metabolic rate, 207
   Benedict, 153, 180
   benzidine, 247
   biuret, 202
   concentration, 247–248
   dilution, 247
   Heller's, 202
   Millon's, 202
   phenolsulphonthalein, 172
   PSP, 172
   urea clearance, 247
Testosterone, 221, 224
Theelin, 221, 224
Theobromine, 170
Thiamine, 165, 225–226
Thioether, 171–172
Thiol, 171
Thiopental, 169
Thoraseal, 68, 69
Thrombin, 237
Thromboplastin, 237
Thrombus, 238
Thymine, 168
Thyrocalcitonin, 218
Thyroglobulin, 216

Thyroid, 216–217
   diseases of, 217–218
Thyrotropic hormone, 223
Thyroxine, 216
Tidal drainage, 58–60
   figure, 61
Tincture, 89
Tolbutamide, 193, 219
Toluene, 144, 150, 156
Torque, 252
Traction, 254–255
Transferase, 210
Transistor, 48
Transverse waves, 259
Tribromoethanol, 149
Tricarboxylic acid cycle, 216
Trichloracetaldehyde, 153
Triose, 177
Triprotic acid, 108
Trisubstitution, 146
Trypsin, 215
Trypsinogen, 213

Ultrasound, 260–261
Ultratones, 260
Ultraviolet rays, 262
Universal gas law, 50
Unsaturated compounds, 18, 141–143, 186–187
Unsaturated solutions, 96
Uracil, 168
Urea, 165, 245
Urea clearance test, 247
Urease, 208
Uric acid, 245
Urine, 242–247
   abnormal constituents of, 246–247
   appearance of, 244
   characteristics of, 244
   composition of, 244–246
   fermentation of, 247
   pH of, 244
   specific gravity of, 244
   sulfur in, 245, 246
   total solids in, 244–245
   volume of, 242
Urinometer, 52–53
Urobilin, 249

Urokon, 270
Urotropine, 164

Valence, 16–20
  dual, 19
  electro-, 16–17
  of middle element, 19
  summary of, 20
Valence numbers, 16, 21
Vaporization, heat of, 24
Vasopressin, 223
Vectors, 253–255
  addition of, 255
    figure, 21–22
  quantities, 253
    figure, 21–22
Venturi tube, 64
Veronal, 169
Vesicant, 172
Viscosity, 56
Visual purple, 228
Vitamins, 225–232
  A, 228
  B_1, 225, 226
  B_2, 226
  B_6, 227
  B_{12}, 227
  C, 228
  D, 229–231
  E, 231
  K, 232
  positive action of, 231
  results of deficiency, 231
  sources of, 231
Volt, 45
Voltage, 270
Voltmeter, 45
Von Bayer, Adolf, 169

Wangensteen suction, 61–64
Wastes, metabolic, 241–249

Water, 83–89
  action with active metals, 86–87
    with metal oxides, 87
    with nonmetal oxides, 87
  chemical properties of, 86–87
  of crystallization, 87
  density of, 83–84
  electrolysis of, 86
  filtration of, 84
  hard, 85–86
  heat capacity, 95
  ionization of, 110
  physical properties of, 83
  physiological importance of, 89
  purification of, 84
  retention of, 133
  softening of, 85–86
  specific heat of, 84
Water soluble vitamins, 225–228
Waves, 259–267
  compression, 259
  light, 262
  short, 260
  sound, 259
  transverse, 259
  ultrasonic, 260–261
Wax, 189
Weight, 2
White corpuscles, 237
Wohler, Fredrick, 137
Work, 28

X ray, 269–271
  deep, 270
  hard, 270, 275
  soft, 270
Xanthoproteic test, 202
Xerophthalmia, 228

Zinc, radioactive, 283
Zwitterion, 166
Zymase, 180

PERIODIC TABLE OF THE ELEMENTS

This table arranges the elements by atomic numbers (above the chemical symbols). The chemical atomic weights (below the symbols) were adopted by the International Commission on Atomic Weights in 1957 and those in parentheses represent the mass numbers of the currently known longest-lived nuclides for elements having no known stable nuclides. The lanthanides (57–71) and actinides (89–103) are placed below the body of the table for compactness.

IA	IIA		IIIB	IVB	VB	VIB	VIIB	VIII		
										1 H 1.008
3 Li 6.940	4 Be 9.013									
11 Na 22.991	12 Mg 24.32									
19 K 39.100	20 Ca 40.08		21 Sc 44.96	22 Ti 47.90	23 V 50.95	24 Cr 52.01	25 Mn 54.95	26 Fe 55.85	27 Co 58.9	
37 Rb 85.48	38 Sr 87.63		39 Y 88.92	40 Zr 91.22	41 Nb 92.91	42 Mo 95.95	43 Tc (97)	44 Ru 101.1	45 Rh 102.9	
55 Cs 132.91	56 Ba 137.36		57 La* 138.92	72 Hf 178.50	73 Ta 180.95	74 W 183.86	75 Re 186.22	76 Os 190.2	77 Ir 192.	
87 Fr (223)	88 Ra (226)		89 Ac† (227)							

*Lanthanide series

58 Ce 140.13	59 Pr 140.92	60 Nd 144.27	61 Pm (145)	62 Sm 150.3
90 Th 232.05	91 Pa (231)	92 U 238.07	93 Np (237)	94 Pu (239

†Actinide series